PHYSICS, COSMOLOGY AND ASTRONOMY, 1300—1700:
TENSION AND ACCOMMODATION

BOSTON STUDIES IN THE PHILOSOPHY OF SCIENCE

Editor

ROBERT S. COHEN, *Boston University*

Editorial Advisory Board

ADOLF GRÜNBAUM, *University of Pittsburgh*
SYLVAN S. SCHWEBER, *Brandeis University*
JOHN J. STACHEL, *Boston University*
MARX W. WARTOFSKY, *Baruch College of the City University of New York*

VOLUME 126

PHYSICS, COSMOLOGY AND ASTRONOMY, 1300–1700: TENSION AND ACCOMMODATION

Edited by

SABETAI UNGURU
Tel-Aviv University, Tel-Aviv, Israel

KLUWER ACADEMIC PUBLISHERS
DORDRECHT / BOSTON / LONDON

Library of Congress Cataloging-in-Publication Data

```
Physics, cosmology, and astronomy, 1300-1700 : tension and
  accommodation / edited by Sabetai Unguru.
      p.   cm. -- (Boston studies in the philosophy of science ; v.
  126)
    Papers from "a spring 1984 international workshop held, under the
  auspices of the Israel Academy of Sciences and Humanities, by the
  Institute for the History and Philosophy of Science and Ideas of Tel
  -Aviv University in cooperation with the Van Leer Jerusalem
  Foundation"--Introd.
    ISBN 0-7923-1022-5 (alk. paper)
    1. Science, Renaissance--Congresses.  2. Physics--History-
  -Congresses.  3. Astronomy--History--Congresses.  4. Cosmology-
  -History.  I. Unguru, Sabetai.  II. Aḳademyah ha-le'umit ha
  -Yiśre'elit le-mada'im.  III. Universiṭat Tel-Aviv.  Institute for
  the History and Philosophy of Science and Ideas.  IV. Mosad Van Lir
  bi-Yerushalayim.  V. Series.
  Q174.B67  vol. 126
  [Q125.2]
  001'.01 s--dc20
  [509'.4'0902]                                          90-48538
ISBN 0-7923-1022-5
```

Published by Kluwer Academic Publishers,
P.O. Box 17, 3300 AA Dordrecht, The Netherlands.

Kluwer Academic Publishers incorporates the publishing
programmes of D. Reidel, Martinus Nijhoff,
Dr W. Junk and MTP Press.

Sold and distributed in the U.S.A. and Canada
by Kluwer Academic Publishers,
101 Philip Drive, Norwell, MA 02061, U.S.A.

In all other countries, sold and distributed
by Kluwer Academic Publishers Group,
P.O. Box 322, 3300 AH Dordrecht, The Netherlands.

Printed on acid-free paper

All Rights Reserved
© 1991 Kluwer Academic Publishers
No part of the material protected by this copyright notice may be reproduced or
utilized in any form or by any means, electronic or mechanical,
including photocopying, recording or by any information storage and
retrieval system, without written permission from the copyright owner.

Printed in the Netherlands

TABLE OF CONTENTS

SABETAI UNGURU / Introduction   vii

PART ONE
ANCIENT BACKGROUND

JOHN GLUCKER / Images of Plato in Late Antiquity   3
ABRAHAM WASSERSTEIN / Hunches that did not come off:
   Some Problems in Greek Science   19

PART TWO
ISLAMIC AND JEWISH CONTRIBUTIONS

GAD FREUDENTHAL / (Al-)Chemical Foundations for Cosmo-
   logical Ideas: Ibn Sînâ on the Geology of an Eternal World   47
BERNARD R. GOLDSTEIN / Levi ben Gerson: On Astronomy
   and Physical Experiments   75
Y. TZVI LANGERMANN / The Astronomy of Rabbi Moses
   Isserles   83

PART THREE
MEDIEVAL COSMOLOGY, NATURAL PHILOSOPHY, AND OPTICS

EDWARD GRANT / Celestial Incorruptibility in Medieval Cos-
   mology 1200–1687   101
EDITH D. SYLLA / The Oxford Calculators and Mathematical
   Physics: John Dumbleton's *Summa Logicae et Philosophiae
   Naturalis*, Parts II and III   129
SABETAI UNGURU / Experiment in Medieval Optics   163

TABLE OF CONTENTS

PART FOUR
KEPLER: COSMOLOGY, ASTRONOMY, AND LIGHT

FRITZ KRAFFT / The New Celestial Physics of Johannes Kepler	185
DAVID C. LINDBERG / Kepler and the Incorporeality of Light	229

PART FIVE
SCIENCE, RELIGION, AND POLITICAL POWER

JOHN D. NORTH / One Truth or More? Demarcation in the Universe of Discourse	253
MICHAEL SEGRE / Science at the Tuscan Court, 1642—1667	295
NOTES ON CONTRIBUTORS	309
NAME INDEX	313

SABETAI UNGURU

INTRODUCTION

Habent sua fata colloquia. The present volume has its origins in a spring 1984 international workshop held, under the auspices of the Israel Academy of Sciences and Humanities, by The Institute for the History and Philosophy of Science and Ideas of Tel-Aviv University in cooperation with The Van Leer Jerusalem Foundation.

It contains twelve of the twenty papers presented at the workshop by the twenty-six participants. As Proceedings of conferences go, it is a good representative of the genre, sharing in the main characteristics of its ilk. It may even be one of the rare instances of a book of Proceedings whose descriptive title applies equally well to the workshop's topic and to the interrelations between the various papers it includes. Tension and Accommodation are the key words.

Thus, while John Glucker's paper, 'Images of Plato in Late Antiquity,' raises, by means of the Platonic example, the problem of interpretation of ancient texts, suggesting the assignment of proper weight to the creator of the tradition and not only to his many later interpreters in assessing the proper relationship between originator and commentators, Abraham Wasserstein's 'Hunches that did not come off: Some Problems in Greek Science' illustrates the long-lived Whiggish tradition in the history of science and mathematics. As those familiar with my work will undoubtedly note, Wasserstein's position is far removed from my stance on ancient Greek mathematics. I chose, therefore, to interfere editorially as little as possible with its conclusions and approach, appealing as they do to many workers in the field, and the paper stands basically as given in its pristine allure. Another hunch that did not come off, one might say.

Bernard Goldstein's paper on Levi ben Gerson shows us a man in the fourteenth century who was not reluctant to use a hands-on 'experimental' approach in the study of astronomy and who, in doing so, preceded in his understanding and results some of Tycho Brahe's and Johannes Kepler's achievements in the measurement of eclipses and on pinhole images. Gad Freudenthal shows tellingly the inherent elasticity of theory in saving a cherished world view, specifically the eternity of the world, that seemed to be inconsistent with the geological features of the earth, as accepted by the theory. By postulating generative geological processes compensating for erosion, and by drawing on alchemical and chemical auxiliary concepts, Ibn Sînâ was able to save Aristotelian natural philosophy from adverse criticism.

While Y. Tzvi Langermann deals with the astronomical views of Rabbi Moses Isserles, a sixteenth-century famous legal scholar in Cracow, during a period that witnessed a strengthening of rationalist undercurrents in Jewish intellectual life, Edward Grant writes on 'Celestial Incorruptibility' and shows how a few significant cracks developed within medieval cosmology, leading, eventually, to the abandonment of the sacrosanct Aristotelian principle of celestial incorruptibility. Edith Sylla uses Parts II and III of Dumbleton's *Summa logicae et philosophiae naturalis* to show how mathematics could also serve in the hands of the Calculators in handmaiden status, at the pleasure of physics, to deal with physical reality. It is this natural philosophical tradition, striving to achieve an accurate knowledge of physical reality by exploiting the fruitfulness of mathematics, rather than the logico-sophismatical tradition that makes the Calculators into an important link between medieval qualitative physics and the new quantitative physics of the seventeenth century.

'Experiment in Medieval Optics' discusses the role and limitations of 'experiment' in Witelo's *Perspectiva*, while Fritz Krafft's paper focuses of the crucial role of *physics* in Kepler's new astronomy. David C. Lindberg explores Kepler's ideas on the nature on light and establishes their roots in the ancient and medieval optical tradition. Finally John D. North tackles the problem of intellectual demarcation in connection with a discussion of the issues of creation and eternity of the world, showing the importance, profundity, subtlety, and intellectual pyrotechnics of this medieval strategy of argument; and Michael Segre treats the question of the relation of science and political power at the Tuscan Court after Galileo's death.

These are, then, the contents of a volume that is meant to provide the reader with a nontrivial sampling of the variety of interrelations amongst physics, cosmology, and astronomy between, roughly, 1300 and 1700, of their tension, and eventual resolution in the new Newtonian science which supplanted medieval natural philosophy.

Tel-Aviv University
Tel-Aviv, Israel

PART ONE

Ancient Background

JOHN GLUCKER

IMAGES OF PLATO IN LATE ANTIQUITY

I have chosen my title quite deliberately, for reasons both subjective and objective. Not '*The* Image of Plato in Late Antiquity', since it is now a commonplace that at no time has there been anything like one and only kind of Platonism.[1] Various interpretations of the dialogues existed even among Plato's own pupils. Aristotle, his pupil for the best part of twenty years, took the creation-myth of the *Timaeus* quite literally as an event in time. Aristotle's friend Xenocrates — a pupil of Plato 'from his youth', who even 'accompanied him on his trip to Sicily' (Diog. Laert. IV, 6) and was his successor's successor as head of the Academy — believed that the myth was merely an 'analysis for the sake of examination' (what one could call geometrical construction), and that the world of the *Timaeus* was eternal. So, for that matter, did Xenocrates' pupil Crantor — but Crantor disagreed with him on the meaning of the creation of the soul in the *Timaeus*.[2] We shall soon see that in late Antiquity there were at least two main 'images' of Plato which, even though not mutually exclusive, were not quite the same — and there were also relics, at least, of another image, rejected by the upholders of both, but known to be of an ancient and honourable ancestry.

Why, then, not '*The* Images of Plato . . .' — apart from the awkward English? Here my reasons are, as I have said, both objective and subjective. Objective, since our evidence is, in the nature of the case, partial and restricted. If it were not for one manuscript, written in A.D. 925 and miraculously preserved until it reached the library in Vienna, we would have had no knowledge of the anonymous *Prolegomena to Platonic Philosophy*,[3] which is among other things (as we shall soon see) one of our main sources for the lingering on of sceptic images of Plato into late antiquity. And the one specimen of a Middle Platonic commentary on a dialogue which we possess has been found in a papyrus roll.[4] For all we know, there may have been some Stoic or — more probably — sceptic treatises on Plato still extant in late antiquity.[5] But they have not reached us.

My subjective reasons are even simpler. I have neither the equipment nor the preparation for offering a Platonic counterpart to Professor Moreaux's monumental *Der Aristotelismus bei den Griechen* (Berlin:

De Gruyter, 1973; 1984); and, had I been able to produce such an enviable piece of work, a paper read at a conference — and on a subject which is only marginal to the theme of the conference at that — would hardly be the proper occasion for presenting it. What I wish to offer here is more in the nature of an archaeological shaft, giving us some glimpses of an area which is still largely to be explored,[6] assessing in a rather provisional manner what one might expect to find in one or two of the main strata — and, inevitably, comparing ancient images, conceptions and approaches with modern ones. For people have been living on the Platonic *tell* and building new settlements on the debris of old ones ever since the generation of Plato's immediate disciples. If Plato is turning in his grave, he must be used to it by now: he has been doing it for more centuries than he would care to remember.

What, then, are the main images of Plato in late antiquity — and, what is no less important, *whose* images? How do we know of these images and distinguish between their worshippers?

That every educated Greek, from the first or second century A.D. or even somewhat earlier, was supposed to have read some Plato, seems at first sight only too obvious. One just has to look at the large number of quotations and testimonia which could be collected for most of the dialogues.[7] But our sources for these 'fragments' and testimonia — where they are not themselves 'Platonists'[8] of some sort — are men of learning like Quintilian or Eusebius. What did the average educated Greek really know of Plato and about him?

The extant commentaries on various dialogues — apart from the papyrus commentary on *Theaetetus*[9] — are Neo-Platonic, complex, and far too long for the average educated reader. Two examples would suffice. Proclus' commentary on the *First Alcibiades* occupies, by a rough calculation, the equivalent of 140 Stephanus pages to Plato's 32. His commentary on the *Timaeus* occupies three Teubner volumes of about 300 pages each. We all remember the length of Plato's *Timaeus*. And I have said nothing on the *lexis* and *dianoia* of those commentaries. Such works were clearly written *sibi et doctis* — for the teacher, his pupils and colleagues, and serious students of Platonic philosophy. What, then, did the average person read?

The dialogues, I will be told. But *all*, or even most, of them? And how thoroughly? One indicator of this, I believe, could be the scholia — or rather, the quantity of scholia. Take our scholia to the *Apology* as a specimen. They occupy about the space of three Oxford pages to 35

Oxford pages of Plato's text. Now compare them with the scholia to the first Oxford page alone (lines 1—19) of Euripides' *Medea*: the quantity of scholia is about the same, with a slight prejudice — if at all — in favour of the one page of Euripides. But surely, the *Apology* is far easier as a Greek text than any Euripides. Yes — but *all that much* easier for a *Koiné* speaking schoolboy, who may not even be a native speaker of Greek? And is the myth of the *Timaeus* all that easy? Yet all the scholia to the *Timaeus* would occupy only about eight Oxford pages.

If we are not yet convinced, let us take a prose author like Demosthenes. His First Olynthic Oration occupies a little over seven Oxford pages. The scholia, in Professor Dilts' Teubner text, run to 35, and larger, pages. The Second Olynthic, of the same size in the Oxford Text, has 32 pages of scholia. It is true that some Demosthenic speeches have shorter sets of scholia than others. But even 34 pages on *De Corona* of about 100 pages is still a far more generous proportion than any of the Platonic scholia. The scholia on Aeschines in Schultz' old Teubner run to about 100 pages in small print, on 230 pages in large print of the text itself, one third of each page being the large apparatus and commentary.

No, the average educated Greek most probably read very little Plato, just as the average educated Englishman reads very little Hobbes, Locke or Hume. If he wanted to acquire the necessary 'coffee-table' acquaintance with Plato — or even if he merely wished, like Justin Martyr (*Dial. cum Tryph.* 2, 219D) 'to appear wise to himself within a brief space of time, and to hope in his stupidity to see God in no time' — he had access to the usual literature written especially for that purpose: the various biographies, doxographies and mixtures of the two. Fortunately, we do not have to speculate — we have relics of this sort of literature: Albinus' *Isagoge* and *Didaskalikos*; the *Life* ascribed to Olympiodorus; the anonymous *Prolegomena to Platonic Philosophy*; Book III of Diogenes Laertius' *Lives of the Philosophers*; and chapter 19 of Hippolytus' 'Philosophoumena'.[10]

We have, then, roughly two types of readers. First, the serious student of Plato and philosophy, who reads the dialogues themselves — or at least, the ones considered by his teachers to be most central and important — with their large and complicated commentaries. Here, one can assume, the fate of most commentaries befell this kind of literature as well. Later — and therefore (as many of us still tend to think today) better and more up-to-date — commentaries replaced the older ones, and every new generation of students had to contend with harder and bulkier books. Longinus, Numenius and the 'School of Gaius' gave their place to Olympiodorus, Proclus, Hermias and Damascius. The survival of the works of the latter and the loss of the works of the former is evidence enough. The serious student of philosophy in late antiquity, we can say, read his Plato through Neo-Platonic spectacles. He may have read the more popularly written lives and prefaces. But he could also read such Neo-Platonic prefaces as Proclus' *Platonic Theology*[11] — and if he also read his Plotinus (which most serious students did more than once), he would receive there, almost on every page, a Neo-Platonic conception of Plato.

The general educated reader, on the other hand, usually read the shorter and more comprehensible compendia of Plato's life and doctrines. Here, he would find no Neo-Platonic doctrines. Albinus, Diogenes and Hippolytus present us with purely Middle Platonic approaches to Plato; and the anonymous *Prolegomena* — although Westerink has shown that it is connected in all but the biographical section with the school of Proclus[12] — hardly enters into problems of doctrine. What the reader would find here — as far as Plato's doctrine is concerned — is an amalgam, in various quantities, of Platonic, Aristotelian and, here and there, Stoic elements; or, to be more precise, ideas which are mainly Platonic, often translated into — or explained by — Aristotelian concepts or Stoic terms.

It is not my task here to provide a summary of the various Neo-Platonic images of 'true Platonism,' or of the various amalgams — and their different consistencies — which one can find in the Middle Platonic compendia. The general outline of Neo-Platonism, and how it was derived by its propounders from Plato's works, and especially from late dialogues like *Parmenides, Philebus, Sophist* or *Timaeus* — all this is reasonably well known today even to the general reader of prolegomena and doxographies. Some of the main views of the Middle Platonists are now better known to the general reader thanks to Professor Dillon's fascinating work. And, after all, I have only promised a shaft, not a full excavation report. I shall therefore concentrate on a few questions of a more general nature. First, are there any features shared by most or all of the Platonic literature of late antiquity?

The division of the dialogues into trilogies and tetralogies and into themes ('physical, logical, ethical, theological, political' and the like) or types of discussion ('hyphegetic, zetetic, aporetic, peirastic, maieutic' and others of the same ilk), occupies much space both in the popular compendia and in the Neo-Platonic commentaries. It was obviously of crucial importance for Proclus to show that the *Parmenides* was not merely a 'logical, gymnastic' dialogue, but a proper theological one, teaching us some positive doctrines of supreme importance.[13] But the uselessness of Thrasyllus' tetralogies for the proper study of Plato needs only to be mentioned in passing. As for the classification by theme or type of dialogue, they are fluid enough in the various sources.[14] But even had we possessed one 'authoritative' division of all the dialogues into themes and types, done by some respected early commentator like — say — Longinus, would a modern scholar agree that the *Republic*, for instance, is merely a dialogue 'on justice, political,' whereas *Gorgias* is 'on the Good, ethical'? Most of us would agree that 'Plato himself never wrote any important dialogue on a single topic ... he always appears to be following the argument whithersoever it may lead ... He never treats subjects separately'.[15] Even a short early dialogue like *Euthyphro* — officially devoted to a search for the definition of a single concept, τὸ ὅσιον, loosely (and misleadingly) translated as 'piety' — teaches us also some things about logic, literary criticism, religion, law and politics. We notice the ancient obsession with labels and classifications with amusement and dismiss it with pleasure.

One issue which the modern reader might expect to find in the literature of late antiquity — since modern research of the last century and a half have accustomed him to it — is the problem of chronology and development. After all, Plato wrote his dialogues over a period of forty years or more, and — as many modern scholars have claimed and attempted to prove — some of his most fundamental views underwent modifications of greater or lesser significance between his early, middle and late period. Alas, the modern reader will search for such things in vain in the ancient sources. The ancients knew that the *Laws* were the last dialogue, transcribed out of Plato's waxen tablets by Philip of Opus.[16] The author of the anonymous *Prolegomena* can even mention that one possible order of the dialogues, proposed by some unnamed people before his time, is by the 'chronology of Plato' — or, as we call it, by date of composition. But all he can say in detail, apart from *Laws* being posthumous and the latest, is that these people also maintain that *Phaedrus* is the earliest dialogue — 'because there, they say, he raises the question whether one should write books or not: and if he had not made up his mind whether to write or not then, how could he have written another book before?'[17] It is possible that the author's source or sources, most probably at some removes, drew on someone who did try to arrange *all* the dialogues in chronological order. But his reasons for putting *Phaedrus* first can hardly strike us as evidence for a reliable historical tradition of the chronology of Plato's works. In the best case, this was yet another attempt at playing the classification game, based on a sounder principle than most, but carried out with insufficient means and in the speculative fashion so characteristic of much of what passed for literary history in late antiquity (— and only then?). In any case, we have no evidence that any conclusion was ever drawn from this chronological ordering as to the possibility of change and development in Plato's views. I do not believe that commentators and compilers of handbooks in late antiquity were capable of drawing such a conclusion. An author is usually believed by the compilers and commentators of late antiquity to be a single, indivisible unit, and all his works are of one piece. To take an example from another field: if Aeschylus, in his *Oresteia*, has three actors, this is evidence enough for the late, anonymous and utterly confused compiler of the late Hellenistic (or early Byzantine) *Life of Aeschylus* to conclude that it was Aeschylus — *tout court* — who introduced the third actor into tragedy. He can then disagree with a good ancient source — Dicaearchus of Messene, a pupil of Aristotle and a responsible scholar — who had said, most probably on good evidence, that it was Sophocles. Why? But that is simple: because we find three actors 'in Aeschylus'! That Aeschylus in 458 B.C. could have been influenced by Sophocles, who had already been on the stage for ten years since his first victory (on Aeschylus himself), could not even occur to our late compiler. An author — like Rome — is always the same; and, like the Torah, there is no 'earlier' and 'later' in his work.[18]

Little care for chronology, then, and not the slightest suspicion of development. All the dialogues — and this is the picture emerging from all the sources of late antiquity — are parts or facets of one unified and consistent Plato. What has one to do, then, when one dialogue appears to contradict the doctrines of another, or of other dialogues?

A good test-case is the first part of the *Parmenides*. In it, Plato's Parmenides supplies us with some of the best arguments against Plato's own classic Theory of Forms — some of them later used by Aristotle — and his refutations remain unanswered. I have argued elsewhere [19] that for the followers of the sceptical Academy, Plato's *Parmenides* must have been a goldmine. Whatever its date of composition (and I doubt whether even Arcesilaus knew or could care much about that), it shows that Plato himself took the Theory of Forms merely as a 'youthful indiscretion' of a young Socrates, from which he — as he grew older and, naturally, more sceptical — and his sceptical pupil Plato, had long ago recovered.

So far, so good. But our late compilers and commentators could hardly accept such a solution. Some of them knew that there had been attempts in the past to present Plato as a sceptic and a 'New Academic' — but they rejected this view and attempted to refute it.[20] What, then, does one do with that awkward first part of *Parmenides*? We know what Proclus does: two of his writings which deal with this dialogue have reached us.[21] For Proclus, the task is fairly easy; for the One, the Mind, the Soul, the Ideas — they can all be made to fit into his Neo-Platonic hierarchy, and 'throwing the dialogues against each other' (with some considerable help from the hefty volumes of older commentaries), will iron out apparent contradictions. But what about the Middle Platonic compiler? Nothing easier, provided we use the proper labels. *Parmenides* can be classified as a 'logical' or 'elenctic' dialogue, whose aim is refutation and 'gymnastic,' or training, in dialectic for the reader. This, as we have already seen, is after all one of the views which Proclus had to contend against — and quite probably, was derived in the last resort from a sceptic source long forgotten.[22] Classification thus has some value: using the right labels can help us preserve the unity of Plato's dogmata — a unity which, for all our late sources, is itself a sacred dogma.

One should add, to be fair, that the unity of Plato's thought is a thesis not unknown in recent scholarship. Shorey, Cherniss and Crombie — to mention but a few distinguished names — have defended this unity in an age when the relative chronology of three periods has been treated almost as dogmatically as the ancient idea of unity (although the proofs of the three-period chronology have been available even to those who have chosen to take Campbell's and Ritter's word for it), and when theories of development — sometimes fairly detailed developments from one dialogue to another, supported by cross-references and recent literature — have abounded. But the modern scholar has, at least, the option. He can see in the contradictions between, say, a middle and a late dialogue the result of change and development in

Plato himself — even the outcome of criticism and of discussions between Plato and his colleagues and pupils in the Academy. He is not committed to viewing the whole Platonic corpus as one, timeless, motionless, changeless 'unshaken heart of well-rounded truth.' The student or reader of Plato in late antiquity was not offered this option.

For late antiquity, then, Plato is a dogmatist, and all his various dialogues are expressions of one and the same set of dogmata. The nature of the Platonic dialogue as drama is discussed in our compendia in a rather formal and schematic fashion;[23] and when we are told that 'he reveals his view through four persons: Socrates, Timaeus, the Athenian stranger and the Eleatic stranger,'[24] we know that we can lay aside any hope for an understanding of the dialogue as a literary form similar to tragedy and comedy, which were still flourishing in Plato's time.[25] After all, an understanding of tragedy and comedy themselves as real dramatic forms was not a very strong point with most scholars and readers in late antiquity.

But if the whole Platonic corpus expounds, in various ways and forms, one and the same set of dogmata — what are these dogmata, and what image of Plato do they present?

Once again, this is not the place to enter into details. They are plentifully available in the compendia and commentaries. Here, one can only deal with the 'images' promised in my title and explained in the opening paragraphs. My impression is that the images are, by and large, also two — again, divided roughly by the kind of books and the sort of readers they are addressed to.

For the Neo-Platonist, Plato is, of course, 'one of us', and the various dialogues are interpreted to fit his scheme of things human and divine: the One, the Mind, the Soul, the emanations, the various kinds and degrees of being and knowledge, and what is above being and knowledge and is the origin and source of both. But what about our general reader, the educated Greek who derives most, if not all, of his knowledge of Plato from the prolegomena, doxographies and 'lives and letters'?

Here we reach, perhaps, the one point which may be of some interest to the historian of science. But being a Classical scholar, I propose to present it in what may appear to the more straightforward historian of science as a somewhat roundabout fashion. Plato was never averse to playing with hypotheses: let us play with one on Plato himself.

Imagine, ὑποθέσεως χάριν, that all of Plato's writings had been lost before the Neo-Platonists came about and began to reinterpret him and to write their bulky commentaries. All the modern reader would have had at his disposal would be some quotations and testimonia in earlier writers — mainly such as Aristotle and Cicero — and our Middle Platonic compendia of late antiquity, which would be the major source of testimonia, if only because of their bulk and their concentration on Plato. The general image of Plato to emerge from these compendia would be that of a classical φυσιολόγος in the best Pre-Socratic tradition. Look, for example, at the ἀρέσκοντα section of Diogenes Laertius' *Life of Plato* (III, 67—80). Virtually all of it is a somewhat confused summary of the *Timaeus*. Now look at Albinus' *Didaskalikos*: four chapters on 'dialectic' (III—VI); eighteen on ethics and politics (XXVII—XLIX) — but still twenty full chapters (VII—XXVI) between the two — virtually as much as both of them put together — on 'theology, physics and mathematics,' all devoted to the exposition of the *Timaeus* — which, we remember, far from occupying half of Plato's literary output, is only about a quarter of one out of five Oxford volumes of Plato. A similar picture emerges from Hippolytus' chapter on Plato: thirteen paragraphs on God, the Universe, Matter, the Soul, and other 'physiological' subjects (1—13), as against nine (15—23) on the Goods, the Virtues, the *Summum Bonum* and the like. A near enough proportion could be found in Apuleius — and, I suspect, in any other compendium of the period which may still turn up.

Our hypothetical editor of *Die Fragmente der Voraristoteliker* would be able to make up an almost complete list of the dialogues, mainly from the partial lists in chapters of our compendia dealing with labels and classifications. Aristotle's *Ethics* and *Politics* would supply some fragments and testimonia on issues other than 'theological, physical and mathematical.' But try to reconstruct the *Republic* as we have it mainly out of Book II of Aristotle's *Politics* — or the Theory of Forms (not to mention the various critiques of it in the later dialogues) out of Aristotle and the compendia of late antiquity. The Plato of our *Die Fragmente der Voraristoteliker* would emerge most clearly as a philosopher with preponderantly cosmological interests, and the one dialogue whose doctrines we would be able to reconstruct in some considerable detail (even if we were not to know that all these details came from it — how much more of an incentive for believing that these were doctrines commonly found in many dialogues) would be *Timaeus*.

True, our reconstructed Plato the Cosmologist would still be very different from most of his Pre-Socratic forbears. Indeed, his cosmos is almost the exact opposite of the eternal universe governed by mechanical laws of matter and various forces presented by most of his predecessors, and this despite all his many borrowings from them which he incorporates in his *Timaeus*. Professor Vlastos has made this point very clearly and forcefully in his *Plato's Universe*. But our main point is that our hypothetical Plato, thus reconstructed mainly from the evidence of what the educated man read about him in late antiquity, would still be first and foremost a cosmologist. His 'dialectic' and 'ethics' could be interpreted without much difficulty as offshoots of his theories of the soul, matter, God, and other elements in the cosmos of the *Timaeus*.

I may be asked what is wrong with this. After all, 'Later tradition, from the time of Plato's immediate successors down to the end of the Academy, is unanimous in regarding the *Timaeus* as a key dialogue for the understanding of Plato's philosophy.'[26] This may be true, at least, for some of Plato's immediate pupils and of some of the 'Platonists'[27] of the first century A.D. onwards. And the *Timaeus* is such a magnificent work that one would, somehow, like to be able to regard it as a 'key dialogue.' But did Plato regard it as such? Fortunately, the dialogue is extant, and I doubt whether its text would justify such a claim.

First, the drama itself. Socrates, Critias, Timaeus and Hermocrates are having a long συνουσία. Yesterday, it was Socrates' turn to 'entertain,' and he expounded his theory of the Ideal Republic. Now it is the turn of the others. Timaeus the Pythagorean of Locri, being good at his astronomy and the study of nature in general, will tell us today about the creation of the world and of man. Critias, who gives us a foretaste in the opening section of the dialogue of what he is about to tell us tomorrow (and in the dialogue *Critias*), will expatiate on an episode from the early history of man — the story of Atlantis. Socrates, one remembers, has discoursed about a subject close to his heart (and Plato's) — the ideal state (in the basic form of *Republic* II—IV, still without philosopher-kings and Ideas). After all, even the Socrates of Xenophon and Aristotle (Xen. *Mem.* I, 1, 11—16; Ar. *Metaph.* A, 6, 987b1—2) has rejected all speculation on the physical universe as inconclusive and pointless, and centered his attention on ethical and political problems. The nature of the physical world and the early history of man are left by Plato to the Pythagorean Timaeus and the statesman and aristocrat Critias. Deliberately? But how could it be otherwise? Is Plato ever careless or casual in his choice of characters?

But if Timaeus is the speaker, Plato is the author; and the mere choice of characters is hardly sufficient. Almost at the outset of his discourse (*Tim.* 29b—d), Plato's Timaeus warns his friends that, speaking of 'becoming' and not of 'being', he can provide them only with 'belief' and not with 'truth'. The word for 'belief' is πίστις, and we remember its connotation in contrast to ἐπιστήμη in *Gorgias* 454c ff. All that Timaeus can tell us about the physical world would be only a 'likely story' — εἰκὸς μῦθος (*Tim.* 29d2). The physical world is a matter of opinion (δοξαστόν, 27e—28a), and of sense-perception (28b8—c1). The distinction between truth as grasped by the intellect and appearance or opinion is repeated and accepted by Timaeus in 51a—52d, with the addition of the famous 'bastard reasoning' of 52b2 for perceiving the *chora*. Plato has clearly not given up his classic distinction between knowledge and all other kinds of perception — knowledge of eternal, unchanging truth and being as against sensation or opinion of transient, mutable things — a distinction he has held ever since his early dialogues.

But, we are told by Vlastos,[28] 'once you renounce hope of attaining knowledge in your theories about the natural universe, would you still have good reason to engage in such theorizing?' I do not see why not, provided you have made it clear from the outset that, in a field like this, one can achieve only an approximation and no certainty — and Plato does exactly that. After all, the sceptical Xenophanes did not refrain from giving us his own theories about God and the physical universe — as long as he also warned us that 'these things should count as opinion, approximating truth' (B35DK); and Parmenides, whose goddess denied to mortal opinion any 'true credence' (notice the words: πίστις ἀληθής B1, 30DK), was still taught by the same goddess also the 'way of opinion' about the structure of the physical world. A modern scientist (or philosopher of science) would hardly claim for any of his theories the status of Platonic ἐπιστήμη. And yet like Plato, he could also hardly resist the temptation to offer some explanation, however provisional and uncertain, to the world of phenomena (or theories) around him.

I have been conjuring up images. Before I take my leave of you, I wish to recall one more image — this time, in the etymological sense of the word: *imago*. Plato and Aristotle occupy the centre of Raphael's *School of Athens*. Plato, pointing upwards, holds in his other hand the *Timaeus*; Aristotle points downwards and holds the *Nicomachean Ethics*. The *Timaeus* in Plato's hand is a testimony to the central place occupied by this dialogue in many periods of antiquity and the Middle Ages — including the Latin west, one of whose few Platonic texts available continuously in Latin was this dialogue in Chalcidius' version. The central position of the *Nicomachean Ethics* is more problematic. What one should note, however, is that this popular representation of the two philosophers is largely misconceived. Not just on purely philosophical grounds. True: Plato, with all his 'upward drive', did not entirely reject 'the moving image of eternity', and even his *chora* is in some respects more active than Aristotle's 'place' or 'matter'. Aristotle, too, did not advocate a total rejection of all things divine: his God is 'thought thinking itself' (which, in a way, is even more 'intellectual' and less 'banausic' than Plato's 'Artisan-God'); and if the highest good for man (as we are told in the *Nicomachean Ethics* themselves) is the life of contemplation, one important reason for it is that this is the fulfillment of that part of his nature which he shares with the Divine.

But my main point is that what we have in this painting is an unclaimed — or even a rejected — reputation. Aristotle, given the choice of one of his books, would hardly have chosen to be represented by the *Nicomachean Ethics*. After all, we are told there, almost at the outset of Book I (ch. 3; 1094a12—28) that ethics, with all due respect to its importance, is a matter of approximation, not of precise knowledge. The study of the soul, for example, hardly less important than ethics, is far more precise (*De Anima* I, 1; 402a1—7). I have tried to show that Plato, given the same choice, would hardly have selected the *Timaeus* — and for surprisingly similar reasons.

Yet this was the verdict of much of late antiquity, the Middle Ages and the early Renaissance. Their images of Plato are hardly like any of the images of him current today — whether that of Plato the analytic philosopher (usually, *malgré lui*); or of the open-ended, developing philosopher, who ended up, perhaps, with more questions than answers; or of the one, unchanging, systematic philosopher — who was, however, at least as interested in dialectic, ontology, ethics and politics as he was interested in the nature of the physical world. Even the Plato of the 'ungeschriebene Lehre' is hardly that of the Neo-Platonists, with all the similarities, and surely not that of our Middle Platonist compendia. Our distance from Plato and his ancient followers, and our modern philological, historical and literary disciplines, may have made us more aware of Plato against his own literary, historical and philosophical background. Yet there are various, and not always reconcilable, images of Plato current today, and perhaps the one thing most of their holders share is their rejection of most of the previous images. Are images of Plato, then, no more than Popperian scientific theories, to be discarded in due time in favour of other — but no less hypothetical and transient — images? Or could we, if we were to broaden our excavation at the proper stratum, reach the ancient city at the bottom of the *tell*? After all, our hypothetical Plato of *Die Fragmente der Voraristoteliker* is — praise be to the Demiurge, the World Soul and the manuscript tradition — only a hypothesis. The complete writings of Plato are still in our hands. Perhaps we should concentrate on broadening our excavation of this most ancient stratum, and take the upper strata for what they are: new settlements built on the rubble of each other, not without their own measure of interest and fascination, but hardly representative of the ancient city itself.

Tel-Aviv University
Tel-Aviv, Israel

NOTES

1. See the cogent remarks of Paul Oskar Kristeller, *Renaissance Philosophy* (New York, 1961), pp. 48—9.
2. Plutarch, *De An. Procr.* 1—3 (*Moralia* 1012D—1013A). See also the excellent discussion of John Dillon, *The Middle Platonists*, (London, 1977), pp. 10—11.
3. L. G. Westerink, *Anonymous Prolegomena to Platonic Philosophy* (Amsterdam, 1962), p. L and n. 142.
4. *Anonymer Kommentar zu Platos Theaetetus*, ed. Hermann Diels and Wilhelm Schubart, *Berliner Klassikertexte, II* (1905).
5. In my *Antiochus and the Late Academy* (Göttingen, 1978), p. 314, I stated (with no supporting evidence) that Proclus' knowledge of the Academic Sceptics 'could not have been very profound.' A relevant piece of evidence seems to me to be Proclus' commentary on Plato's *Parmenides*, Book I, pp. 630—33 Cousin (*Procli Opera* (Paris, 1864)). This section deals with 'some people — and there have been some also in the past ' (Εἰσι δέ τινες καὶ γεγόνασι τῶν ἔμπροσθεν) who claim that the sole object of the *Parmenides* is a refutation of Zeno's doctrine. The expression ἰδόντες ἐφ' ἑκάτερα χρονομένους τοὺς λόγους, echoed by τὴν εἰς ἑκάτερα γυμνασίαν of *Theol. Plat.* I, 8, is reminiscent of ἡ εἰς ἑκάτερα ἐπιχείρησις (*in utramque partem disputatio*) of the sceptical Academy — see *Antiochus*, p. 33 n. 78; p. 291. This is not to say that Proclus was directly familiar with writings of the sceptical Academy — only that his sources (most probably Middle Platonists and early Neo-Platonists — see *Antiochus* p.216) had made use

of such writings and that Proclus could see from their works that this was an ancient position. I take it that εἰσὶ ... τινες of pp. 630 and 631 Cousin refers most probably not to contemporaries of Proclus or his sources but merely to commentaries which were still available to him or, most likely, to them. But this would imply at least that at the time of Proclus, there were still some works which could treat a dialogue of such crucial importance to the Neo-Platonic image of Plato as the *Parmenides* as purely aporetic.

6. The literature on Neo-Platonism is, of course, vast, and the Middle Platonic industry is hardly lagging behind. But most of this literature is concerned with the particular kind of Platonic or Platonizing synthesis made by each of these later philosophers, rather than with their overall and selective view of what is of primary, secondary, of marginal or of no importance in the Platonic corpus.

 One exception known to me is D. T. Runia's doctoral dissertation, *Philo of Alexandria and the Timaeus* of Plato, 2 vols. (Amsterdam, 1983). On p. 36 of vol. I, Dr. Runia notes the importance of the *Timaeus* in the 'dogmatic' part of Albinus' *Didaskalikos*. He writes: 'Not only does the section on Physics ... occupy more than half the work, but its contents are dominated by the doctrines of the *Timaeus*.' In Note 99 to this page (vol. II, p. 468), he does the same exercise for the *Placita* section of Diogenes Laertius' *Life of Plato*. I did the same exercise in a later section of this paper before I knew of Dr. Runia's book. I am grateful to him for sending me a copy of his dissertation, and pleased to see that we have reached similar conclusions independently.

7. This, like much else in this area, is still a task largely to be accomplished. The 'apparatus of testimonia,' placed between the text and the critical apparatus, is a fairly recent practice among editors which should become more general. Its importance is not merely for the establishment of the text (although what is wrong with that?), but also for the history of the popularity of that particular text in antiquity. For Plato, the best of its kind I know of is still the apparatus of testimonia in E. R. Dodds' edition of the *Gorgias* (Oxford, 1959 and reprints). See also his 'Introduction,' pp. 62—66. A full apparatus of this kind for the whole of the Platonic corpus would give us some indication of the popularity or otherwise of the various dialogues in different periods of the ancient world. Compare, for example, Eusebius, *Praep.* XIII, 5 — the only ancient testimonium I know of for *Euthyphro* — with the wealth of testimonia collected by Dodds in his *Gorgias*.

8. Either followers of a Platonic 'school of thought', or commentators on Plato or students of his philosophy (see *Antiochus*, p. 206ff.).

9. Note 4 above.

10. For the first four, the most convenient edition is still vol. VI of Carl Friedrich Hermann's Teubner *Plato*, pp. 152—222. For the fourth, Westerink's edition (n.3 above) is fundamental, not only for its text and translation, but for its excellent Introduction. Diogenes Laertius is available at last in a critical edition: H. L. Long's Oxford Text. The text of Hippolytus is most easily accessible in Diels, *Doxographi*, pp. 567—570. Of the secondary literature I shall only refer to R. E. Witt's *Albinus and the History of Middle Platonism* (Cambridge, 1937 and reprints), and to Dillon (n.2 above), pp. 267—306 and 408—414. I shall not deal with Apuleius, since I wish to concentrate on the Greek world. Apuleius adds little or nothing that

is very original to what we already know from Greek sources of the popular image of Plato and Platonism.
11. On the popularity of which among students of philosophy in late antiquity, see H. D. Saffrey and L. G. Westerink (eds.), *Proclus, Théologie platonicienne*, Livre I, (Paris, 1968), pp. CL—CLIV. It is true that this work almost disappeared in the Middle ages — on which and the likely reasons, see *ibid*, pp. CLIV—CLX. Saffrey and Westerink's stemma, p. CLI, makes all the extant MSS depend on two copies made in the thirteenth and fourteenth centuries. But we are only concerned with late antiquity.
12. Westerink (n.3 above), pp. XXXIV—XLI.
13. Proclus, *Comm. on Parmen.*, p. 630ff. Cousin; *Theol. Plat.* I, 8ff. The *Parmenides* is classed as 'elenctic' — together with the *Protagoras* — in Albinus' *Isagoge* III, p. 148 Hermann. Diogenes Laertius III, 50, classifies it as 'logical', together with *Statesman, Cratylus* and *Sophist*.
14. See examples in sources cited in last note. What use is it to be told (Albinus, *ibid*.) that *Euthyphro, Meno, Io* and *Charmides* are 'peirastic' dialogues, whereas *Protagoras* is 'elenctic' — or (Diog. Laert. *ibid*.) that *Laches* and *Lysis* are 'maieutic,' whereas *Protagoras* is 'endictic'?
15. G. M. A. Grube, *Plato's Thought*, (London, 1935), p. viii.
16. Diog. Laert. III, 37; *Anon. Proleg.* 24, p. 218 Hermann = 45, 10—15 Westerink. The author of *Anon. Proleg.* makes Philip of Opus Plato's successor in the school : so much for historical knowledge and interest amongst authors of our compendia.
17. *Anon. Proleg.* 24, p. 217 Hermann = 45, 4—8 Westerink (I give Westerink's translation, p. 44). Their second 'reason' is that in the *Phaedrus*, Plato still employs the 'dithyrambic' style — no more a piece of historical evidence than the first one, although I doubt our Anonymus understood what he was copying here.
18. See my 'Aeschylus and the Third Actor', *Classica et Mediaevalia*, XXX, 1—2 (1969), pp. 56—77. The whole conception of change and development in an author seems to be fairly recent, and is perhaps the result of dates appearing on the various printed works of modern authors. One wonders how those ancient readers, to whom both the lost ('Exoteric') Aristotle and the acroamatic writings were available, bridged the gaps between the two.
19. *Antiochus*, pp. 40—47.
20. To the two sources mentioned in *Antiochus*, pp. 38—9, add Diog. Laert. 51—2 (where the compiler has obviously misunderstood what he was abbreviating) and the references from Proclus, notes 5 and 13 above.
21. Notes 5 and 13 above.
22. Note 13 above and text. Some modern authors would agree with such a 'gymnastic' view of *Parmenides* — see, *e.g., I. M. Crombie, An Examination of Plato's Doctrines*, II (London, 1963), p. 325: 'This [the method of *Parmenides* — J. G.] is to throw at the reader a piece of obviously tangled and fallacious argument, and leave it to him to untangle it'. For what seems to me to be a more realistic view, see Grube (n. 15 above), pp. 34—5.
23. E.g., Albinus, *Isagoge* I—II, pp. 147—8 Hermann; *Anon.Proleg.* XIV—XV, pp. 208—210 Hermann = 27—31 Westerink.
24. Diog. Laert. III, 52; *cf.* Sextus, *P. H.* I, 221.

25. See Alexandre Koyré, *Introduction à la lecture de Platon*, 2ème ed. (Paris, 1962), pp. 17—19.
26. G. C. Field, *The Philosophy of Plato* (Oxford, 1949), p. 126.
27. Note 8 above.
28. Gregory Vlastos, *Plato's Universe* (Seattle, 1975), pp. 95—6. I have not even attempted to cite secondary literature on the *Timaeus*, on Platonic myth or cosmology, or on Plato in general. This would merely double the size of the notes in an article which is not mainly about Plato. The expert will find numerous modern discussions without my assistance. No amount of modern discussion is likely to affect the meaning of things which are clearly and repeatedly said by Plato himself.

    I shall, however, make one exception. E. N. Tigerstedt's *Interpreting Plato* (Stockholm, 1977), is by far the finest and most intelligent book on the whole problem and history of Platonic exegesis that I have seen. Tigerstedt's conclusions (Ch. VII, *The Fair Risk*, pp. 92—108) are not unlike mine. His excellent advice (p. 107), 'It is Plato's voice we should listen to, not our own, disguised as Plato's', should be inscribed on the door of every seminar room where Plato is seriously studied — and, of course, in Greek.

ABRAHAM WASSERSTEIN

HUNCHES THAT DID NOT COME OFF: SOME PROBLEMS IN GREEK SCIENCE

European science has its origin in Greece. Rationalism, openness to new ideas, the willingness to learn from others, freedom from religious constraints and superstitious fears, flexibility of mind, critical attitudes to tradition, — here is a random list of characteristics said to be common to Greek science and to ours. We have much to thank the Greeks for. Still: the inheritance of Greece has also been a heavy burden on European science. For some of its important legacies to posterity had for many centuries a potently inhibitory effect on scientific thought and enquiry. I need mention only the survival throughout antiquity and the middle ages (almost, though not entirely, without opposition) of such powerful models in astronomy as that of the geocentric universe, or of the circular motions of the heavenly bodies; or of teleology in the biological sciences; or of certain Greek attitudes to work and to the translation of science into technology.

It cannot be said that there were no other voices. Though the geocentric universe was the dominant astronomical model down to the Renaissance, we hear now and again of other models, too. For instance, the Pythagorean universe is not geocentric; and Aristarchus was capable of conceiving of a heliocentric system.[1] Circular motion indeed was almost universally accepted as the only possible way in which heavenly bodies could be moving; but even here it was possible as early as the sixth century B.C. for Xenophanes to speak of the sun as moving not in a circle but in a straight line unto infinity.[2] The same Xenophanes is said to have cited fossils in support of what looks like a theory of geological evolution;[3] and a theory of biological evolution is reported to have been taught by his fellow Ionian Anaximander.[4] Thus, teleology did not have it all its own way in biological thought.[5] Even in considering what looks like the most serious lacuna in Greek science, its tendency to be divorced from practical application, we must not forget some steps in the direction of translating theory into practice, such as a good deal of medical procedure, some astronomy, even some engineering.[6]

Hunches will occasionally arise from the sudden determination to free oneself from the burden of received doctrine, a determination exemplified by the unexpected turning away from circular motion in the suggestion of rectilinear movement put forward by Xenophanes; or from an equally sudden determination to reshuffle the data of a problem or a theory and thus to transform the whole picture of the world: an instructive instance is provided by the fifth century atomists' sudden hunch that it might be useful to look upon motion as no less natural a state of matter than rest;[7] and the equally innovative assertion by the same thinkers that there can be, and that there is, a void.[8]

Thus, occasionally, some Greeks had a hunch that things were different from the way they appeared in traditional models. These hunches, however, or most of them, did not lead to further developments or, in any case, they did not cause Greek science to go in a new direction. Why was this so?

A hunch is not quite the same thing as a guess. The latter is based on nothing more than an arbitrary, unreasoning choice between possibilities present to our mind, with nothing at all indicating that one choice is preferable to the others. Such a choice is not only arbitrary and unreasoning; it is also unsupported by that state of mind that we call certitude. Hunches seem to be different: we may have no certitude concerning them — else they would be more than hunches. But we have a feeling, an intuition, a presentiment, that something may be the case, or that something may turn out in a certain way. Though we may have no certitude about the truth of the proposition which we happen to be considering or enunciating, yet, for reasons which we have neither analysed nor argued to ourselves, we seem to feel somehow more strongly (than when we merely guess) that so-and-so may be the case or that such-and-such may be the outcome of a process or an investigation. Hunches may be supported, without our knowing it, by some sort of unconscious or unformulated interpretation of observation, by some sort of unsystematic, almost automatic, ordering of experience. That such hunches often initiate a psychological process leading to scientific hypothesis, to theory, even to demonstration in mathematics, is well known; that they sometimes remain themselves very similar to guesses, namely unreasoning and arbitrary, and, what is more important, that they sometimes fail to lead to the attempt to integrate them into a scientific framework, or to planned and controlled methodical observation and experimentation, is also true. Even so, the significance of such abortive hunches it not negligible. They may have no direct traceable posterity in the history of thought in a particular period, and to that extent they may have been ineffective. Nevertheless, for the historian of scientific ideas they have a certain importance and indeed a certain charm. Not, let me say this most emphatically, because they may be anticipations of later theories or discoveries (atomism for instance) but rather because in spite of their ineffectiveness they help to characterise the world of ideas in which they occur. Further: their very ineffectiveness may raise questions which will help us understand better the scientific world picture in which they did not find a place, in terms both of imagination and of method as well as of the aims of scientific speculation.

Again: hunches may be effective, and sometimes powerfully effective, in scientific development even when they are supported only by mistaken interpretation of experience, or by no more than the arbitrary choice between two equally possible alternatives, unsupported by any observation or experiment. Let us consider two examples. They are very different from each other.

The first is the fifth century hunch that matter may be atomic. This theory commends itself certainly no more and probably rather less to our intuition than its contradictory alternative, namely that matter is infinitely divisible. Indeed it ought to be clear that we are here dealing not with an anticipation of modern physical doctrine but with something typically Greek: you simply take a hunch (*e.g.*, about the nature and divisibility of matter) and see where it leads you. Or, equally typical of ancient science, you accept such a doctrine not because it is intuitively plausible, nor because it is suggested by observation or experiment, nor because it possesses better explanatory powers than its competitors, but because you want it for other reasons. I shall come back to this point later.

Let me repeat here only that it is irrelevant that many centuries later it happens to turn out that atomicity of matter is in fact better supported by what we have learned of nature than infinite divisibility. For it must be clear that in antiquity no one had any serious observational or experimental reasons for believing that matter was atomic, or for believing that atoms moved randomly in the void. (Likening the behaviour of atoms in the void to that of motes in a beam of sunlight is, of course, no more than giving an illustration of what one means; it is not an argument or a demonstration).[9] Indeed it is significant that in the history of scientific thought atomism was not really, until modern times, either safely acceptable or, what is more important, generally, or even largely, accepted. Its fame rests not on its inclusion into the physical world picture of antiquity or the middle ages; on the contrary, it was *not* part of that world picture which from Aristotle onwards (but also, and particularly, in the posterity of Stoic thought) on the whole treated matter, analogously to mathematical magnitudes, as continuously divisible. The fame of atomism and its importance in the history of European thought have little to do with physics: they are due to the circumstance that atomism is part and parcel of an influential and interesting *ethical* system,[10] Epicureanism, whose interest for posterity depends less on its implications for physics than on the mental prurience aroused in intellectual circles by the theological hostility of its Jewish and Christian critics in antiquity, and on the aesthetic attraction of its most influential literary exposition, the *de rerum natura* of Lucretius.[11] Now, was this a hunch that came off? Or not? In any case, let us remember that there was in antiquity no empirical reason whatsoever for preferring atomism to a theory such as that of Anaxagoras or that of the Stoics, who regarded matter as continuous, infinitely divisible.

We now come to our second example: circularity of heavenly motion. This is, surely, as far as intuition is concerned, far more securely founded than atomism, or than 'evolution.' After all, we can see, or we think that we can see, that practically all the bodies in the heavens seem to move around us in circles either diurnal or annual: sun and fixed stars move once every twenty four hours, from east to west, and the sun also once in the course of a year, against the background of the stars, from west to east. Here intuition in the most literal sense of the word is almost coextensive with, and certainly seems to be supported by, observation. And indeed, the circular motions of the heavens seemed to the Greeks to be such impressive evidence both of the regularity of nature and of the aesthetic and mathematical perfection of circular orbits, that out of what began as an empirically based theory there grew the hunch that *all* heavenly motion must be circular; not only the movements that really seemed to be circular, but others, too, and in particular the movements of the planets (including, for our purposes, the sun and the moon).

Thus, for many centuries the Greek astronomers concentrated their best efforts on the attempt to account for the apparently irregular, or apparently non-circular, planetary motions, by the assumption of underlying circular orbits; spheres were built around spheres, epicycles constructed upon epicycles, not only to save the phenomena, but, just as importantly, to preserve the circularity of all heavenly motion. Here at once you see more than a hunch: you see the systematic attempt to 'save the phenomena.' Let me say here, only by the way, that this famous slogan of ancient astronomy does not exhort the philosopher or scientist to devote his efforts to explaining the phenomena by means of *true* assumptions, but rather to fit them, or some of them, into a predetermined theory.[12] What is to be saved, really, by 'explaining' it, by integrating it into a larger framework of natural law, is not the fact presented to us by nature but rather what we might regard as a metaphysical prejudice or a purely speculative hypothesis, even an amusing fiction — these are to be saved by arranging the facts of nature.

Let me explain this a little more clearly: if you look at the sky you see a number of bodies that seem to move in a circle around the earth; the apparent daily westward motion of the sun (due to the earth's rotation) and its annual motion eastwards against the background of the stars at a rate of about 1° per day are both circular. So are the motions, or apparent motions, of the moon and of the sphere of the fixed stars. But among these heavenly bodies there move some whose movement does not so easily seem to fit into the circular scheme of things: the planets appeared to the Greeks to 'wander erratically' about the sky (that is what the Greek word πλάνης or πλανήτης, planet, means). Rising and setting nightly, like the other stars and with them, from east to west, they also like the sun, move against the background of the stars; and what makes their movement so odd is the fact that it does not seem to be either uniform or circular. As the planets circle the sky against the background of the stars they sometimes seem to slow down, to stop, to reverse the direction of their motion, to stop again, and then to reverse their motion again. Now, looking at the path of a planet one does not, necessarily, think at once of circular motion at all; on the contrary, one might easily accept a loop or some complicated knot as an approximately true description of the real path. But not so the Greeks of the classical age. To them all heavenly motion had to be uniform and circular; thus we are told of Plato that he set the following problem for students of astronomy:

By the assumption of what uniform, circular and ordered motions can the apparent motions of the planets be accounted for? (Simplicius, *in Arist. De Caelo*, p. 493. 2—4, Heiberg).

To this question, which became, it would be no exaggeration to say, the fundamental question of ancient astronomy, a number of answers were propounded differing in many important respects from each other, but all sharing one characteristic: they were all faithfully attentive to the exact formulation of Plato's question in that they were all based on the assumption that the motion, or apparent motion, of the planets could and should be described as the resultant of a number of other motions *all of them circular*. Thus we find Eudoxus explaining planetary motions as the resultant of the varying motions of a number of spheres of different sizes, all moving around the same centre, the earth, though on different axes, and with different but uniform speeds, and in different directions.[13]

Or take other answers: such as the attempt to describe the motions of heavenly bodies by means of eccentric circles; the sun was conceived of by Ptolemy as moving in a circle (see figure 1) around the centre C eccentric to Earth E.

It is interesting to note that Ptolemy who knew of the equivalence of this sytem of representation with that of representation by epicyclic motions chose the eccentric system for the theory of the sun because it was the more *economical* one involving only one circle.[14]

In the epicyclic system (see figure 2 below) the planet is conceived as a point moving in a circle (the epicycle) the centre of which is itself moving in another circle (the deferent). This is really not very difficult to visualise: we need think only of the motion of the moon *as seen from the sun* and we have a perfectly good epicyclic system: the moon moves in a circle round a centre (the earth) which itself moves in a circle

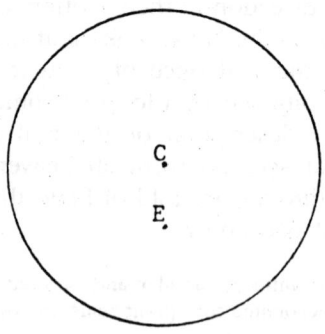

Fig. 1.

around another centre (the sun). It does not matter whether these centres are real or notional.

The epicyclic system (sometimes combined with eccentrics) often involved a great number of complicated adjustments. When necessary, one epicycle was put on another and another one on that; but no matter how complicated the system might become, uniformity and circularity were always 'saved.' The legacy of the Greeks in this respect lasted for a long time: heavenly motion had to be circular practically by definition. It took a Kepler to do away with this.

It is clear from one of the examples I have cited above, Ptolemy's preference for eccentrics over epicycles in his solar theory, that even this great astronomer whose work was destined to be for many centuries the basis not only for astronomical theory but of practical astronomy, too, was capable of determining a question about physical fact by an appeal to a metaphysical prejudice, namely that nature is economical. Still, in spite of that, he at least fits, even here, his theory to the phenomena, that is to say: even though his theory of the sun is arbitrary it fits the solar observations, it does not contradict any of them.

Greek astronomers are sometimes capable of being less empirical than this.

In the *Republic* Plato puts forward a working programme for his ideal astronomer, who must look at the heavens not with his eyes but with his mind; he must look at what is real and invisible and hence

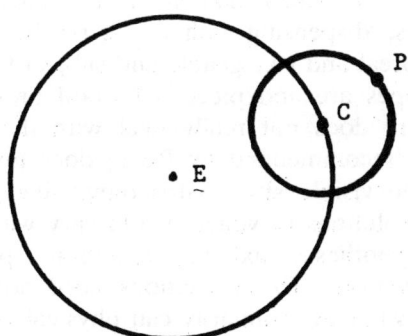

Fig. 2. Epicycles: P is a bodys moving in a circle (the epicycle) around a point C which itself moves on the deferent circle around E.

accessible to thought but not to sense, rather than at that which is perceivable by sense but, being changeable and perishable, is not a proper object of contemplation by the mind, which is concerned with unchangeable stable reality.

We should use the phenomena [he calls them 'broideries in the heavens'] as *illustrations* to facilitate the study which aims at those higher objects, just as we might employ ... diagrams in geometry; anyone acquainted with geometry who saw such diagrams ... would regard it as ridiculous to study *them* seriously in the hope of gathering from them the truth about equality, doubleness and so on. So also the true astronomer will have the same feeling when he looks at the movement of the stars; he will regard as absurd the man who supposes these things which are corporeal and visible to be changeless and subject to no aberrations of any kind; he will hold it absurd to exhaust every possible effort to apprehend their true condition. ... Hence we shall pursue astronomy as we do geometry, by means of problems, and we shall dispense with the starry heavens if we propose to obtain a real knowledge of astronomy.[15]

True, the abstract 'astronomy' recommended by Plato has not much in common with that practised by Ptolemy. But it has its place in our discussion. For Plato's programme as proposed here exhibits what seems to me a characteristic of much of Greek astronomy: namely that astronomy is a mathematical science in a peculiar sense of that term. The Platonic astronomer, like the geometer, is not really concerned with physical reality in the rough; like the geometer he deals with problems the constituent elements of which are 'purer' than, and ultimately independent of, what goes on 'up there.' This astronomer studies mathematical problems, and thus the science is treated as if it were a collection of purely mathematical problems, unconnected with the appearances, 'dispensing with the starry heavens.' For the visible heaven is corporeal and changeable and subject to aberrations of every kind just as ropes are and pieces of wood or stone, and fields and rivers. Geometry does not really deal with the latter; and similarly 'astronomy' (as recommended by Plato) does not deal with what our eyes see on the visible sky. 'Astronomy,' like geometry, deals with 'problems' the solutions of which would vary with the variation in the assumptions, hypotheses, axioms, definitions, postulates, data, phenomena, observations. These solutions need not have any necessary relationship to what we ordinarily call physical reality, or indeed even to observed phenomena. There are documented cases which illustrate this attitude.

In Simplicius' commentary on Aristotle's *de caelo* we read of a criticism directed by Sosigenes[16] at Eudoxus' attempt to give an answer to the problem proposed by Plato, namely to account for the apparently non-circular planetary motions on the assumption of uniform, circular and ordered motions:

> The theories of Eudoxus do not really save the phenomena; not only do they fail to save the phenomena that were observed at a later date; no, they do not even save the phenomena that were known before, that were known and accepted by Eudoxus himself.[17]

Sosigenes criticises Eudoxus not for mistakes due to ignorance of facts, but for ignoring facts known to him; he criticises him for not taking account, in building his theory, of phenomena and observations known to him and indeed already to his predecessors.

From the point of view of the empirical astronomer there is a fatal flaw in Eudoxus' theory of concentric spheres. In such a system the planets are represented as points moving on spheres revolving around a common centre but on different axes. By judicious variation of speeds, radii and direction of motion any, even the most complicated, planetary orbit could be represented as the resultant of a number of circular motions. But in such a system the distances of the planets (points on spheres) from the common centre are invariable. Now the planets are observed to be brighter at some times than at others. The sun and the moon seem to vary in size (in their angular diameters). How is one to explain this? Eudoxus does not only fail to help us here: If we explain, as we feel bound to do, varying brightness of the planets and variations in the apparent size of sun and moon as a function of varying distance, then we must accept that these phenomena are not only left unexplained, but that they contradict Eudoxus' theory. Now this in itself would not be important. What is truly remarkable is that Eudoxus knew of fatal objections to the theory of concentric spheres and yet did not seem to be worried by them.

What this means is clear: The 'solution' to Plato's problem proposed by Eudoxus was not concerned with presenting a 'true' account of planetary motion. It was an exercise in spherical geometry. What Eudoxus was offering was a geometrical construction accounting for *an arbitrarily selected set* of phenomena. This, of course, implies that another selection (or the consideration of data not hitherto taken into account) would impose another solution. Eudoxus was not describing the world as it is, or as we see it, but illustrating a method of solving certain kinds of geometrical problems. The problem and its solution could not be claimed to be 'true' in the sense of telling us what really happens in the world. The theory could not make such a claim precisely because it did not pretend to take into account (and that means: to account for) *all* phenomena. A mathematical problem, unrelated in detail to physical reality, is presented to the mathematician; the solution that he offers is related only to the terms of the problem as presented to him. Not all the phenomena were accounted for, because Eudoxus was not dealing with a real problem but with a purely geometrical exercise.[18]

All this illustrates an important fact about Greek science: its aim was not always, as starry-eyed admirers of the Greeks sometimes assert, the acquisition of knowledge, knowege for its own sake. On the contrary, there are cases, as we have just seen, in which the Greek scientist builds a theory, not in the hope that it will correspond to physical fact, but simply as an exercise in mathematical problem solving — almost a game. The Greek scientist will say: let us take this hunch and think out all the logical consequences of positing it (or its contradiction); the resulting structure is often all that interests him — independently of its agreement with the deliverances of experience. The modern scientist means to do something quite different: he follows his hunch to see what happens if he tests it through experiment in the laboratory or by observation of natural events; and he is as ready to abandon his theory, if he has succeeded in falsifying it, as to uphold it when he has established — at least *pro tempore* — its tenability. The Greek scientist is not always interested in the truth value of his theory but rather in its internal coherence even when it can be shown by appeal to further data, extrinsic to the problem as it was presented to him and which he has therefore not considered, that his theory is not 'true'. Hence the ineffectiveness of some, indeed many, Greek hunches.

Unlike the linear motion of the sun to infinity adumbrated by Xenophanes, circular motion of the heavenly bodies fits into an elegant picture of the universe and lends itself to playing about with interesting and complicated mathematical constructions. It also, and not just incidentally, fits an aesthetic view of the world that is metaphysically determined.

Here are two further examples of hunches that *did* come off.
  1. How many prime numbers are there? Certainly very many, for no

HUNCHES THAT DID NOT COME OFF   29

matter how far we go we shall still find prime numbers along the way. But there is nothing that indicates how many there are. Intuitively, it might seem entirely plausible that, as we continue counting, they become fewer and fewer, ever more distantly located from each other, and that at some point, no matter how far away, we reach the greatest, i.e., the last, prime number; there simply are no more primes to be found along the way, or, in other words: the number of primes is finite. But intuition here does not help; indeed it may be torn in different directions and may not be averse from conceiving of the possibility that the number of primes is infinite, just like the number of natural numbers. How are we to decide? Is the number of primes finite? Or is it infinite? There simply was no direct proof for either proposition. Now, in the absence of proof, someone has a hunch: there is an infinity of primes. Since he cannot directly prove that to be true, he asks himself: what would happen it it were not true? What happens if we assume that the number of primes is finite? Or, in other words, if we assume that there is a prime number which is the greatest of all prime numbers? On this assumption we can do the following: Take the product of all primes $p_1, p_2, p_3, p_4, p_5 \ldots p_n$. Call the product, say, m. Consider the number $m + 1$. It is not divisible by p_1 or by p_2, or by $p_3 \ldots$, or by p_n; such division would always leave a remainder 1. If divisible at all, it must be divisible by a prime greater than p_n, contradicting the assumption that p_n is the greatest prime. Or it is not divisible at all: again contradicting the same assumption, for it would mean that it is itself a prime. Since we can continue that process without end, we have shown that there is no greatest prime, that the number of primes is infinite.[19]

But we have not proved this directly. We have used (perhaps invented?) a logically suspect kind of argument, namely *reductio ad absurdum*.

2. The irrationality of square roots of non-square integers. Here is a simplified proof of the irrationality of the square root of 2, known in the fifth century and extended not much later to the square roots of all other non-square integers; it was generalised.[20] For the sake of simplicity I give the proof here in terms of arithmetic, though it is most important to remember that the proof, as alluded to by Aristotle in the *Anal. prior.* I. 23, 41a 23-27, is geometrical, dealing not with irrational numbers but with incommensurable magnitudes. I shall also confine myself here to the case of $\sqrt{2}$.

What is the ratio of the diagonal of the square to its side? Clearly, it ought to be of the form m/n, the terms m and n having no common factor. Then, if m/n = $\sqrt{2}$, it follows that $m^2/n^2 = 2$; and thus $m^2 = 2n^2$. Hence: m^2 is even. Similarly, if m^2 is even, it follows that m is even, that is to say, it is exactly twice as large as some other number. Call that other number p. Then m = 2p; and $m^2 = 4p^2$. But $m^2 = 2n^2$. Thus $2n^2 = 4p^2$. Hence: $n^2 = 2p^2$. Thus n^2 is even, and, of course, n is even, too. But we started from the assumption that the ratio m/n subsisted between terms which have no factor in common (if they had, we could have reduced the ratio to its lowest terms and started from there). We now find that from our initial assumptions (which included the one concerning the fact that the terms of the ratio have no factors in common) it follows that the ratio of the diagonal of the square to its side, if expressed in numbers, would be, would have to be, expressed in terms both of which are even, thus contradicting one of the premises of the argument; if m/n is in its lowest terms then either m or n or both must be odd; if it turns out that both are shown to be even then we are faced by a situation which as Aristotle (*cf.* also 50^{a}37) puts it in alluding to this proof, would oblige us to accept the conclusion that an odd number can also be even. We note, though only by the way, that we have again used a *reductio ad absurdum* proof.

Both our hunches have come off successfully. They have strengthened our grasp of the field of numbers, extending it at the same time. But let us see what they have done to us on the way. We have had to acquiesce in a method of proof which seems to lack in rigour, *reductio ad absurdum*, because we had no direct proof. And secondly we may feel a little uncomfortable about jumping the gun: for though we seem to have extended the field of numbers to include irrationals, we have done so — if at all — before we have become acquainted with rational fractions, any fractions. What we have really done, however, is *not* to prove the existence of irrational numbers but of incommensurable geometrical magnitudes.[21] Our hunches have indeed paid off; but we have paid a price, a double price: 1. Lack of rigour. 2. By not proceeding step by step we have missed the opportunity of really extending the field of numbers. For it is important to remember this: Greek arithmetic did not really go beyond the natural numbers. The Greeks did not extend the field of numbers to include negative numbers, or zero; indeed they were not clear that 1 or even 2 were numbers;[22] they did not think of fractions as numbers; they could conceive of them only as ratios between integers, and thus, when they came to think of what to us look like irrationals they formulated their thought in terms of incommensurable *geometrical magnitudes* rather than in terms of irrational *numbers*; hence the overly geometrical character of their mathematics, hence also their failure to extend the field of numbers in an orderly systematic way.

We have seen that the Greeks had a price to pay sometimes for listening to their hunches, and at other times for neglecting them. Incidentally, it must not be forgotten that such successes as they had in the field of the theory of numbers may well be due to the fact that the problems that they confronted there were mathematical; the hunches they neglected, or in the case of which they were less successful, were physical (in the widest sense of the word, including, of course, both biology and astronomy).

In a way, it is true to say that the Greeks bequeathed science to us. Yet, ultimately, their science was different from ours; and not only because it was less successful — though this factor, too, requires explanation.

By and large, with some exceptions, Greek scientists often aimed neither at understanding the world nor at changing it. Their scientific theories were not, like those of our scientists, always meant to explain what goes on in the universe, to enable us to predict events and thus to control them or their effects. They reached their theories not by patient observation and experiment but by listening to hunches (which they often left untested); or by 'Gedankenspiele' including mental experiments; or by arbitrary postulation or acceptance of axioms; or by giving answers to questions that were asked only in order to elicit predetermined answers.

Greek thinkers are much given to excessive and premature generalisation: 'All is water', 'All is air', 'All is fire', 'All is number', 'All is one' may be a formula which satisfies our aesthetic longing for all-embracing order; it prevents us from seeing and trying to explain the infinite variety and not infrequent disorderliness of nature. When monism is abandoned by Greek thinkers it is replaced by an equally tidy (and equally ineffective) pluralism which imposes on the world a pattern of, e.g., the four elements in cosmology and physics, or the humoral theory in medicine. Even atomistic pluralism is fundamentally the fruit of the same kind of mentality: there are many atoms, but each one of them is really like the single One.[24]

Greek scientists were rationalists. They are rightly praised for framing their descriptions of the world 'without bringing in Marduk'.[25] They describe, explain and try to heal our illnesses not by incubation, incantation, prayer but in rationally thought-out ways, such as reference to climatic and ecological conditions, diet, exercise. There is no miracle-mongering, no superstition, no prophet and no priest. Greek science, indeed Greek civilisation as a whole, existed for many centuries without sacred scriptures, without being dominated by institutionalised religion.

But superstition is not the only danger that threatens science. That very rationalism which is, rightly, regarded as the glory of Greek philosophy and science may represent no less insidious a threat — partly because it is sometimes confused with the scientific attitude.

Rationalism is not identical with science. Rational procedure is not necessarily the same as scientific procedure. Rationalism does not guarantee the scientific character of a description of the world, or of a theory in astronomy or medicine. An excess of rationalism can be as fatal to the proper development of science as magic or superstition. Greek rationalism often despises, or at least neglects, the inductive procedures of natural science, the worry about particular facts. This neglect of induction is accompanied by its obverse: an obsessive preference for deductive methods modelled on the mathematical sciences, in which our starting points are fundamental general principles from which we rigorously deduce particular detailed theorems; in modern natural science we start from particular facts and go back to general principles. It is, of course, true that there is always a danger in scientific work that one may be dragged along by theory further than one would wish to go. But there are precautions one can take, and indeed science is powerfully protected against the danger of jumping to hasty conclusions by constant recourse to observation and experiment and stubborn insistence on the fundamental method of modern science: systematic confrontation of theory with fact, testing theory by fact; willingness to abandon theory if the facts so demand; unwillingness to disregard facts if they do not fit the theory. This attitude has been so fundamental to all scientific work since the Renaissance that it is sometimes difficult to conceive of any scientific work that obeys different rules. Modern science has very largely freed itself not only from *a priori* presuppositions of a religious nature but also from those offered by philosophy. If there are philosophical problems connected with modern science, it is the latter which poses them to the philosopher, and not the philosopher to the scientist. This is not so in Greek antiquity. The philosophers presented the programme, the model, the problem to the scientist. Often a fundamental scientific theory appeared not as the result of minute examination of facts, observations, experiments but as an *a priori* doctrine furnished to the scientist by the philosopher. This can be illustrated not only from the mathematical science of astronomy, as we have seen above, but also from medicine, and not only in isolated cases, but throughout many centuries of the history of scientific medicine. We need mention here only in passing the influence discernible in various parts of the Hippocratic corpus of doctrines of one philosophical school or another, of Anaxagoras, for instance, or of Heraclitus, or of Empedocles. More interesting are the warnings repeated again and again in medical writings concerning the use of deductive methods in medical thought and practice. These warnings are found as early as the fifth century B.C. and as late as the second century A.D. Here is a passage from the Hippocratic treatise *de vetere medicina* (chapter I):

> All who, on attempting to speak or to write on medicine, have assumed for themselves a postulate as a basis for their discussion — heat, cold, moisture, dryness or anything else that they may fancy, — who narrow down the causal principle of diseases and of death among men, and make it the same in all cases, postulating one thing or two, all these obviously blunder in many points even of their statements, but they are most open to censure because they blunder in what is an art, and one which all men use on the most important occasions, and give the greatest honours to the good craftsmen and practitioners in it. Some practitioners are poor, others very excellent. This would not be the case if an art of medicine did not exist at all, and had not been the subject of any research and discovery, but all would be equally inexperienced and unlearned therein, and the treatment of the sick would be in all respects haphazard. But it is not so; just as in all other arts the workers vary much in skill and in knowledge, so also is it in the case of medicine. Wherefore I have deemed that it has no need of an empty postulate, as do insoluble mysteries, about which any exponent must use a postulate, for example, things in the sky or below the earth. If a man were to learn and declare the state of these, neither to the speaker himself nor to his audience would it be clear whether his statements were true or not. For there is no test the application of which would give certainty.[26]

There is another passage from an obscure author writing six or seven centuries later (Cassius Iatrosophistes, in Ideler, *Physici et Medici Graeci Minores I*, p. 144) in which the followers of Herophilus are criticised for using what the author calls 'geometrical methods of demonstration', that is to say, deductive reasoning, in their medical work (γεωμετρικῇ χρώμενοι ἀποδείξει). Justified or not, this criticism illustrates the extraordinary persistence, even in medicine, of *a priori* deductive reasoning. What is still more interesting is this: the criticism is directed against the followers of Herophilus. He lived in the last third of the fourth century and at the beginning of the third. He was reputed to be one of the greatest doctors of antiquity; he was compared to Hippocrates and (later) to Galen. He is best known for having turned anatomy into a scientific discipline. He was apparently the first doctor to practise systematically dissection of human bodies. Galen (who, incidentally, calls him διαλεκτικός: X. 28 Kühn, repr. Hildesheim 1964—65) says about him that he was more interested in observation and experiment than in logical method (IX. 278, Kühn). He is said to have substituted empirical method for dogmatism (Gossen, *s.v.* Herophilos, *Paulys Real-Encyclopädie der classischen Altertumswissenschaft*, Neue Bearbeitung, vol. 15 (1912), pp. 1104—1110). His work gave the impulse to the growth of the empiricist school. And yet, here we find him (or his school) accused, rightly or wrongly, of using 'geometrical', i.e., deductive, methods of demonstration. After centuries of empiricist propaganda it seems still to have been conceivable that medical writers could use such methods — otherwise there would be no point to Cassius' criticism. Similar criticisms are found in the works of Galen.

Medicine is not, of course, the only field in which the dangerous effects of deductivism can be observed. Let us look at an amusing example in one of Aristotle's arguments on the borderland between physics and astronomy.

That the earth is spherical is not immediately obvious to the ordinary man, and Aristotle rightly thinks that this proposition stands in need of demonstration. Such a demonstration composed of a number of arguments is offered in the second book of *de caelo*. Having established that bodies naturally move towards the centre of the universe and that the earth is situated at that centre, and that it is immobile, he now offers arguments to show that it is spherical. Among these there is the following:

> Here is another argument for the sphericity of the earth: all falling bodies hit the ground at right angles; yet the straight lines of their fall are not parallel to each other. This can be accounted for only on the assumption that the earth is spherical.[27]

This argument has all the appearances of being an inductive argument. It appeals to fact; it appears to reason backward from an observation, from a particular, known, fact, to a general hypothesis that explains it, from which, had we but known it, we could have deduced the fact. It is the apparently known fact or observation of the non-parallelity of the lines of fall that are yet perpendicular to the plane of impact that establishes the hitherto unproven hypothesis that the earth is spherical.[28]

But here we are faced with a difficulty: are we really reasoning from a known fact or observation? Observation certainly suggests that the lines of fall are perpendicular to the plane of impact. But does observation also suggest that these lines are not parallel? Ordinary everyday observation suggests nothing like this. On the contrary: common observation, rightly or wrongly, suggests the exact opposite, namely that the lines of fall of sufficiently[29] heavy bodies *are* parallel and *not* at an angle to each other. Nor can one easily think of a controlled experiment that Aristotle or anybody else could have performed in order to measure or even establish the deviation from parallelity of the lines of fall.

How then did Aristotle know that these lines are not parallel? The question need only be asked for the answer to be clear. He knew it because he *deduced* it from the theory that the earth is spherical. This *petitio principii* does not stand alone in the works of Aristotle. I cite it here not in order to show how even Aristotle could go wrong (it is hardly necessary to add that he also knows some valid arguments for the sphericity of the earth) and be liable to logical error; but rather because the error is instructive. It illustrates what is a typical feature not only of Greek mathematics but of Greek physics, too: the deductive character of Greek scientific thought. Aristotle imagines that he is working upwards from facts to theories; but he slips (here) into an argument that works the other way: downwards from the theory to the particular fact. That he does so unwittingly and mistakenly strengthens our suspicion that he does so characteristically. He had a hunch that the earth is spherical; the hunch happens to be true. More: Aristotle knew good arguments in favour of it, e.g. that the shadow of the earth on the moon during a lunar eclipse always has a curved outline, a fact that fits in well (even though not decisively) with the assumption that the body casting the shadow is spherical. But when he gives us *this* 'proof' (i.e., that based on the non-parallelity of the lines of fall), mistakenly and wrongly, he illustrates both the tendency to deduction and the readiness of Greek scientists to be content with *Gedankenexperimente* instead of laboratory work, with *Gedankenspiele* instead of scientific theory solidly based on research, observation, experiment.

The well known fact that Aristotle (like many Greek doctors) often argues for, and relies upon, empirical work and argument must not obscure such other non-empiricist elements in Aristotle and in the Hippocratic corpus. Much of Aristotle's science is based on metaphysical presuppositions, not only in physics and in astronomy, but also in biology. And we must not imagine the writers of the works collected in the Hippocratic corpus to be consistently paragons of empiricist virtue. The author of *de vetere medicina*, whose warning against relying on philosophical postulates in medicine has been quoted above, develops in the same treatise the humoral theory.

It is true that in the above example Aristotle is slipping from intended induction into unconscious deduction. Still, the hunch could in any case not easily have been checked by observation or experiment. But there are other alleged empirical facts, reported by Aristotle as if they were well-known, which he could easily have tested. But he did not. Thus, he thinks that it is an empirically established fact that a cask which holds a certain quantity of wine not in skins will hold the same quantity in skins; or that a vessel filled with ashes will hold as much water as the same vessel when emptied of ashes (*Physics*, 213 b 16—22). It does not matter in these cases whether the attempted explanations of these alleged facts are interesting and rational or not; what is important is that the alleged facts are simply accepted as empirical facts, which, of course, they are not. This cavalier attitude to empirically verifiable fact, in the case of a thinker so conscious of the need for inductive method in natural science as Aristotle was, is to my mind no less significant than what is, no doubt, as a matter of methodological principle, of greater import, namely the tendency to prefer deduction to induction.

The scientist asks questions because he has a problem: he may be puzzled by a phenomenon of nature or he may be faced with a practical task which scientific investigation may give him the knowledge, and thus the means, to perform. He may or may not find the answer he seeks. In either case: the question comes first, not only chronologically but logically. Not so in some Greek scientific work. Here is an example from the history of ancient atomism. It illustrates my point particularly well since we can observe in it the line of development between Democritean atomism and that of Epicurus. Atomist doctrine in reference to the motion of atoms was altered for reasons which a modern scientist would find unintelligible, or at least unacceptable.

We have seen that Greek scientists sometimes adopted scientific theories for extra-scientific reasons; and that they adopted them sometimes even when the phenomena contradicted them. Some ancient thinkers went even further: they were arbitrary not only in the choice of solutions but in the *deliberate creation* of problems to fit in with pre-existing solutions, the assertion of which seemed desirable to them for reasons which have nothing to do with science or observation. A famous case is that of Epicurus, in whose physical theory the so-called clinamen (παρέγχλισις = swerve) plays an important part.[30] Epicurus had inherited from his predecessors in the classical age, from Democritus in particular, a physical theory which he felt bound to modify in some essential points. For the earlier atomists, and especially for Democritus, atoms move at random in all directions in empty space. Their motion is uncaused and need not be accounted for, any more than rest seems to be in need of explanation in other systems. Since the atoms move randomly some of them will randomly collide. It is worth repeating that like motion itself these collisions are uncaused. The collisions lead to combinations of atoms into larger agglomerations. Thus objects and worlds come into being. This inherited scheme is fundamentally changed by Epicurus. His atoms move, not at random, in all directions, but perpendicularly downwards, in parallel lines, at equal speeds. It follows from this that Epicurus finds himself in a difficulty: if such indeed is the proper description of atomic movement, then it does not account for collisions and hence not for combinations of atoms and the coming into being of this world. Epicurus proposes a solution to the problem. True, atomic motion normally is perpendicular, downward, in parallel lines. But occasionally, without any cause, an atom will deviate slightly from its path and, swerving aside, will cross the path of other atoms. Hence collisions and further collisions (like those of billiard balls) and combinations and the coming into being of objects and of worlds. Epicurus was criticised in antiquity[31] for his arbitrariness in postulating an *uncaused* swerve. But what interests us here is another question. Why did Epicurus need this arbitrary solution? He needed it only because he had changed the classical Democritean theory concerning atomic motion. Atoms moving at random will collide at random; their collisions need not be explained. Epicurean atoms do not move randomly but in such a way (perpendicularly downwards, at equal speeds, in parallel lines) as to make it impossible for them to meet except if a new mechanism, namely the swerve, is introduced. But why make them move in such a way in the first place? There was no obvious need for Epicurus to change the classical atomist theory of motion. There were no observations and no logical considerations that necessitated such a change. It is sometimes said that the attribution of weight (= perpendicular downward movement) is due to Aristotelian influence in Epicurean physics. This is not a satisfactory explanation. Epicurus rejected much in Aristotelian physical doctrine, and it is difficult to see why he should choose for adoption precisely that part of it which makes it necessary for him to adopt so far-reaching and 'unscientific' a modification of Democritean atomism as the uncaused swerve. In any case, the argument alleging Aristotelian influence does not address itself to what is the real difficulty: even if we granted that Epicurus had some good Aristotelian reason for postulating weight (i.e., downward motion) of atoms, we would still have to ask ourselves why he added the further postulate of 'equal speed'. It is the addition of this latter postulate which really creates the difficulty. For without it even atoms moving downwards in parallel lines could have been represented as colliding, combining and thus conglomerating into objects and worlds, without the unscientific monstrosity of the uncaused swerve. 'Equal speed' is not an inheritance from Aristotelian physics. On the contrary, Aristotelian matter falls (or rises) through any one medium (other things being equal) with a velocity proportional to its weight (or lightness). Indeed, no matter how wrong this doctrine may be, it is at any rate suggested by ordinary, untested, experience. There was no reason of experience or of traditional doctrine for Epicurus to postulate 'equal speed' for his atoms moving perpendicularly downwards in parallel lines. Yet by introducing this postulate Epicurus creates the need for the uncaused swerve, *which he would not otherwise need*. The fact is, the swerve is not invented in order to meet a pre-existing difficulty. On the contrary, the difficulty is constructed in order to necessitate the introduction of the swerve. One of the ancient critics of Epicurean physical theory notices this:

Epicurus thinks that the inevitability of fate is avoided by the swerve of the atom ... Epicurus introduced this doctrine because he feared lest, if natural gravity by necessity always carried the atom (downwards), we would have no freedom, since the motion of the mind would then be controlled by the motion of the atoms.[32]

The swerve, precisely because it is uncaused, becomes the physical analogue, illustration, the necessary and sufficient condition, of free will. This is what Epicurus really wants to establish: freedom of will, freedom from the bonds of fate and determinism, from the constraints of divine governance. The swerve must be understood not as fulfilling a function within the framework of scientific explanation but rather as being the translation of an *ethical* postulate into an expression formulated in terms of physics. That means that what we have here is not merely an arbitrary answer to an allegedly existing physical problem ('weight + equal speed = non-collision of atoms. But atomic collisions are needed to account for atomic combinations and for the coming into being of worlds. How are we to account for the collisions?'); but rather: Epicurus creates an arbitrary, unnecessary problem *in order to need* a pre-existing answer. He wishes to be able to preach the doctrine of free will (and its physical analogue, the uncaused swerve of the atoms). He therefore builds the rest of his physical theory, particularly his theory of atomic motion, in such a way as to need the swerve as an element in the system. The motion of atoms in the Epicurean world is constructed precisely in such a way as to make it impossible for them to meet — this is the purpose of giving them the sort of motion they have. That is why Epicurus chooses weight, and that is why he adds equal speed to it. These, chosen arbitrarily and without any scientific need or traditional sanction, in combination, create a problem. It is not right to say 'Epicurus grasped the important fact that differences of weight make no difference to the velocity of bodies falling in a vacuum'.[33] Epicurus grasped no such *fact*. What he did grasp was something quite different, namely that he needed such an arbitrary postulate to create the need for the *clinamen* (the swerve). Neither Epicurus nor anyone else in antiquity had any serious reason for believing that differences in weight make no difference to the velocity of bodies falling in a vacuum.

It is not the system that determines the invention of the *clinamen*, but the wish for the *clinamen* which determines the structure of the system.

Questions are asked not because one wants to know the answer but because one needs a question (a difficulty) in order to be able to use a predetermined answer, an answer which one wishes to give, not for empirical reasons but for some outside purpose, which has nothing to do with physical truth or experience.

In some sense it would be true to say that all Greek science was a hunch that did not come off. There are a good many reasons for this; some of them have been discussed in this paper.

That science was not translated into technological application must be due to a complex of various causes: social, political, cultural. Slavery probably had something to do with it, though perhaps less than is sometimes claimed. Purely technological factors must be taken into account too; the man who designs a steam engine and then wants to build one that really works needs metallurgical techniques which may well not be developed yet; he needs valves which may not exist yet; he needs a comparatively cheap and reliable supply of raw materials and fuel, which again depends on economical transport facilities; and there may be many more factors which would make the technological application of the steam engine difficult or impossible. There probably were many other social, political, cultural, technical, economic difficulties to be overcome. But it seems clear that, over and above all practical problems, there was one that seems to have been decisive: Greek science was different from ours. It is in this difference that we must seek the principal reason for the failure of Greek science, both in what we consider the task of science, namely to make the world more intelligible through an open-ended quest for knowledge about it, and in what in modern times has rightly been thought of as the inseparable companion of the quest for understanding: the opportunity given to mankind, through using science for technology, not only to understand but also to control the world

Greek science was not serious. It was flawed in more than one respect. It lacked both the practical orientation of our science, and also the perhaps naive attitude of the modern scientist who sees his task in looking for real physical truth. Greek attitudes here were inimical to real science. It is true that not all scientists held the same view of their work. It is also true that we can discern here and there signs that other attitudes were also present, at least in germ. But even so; the tendency to believe that the world can be described without recourse to the supernatural is accompanied by an overdeveloped urge to see the world as aesthetically satisfying; by a tendency to rash generalisation; by the inability (one can only call it that) to found science on the firm basis of experience. Hence the comparative lack of experimentation, hence the inability to absorb the argument for induction. That this is connected with the almost pathological, morbid, distrust of the senses propagated by some Greek philosophers is not an accident, even though the connection need not be one of cause and effect. Greek scientists thought they could do without supernatural explanation because they had faith in reason; but because they had faith in reason they sometimes thought they could also do without facts. They were not always, or typically, serious about science: they were not really always after physical truth; science was often a game to them, mental gymnastics or theatre. That they allowed metaphysical presuppositions to an undue degree to govern their thinking about the world favoured their tendency to deductivism; the same metaphysical bent was responsible for their tendency to reify constructs of the mind, e.g. to make substance of number; and for the temptation to confuse substance with value.

The Greek tendency towards deduction in preference to induction explains their comparatively greater success in mathematics than in the natural sciences. It also explains why even in natural sciences they often seem to have been content with solving abstract problems rather than explaining the world. Even where we find experiment it is often 'Gedankenexperiment'; even where we find explicit reference to what looks at first sight like experiment and observation, it is to isolated observation rather than to controlled, repeated and repeatable experiment; it is used for illustration rather than for demonstration; and most importantly (in this field) the Greek scientists failed to bring about that marriage between physics and mathematics that since the Renaissance has been so fruitful in scientific work and that expresses itself very simply in the enunciation of physical laws by means of mathematical formulae. The classical Greeks occasionally measured things; but not many things; and hardly ever in order to formulate quantitative uniformities. In physics I know of only one single law formulated in antiquity before the Hellenistic age in precise mathematical terms: the ratios descriptive of the intervals of the musical scale.

Greek science in all these respects was different from ours. It is a paradox of history that in one way or another, though so unsuccessful itself, it became the progenitor, or one of the progenitors, of modern science.

Hebrew University
Jerusalem, Israel

NOTES

1. It is to be noted that in spite of abandoning geocentricity the Pythagoreans and

others, *e.g.* Aristarchus, held fast to the circular motion of all heavenly bodies. The same is, of course, true of Copernicus.

2. This information is contained in a passage (Aetius II. 24,9 = DK 21 A 41a) in which we are told that according to Xenophanes
   (a) there are many suns and many moons corresponding to the different regions of the earth;
   (b) the sun being temporarily located over a region of the earth not inhabited by us produces an eclipse; and
   (c) 'the same man says' that the sun progresses toward infinity, and that the appearance of the circularity of solar motion is due to the sun's distance.
   It seems clear to me that in (c) Aetius relies on a source different from that to which he owed (a) and (b); and that there is no need to attempt to harmonize the statement concerning linear solar motion to infinity with the doctrine reported by Hippolytus (*Ref.* I. 14,3) that a new sun comes into being each day through the conglomeration of little sparks of fire. Further, there is nothing in Aetius' text to justify a paraphrase of it (G. S. Kirk and J. E. Raven, *The Presocratic Philosophers* (Cambridge, 1962), p. 175) as claiming an indefinite westward movement of the sun. Indeed the whole point of that report is that the sun is not circling the earth but moving away from it. Nor do I see any reason to translate εἰς ἄπειρον = indefinitely (Kirk and Raven, *loc. cit.*; W. K. C. Guthrie, *A History of Greek Philosophy, I* (Cambridge, 1962), p. 393). It is true that our text does not literally and explicitly assert rectilinear movement; but in the context of contrasting the sun's real movement with its apparently circular motion this interpretation imposes itself. See Guthrie, *op. cit.*, p. 394.

3. DK 21 A 33.

4. DK 12A 10, 11 and 30 = [Plut.] *Strom.* 2; Hippol. *Ref.* I. 6,6; Censorinus IV. 7; Plut. *Symp.* VIII. 8,4 p. 730E.

5. The evolutionary theory of Empedocles even includes a doctrine reminiscent of part of Darwinian theory. In its ascription of evolutionary change to a mechanism which relies on accidental mutation and selection, this latter theory shows that accident and selection can be pointed to as performing the same function in the generation of biological features that have greater survival value than others, as that which design has in other explanatory models, and that they would lead to the same result: 'The same is to be said of other organs in [the creation of which] a purpose *seems* to have been at work. In those cases where everything turned out in such a way as it would have done if it had been designed for a purpose, the resultant creatures, being formed through a combination that chanced to be fitting, survived; those creatures that were not [thus fittingly provided for by chance with features that made for survival] perished and still perish, as Empedocles says' (Aristotle, *Physics* II. 8, 198 b 27—32). It is to be noted that we owe our knowledge of this 'Darwinian' argument of Empedocles to Aristotle's discussion of final causes.

6. We need mention here no more than the mechanical achievements of Archimedes and of Heron of Alexandria; the use of astronomical observation for navigation and calendar making; the empirical attitude of some of the writers of the Hippocratic corpus; though in many cases there is a good deal of *a priori* thinking, too.

7. See Guthrie, *op. cit., II* (Cambridge, 1965), pp. 398f.
8. DK 67 A 7.
9. See Aristotle, *de anima* I.2, 403 b 31ff.; Lucretius II. 112ff.; Lactantius, *de ira dei*, 10, 9. Note that Lucretius uses such expressions as *simulacrum* and *imago* (*loc. cit.*). In any case, a good deal of what some modern historians of ancient science describe as remarkable anticipations of modern physical theories is really nothing of the sort. It would be well if historians kept in mind the salutary warning of a recent biographer of Einstein: '. . . the most difficult task in studying past science is to forget temporarily what came afterwards' (A. Pais, *Subtle is the Lord, The Science and the Life of Albert Einstein* (Oxford and New York, 1982), p. 9).
10. See below and also A. Wasserstein, 'Epicurean Science' in *Hermes, 106* (1978), pp. 484–494.
11. The truly remarkable capacity of Epicurean physics to fit into moral theory is paralleled by that of Stoic physics.
12. See A. Wasserstein, 'Greek Scientific Thought' in *Proceedings of the Cambridge Philological Society*, N.S. *8* (1962), pp. 51ff.
13. Simplicius, *in Arist. de caelo*, 492, 31ff.; G. Schiaparelli, 'Le sfere omocentriche di Eudosso, di Callippo e di Aristotele', in *Pubblicazioni del R. Osservatorio di Brera in Milano, IX* (1875); J. L. E. Dreyer, *History of the Planetary Systems from Thales to Kepler* (Cambridge, 1906), pp. 89ff.; O. Neugebauer, *The Exact Sciences in Antiquity* (New York, 1962), pp. 153f; D. R. Dicks, *Early Greek Astronomy to Aristotle* (London, 1970), pp. 176ff. (with more modern literature). The clearest exposition for the non-specialist reader (and the most easily accessible) is still that by Sir Thomas Heath in *Aristarchus of Samos* (Oxford, 1913), (reprinted 1959), pp. 193ff. (based on Schiaparelli and Dreyer).
14. Ptolemy, *Almagest* III. 4, Heiberg I. p. 232.
15. *Republic*, VII. 529–530.
16. A second century A.D. peripatetic philosopher, teacher of Alexander of Aphrodisias; not to be confused with the astronomer associated with Caesar's reform of the calendar.
17. Simplicius, *in Arist. de caelo*, p. 504, 17–20.
18. This seems to me to be the proper account of the matter. We are, however, left with another problem: if the above is a fair discussion, we still have to answer a further question. Why does the theory of concentric spheres persist? Why does Aristotle, for example, accept it or a modified version of it? It is, after all, possible to avoid Sosigenes' criticism without giving up homocentricity. Thus one might postulate the displacement of the observer to a point at a certain distance from the centre of the concentric spheres. This would enable the theory to account for apparently non-circular motions on the continuing assumption of underlying circular motions and without giving up homocentricity; but it would also allow the observer to be at various times at various distances from the planets and thus to observe their apparently varying brightness and apparently varying sizes, both regarded as functions of varying distances. Such an anticipation of eccentricity was not only conceptually possible but actually called for. It would have dealt both with the criticism based on the observation of variation in the brightness of planets and with the even more damaging criticism based on the observation that solar eclipses

are sometimes total and sometimes annular (Sosigenes as quoted by Simplicus *loc. cit.*) This latter observation is more damaging because it cannot be explained easily in any way other than as a function of distance; whereas the observation of varying brightness could have been dealt with by the simple expedient of denying that the degree of brightness is necessarily and exclusively a function of distance. Yet, neither Eudoxus nor others, *e.g.* Callippus or Polemarchus, thought it worth while to try and adapt their theory to observation. That Aristotle, who, after all, modified the hitherto entirely geometric theory of concentric spheres in such a way as to give it the explanatory force of a mechanical system, still did not tinker with the uncompromising homocentricity of inherited doctrine is even more striking.

19. See Euclid IX. 20; and *cf.* A. Wasserstein, *Economy and Elegance* (Leicester, 1961), pp. 14 ff.
20. For the history of the generalisation see A. Wasserstein, 'Theaetetus and the History of the Theory of Numbers', *Classical Quarterly*, N.S. *VIII* (1958), pp. 165ff. For the proof see Euclid, *X*, App. 27 (ed. Heiberg-Stamatis).
21. See Guthrie, *op. cit*, (above, note 2), *I*, p. 265, n. 1.
22. See T. L. Heath, *A History of Greek Mathematics* (Oxford, 1921), (repr. 1960), *I*, p. 71.
23. Proclus, *Hypotyposes* p. 150f (Halma).
24. See, e.g., J. Burnet, *Early Greek Philosophy* (1920), pp. 335f.; Kirk and Raven, *op. cit.*, pp. 306 and 406; Guthrie, *op. cit., II*, p. 392.
25. B. Farrington, *Greek Science* (Harmondsworth, 1944), p. 30.
26. *Hippocrates*, transl. W. H. S. Jones, Vol. 1 (Cambridge, Mass., 1984), Loeb Classical Library, pp. 13—15.
27. 297 b 18f.; *cf.* also 296 b 19f. The above is a paraphrase. A literal translation would present problems the solutions of which are irrelevant to our purpose. See W. K. C. Guthrie, transl., *Aristotle, On the Heavens.* (Cambridge, Mass., 1971), Loeb Classical Library, p. 244 note, and diagram on p. 245.
28. We are not concerned here with the complication that the assumption of other curved surfaces, too, would enable us to account for the 'facts' adduced.
29. We need not consider the special case of objects with a surface large relative to their weight, such as kites and feathers; these, of course, fall in disorderly patterns.
30. For the following see, in greater detail, A. Wasserstein, 'Epicurean Science' in *Hermes, 106* (1978), pp. 484—494; and *cf.* Lucretius II. 216—293; Cicero, *de finibus* I. 6, 19; *id., de fato* 9, 18; 10, 22; 20, 46; *id., de natura deorum* 1, 25, 69; Plut., *de sollert. an.* 964 c; *id., de an. procr. in Tim. Plat.* 1015 c; Aetius 1, 12, 5; *id.*, 1, 23, 4; Diog. Oen, 32, II—III (Chilton); Augustine, *contra academicos* 3, 23.
31. See Cicero, *supra*, note 29.
32. Cicero, *de fato* 10, 22—23; *cf.* also 9. 18 and *id., de natura deorum* 1, 25, 69.
33. A. A. Long, *Hellenistic Philosophy* (London, 1974), p. 36.

PART TWO

Islamic and Jewish Contributions

GAD FREUDENTHAL

(AL-)CHEMICAL FOUNDATIONS FOR COSMOLOGICAL IDEAS: IBN SÎNÂ ON THE GEOLOGY OF AN ETERNAL WORLD

I. INTRODUCTION

Historically, the foremost problem of cosmology is arguably that of cosmo*gony*. Immanuel Kant still considers the question whether the world had a beginning in time to provide one of the four antinomies of pure reason.[1] The problem came to a head with Aristotle. In the ancient Mediterranean societies, the idea that the universe came to be after it had not been was never questioned, of course.[2] Also the Milesian philosophers, followed by Plato, took the coming-to-be of the universe to call for an account. The radically new view according to which the actually existing world may not at all have had a temporal beginning was framed by Aristotle. Not only the underlying matter of the world, he argued, but the world as it *is*, i.e., with its very structure, has existed since all times: the heavens and all *forms* in the sublunary world, notably the species of plants and animals, are eternal. This set going a heated debate that was to continue for more than two millennia.[3]

The upholders of eternity were notably Aristotelians, later joined by the Neoplatonists; the defenders of the thesis of the timely origin of the world were at first Stoic and Atomist philosophers, later those who upheld creationist cosmogonies, which they wanted to be in conformity with both the principles of natural philosophy and the Scriptures. The pros and cons were derived from almost every domain of thought — from the most abstract theological and metaphysical speculations and analyses (e.g. of the nature of God, time, motion, infinity) down to physical considerations drawing on empirical evidence. This paper will be concerned with the latter. We will see how, beginning with Plato and the Stoics, adversaries of the eternity thesis argued that certain observable geological phenomena — namely those processes we today call 'erosion' — cannot be reconciled with the view that the world has existed since ever: erosion, they reasoned, is a unidirectional process and if it had been at work since an infinite time, all accretions on the surface of the earth would have been planed down long ago. In other terms: the observable existence of mountains empirically refutes the eternity postulate. But theories, we know, can all too easily be accommodated even to the apparently most adverse evidence. In fact, the upholders of eternity, beginning with Aristotle and Theophrastus, were not slow to retort: by postulating *generative* geological processes compensating for erosion, they sought to defuse their opponents' criticism.

For many centuries, however, Aristotelians — these are the empirically-minded proponents of the eternity thesis with whom we will mainly be concerned here — were unable to say what these theoretically-postulated generative geological processes were: their theory of matter could not provide an account of how the four sublunary elements *cohere* to form stones and mountains. The difficulty was this. According to the received Aristotelian theoretical notions, cohesion was brought about by the moist component of substances: a clump of matter from which all moisture was eliminated, crumbled into dust. But stones, in as much as they are solid, were construed as 'dry.' How, then, do they hold together? Indeed, a simple empirical fact such as the hardening of clay through heating could not straightforwardly be integrated within the Aristotelian theory of matter. *A fortiori*, Aristotelians were at a loss to explain the formation of stones or mountains. Thus, within Peripatetic philosophy, an account of the formation of mountains remained a desideratum with crucially important metaphysical and theological implications.

This is the context, I will suggest, in which we have to place the petrological and orogenic theory put forward by Ibn Sînâ (Avicenna) at the beginning of the eleventh century. Ibn Sînâ adduced an innovative geology, founded, I will show, on notions deriving from chemistry and alchemy. Specifically, Ibn Sînâ draws on the theoretical notion of *unctuous moisture* — a non-evaporable moisture —, a notion whose origin goes back to the fifth century B.C., but which came to prominence only in the wake of the widespread use of fractional distillation by Arab (al-)chemists. From chemistry, the notion passed to natural philosophy and by the tenth century it had become a well-entrenched theoretical concept with great explanatory import. By definition, unctuous moisture is a moisture capable of conferring cohesion and Ibn Sînâ founds on this notion his account of how stones and mountains can be formed through desiccation.

II. GEOLOGY AND THE ETERNITY OF THE WORLD[4]

The existence of continents and islands separated by masses of water posed a double challenge to Aristotle's postulate that the past of the world extends to infinity: (i) the first has its point of departure in the Aristotelian notion of natural motion, i.e., motion toward the natural place. Aristotle himself tells us that 'some people' — this seems to refer to Plato[5] — have argued that if the world were eternal, each of the four elements should long ago have reached its natural place; the sublunary world should then have consisted of four perfect, immobile concentric spheres of earth, water, air, and fire. The existence of elevated land, reaching out into the spheres of water and air seems to disprove this conclusion. (ii) The second challenge is that, if one accepts, with Aristotle, the idea of exhalations raised by the sun's heat,[6] one would expect, with certain Presocratics, the sea to dry up constantly;[7] hence, if it had existed eternally, it would already have dried up completely.

Aristotle's answer to both arguments is essentially the same. Eternal movement, Aristotle argues, cannot but be *cyclical*. The paradigmatic instance, to be sure, is the movement of the heavenly bodies,[8] but the principle holds of change in the sublunary world too. Thus, in response to the first argument Aristotle adduces the idea that the sun, owing to its double movement (daily and seasonal), causes a constant transformation of the four elements into one another and, moreover, brings about regular generation and corruption of the sublunary substances.[9] Hence, generation and corruption can go on, and indeed has and will go on, eternally. The laws of nature do not imply that in an indefinite time the world must end up in a static state.

Against the second argument, Aristotle makes the point — in fact a corollary of the former — that evaporation is compensated by precipitation: this cycle is eternal too and thus evaporation does not lead up to a static final state either. Specifically, contrary to what the Presocratics had believed, the drying up of the sea is not a unidirectional process, so that the postulate of the eternity of the world does not conflict with the continued existence of seas.[10] Rather, this stretch of land may become submerged in the sea and that part of the sea may dry up, but the process will always follow an order and be *balanced*: land and sea constantly change their respective places, but both are eternal and maintain their original equilibrium.[11] Indeed, the relative quantities of the elements in the world are constant.[12] Now it was an easy task for

Aristotle to account for the submerging of land in water, since he accepted the widespread notion of occasional catastrophic deluges. (This idea, incidentally, also accounts for the fact that culture is of young age, although mankind, as all other species, must be eternal.[13]) The other side of the equation was more difficult to come by: Aristotle at one place[14] suggests that the drying of sea into land may be part of an ageing process (indeed, in Aristotelian terms ageing involves getting dryer[15]) — but does not go into any detail. He nowhere explains concretely how new land is formed in compensation for land that turns into sea.

Theophrastus, Aristotle's pupil and successor as the head of the Lyceum, was confronted with much the same problem. In fact, in the meanwhile the geological argument against eternity had been forcefully and systematically developed by Zenon of Citium, the founder of Stoicism. The Stocis, as is well known, believed that, although the underlying 'matter' of the world is eternal, the actually existing world is perishable: in a succession of 'world periods,' different worlds come into existence and are annihilated again in a conflagration (*ekpyrosis*).[16] This cosmogony supplied a conceptual framework within which unidirectional geological processes could be easily construed. One Stoic argument against the eternity of the world, as reported by Theophrastus (and preserved by Philo), is of particular interest to us:

> If the earth had no beginning in which it came into being, no part of it would still be seen to be elevated above the rest. The mountains would now all be quite low, the hills all on a level with the plain, for with the great rains pouring down from everlasting each year, objects elevated to a height would naturally in some cases have been broken off by winter storms, in others would have subsided into a loose condition and would all of them have been completely planed down. As it is, the constant unevennesses and the great multitude of mountains with their vast heights soaring to heaven are indications that the earth is not from everlasting.[17]

The Stoics, we see, clearly recognized the class of phenomena we call today 'erosion' as constituting a unidirectional geological process. They invoked this process in order to prove that the world could not be eternal.[18] The upholders of the opposite view had therefore to show how the existence of continual destructive processes such as erosion can be reconciled with the postulate of the eternity of the world.

Theophrastus himself indeed sought to counter the Stoic arguments. His rebuttals are of particular interest to us, for they highlight the difficulties inherent in the Aristotelian position. In line with his general metaphysical conception of the world as an organic whole, Theophrastus claims that 'trees and mountains differ not in nature' and just as 'the trees shed their leaves at some seasons and then bloom again at others,' so also 'the mountains, too, have parts broken off but others come as accretions.'[19] In other words: erosion is compensated by the formation of mountains. But how are mountains formed? Theophrastus in fact goes beyond Aristotle and describes a sort of 'volcanic' theory, according to which the element fire, rising upward from the bottom of the earth, 'pulls up with it a large quantity of the earthy stuff,' thus giving rise to a mountain.

Theophrastus' theory, it seems, remained unsatisfactory even to its author. Indeed, although the theory was adduced as giving an account of how the formation of mountains counterbalances erosion, Theophrastus then takes it to *explain away* erosion altogether: fire and earth, he claims are so firmly held together that mountains are not at all destroyed.[20] Perhaps Theophrastus is here unwittingly revealing the soft point of his account: the premises of Aristotelian theory of matter (or 'chemistry') do not at all warrant the claim that fire and earth may hold together, nor, consequently, the idea that the rising fire can pull behind it earthy matter (except as an exhalation). We will in fact see in some detail below that in Aristotelian terms, natural substances require *moisture* in order to cohere. Earth and fire, therefore, cannot form a stone or a mountain. Theophrastus was perhaps not unaware of the difficulty: the subsuming of the formation of mountains under the biological analogy, or metaphor, which nothing links to the would-be physical account, may have been meant to secure the idea of a cyclical formation of land and sea on a conveniently abstract level, for want of something better. At any rate, Theophrastus' account did not find acceptance with subsequent natural philosophers.

A theory of the formation of mountains thus remained an important desideratum in the framework of Peripatetic physics and metaphysics: erosion was an easily observable process, but one rarely witnessed the formation of a mountain. The existence of constructive processes, counterbalancing the destructive ones, was a corollary of the metaphysics of the eternal world, but Peripatetics remained unable to specify their precise nature: why should mountains behave like trees and 'bloom' again after having been eroded? What — in Aristotelian terms — is the efficient cause of mountains? The question was crucial, for the postulate of the eternity of the world hinged on it. Indeed, as will be seen, it came to the fore as soon as creationist philosophies began to attack Aristotelian metaphysics with the tools of its own physics.

The Stoics themselves, at least the later ones, tried to provide a positive account of the existence of the continents. Their main explanandum was this: Why is the surface of the earth uneven? In fact, were it not for the mountains, the sea would have covered the totality of the earth, making life impossible for man and most animals and plants. In answer to this, the Stoics maintained that existence of continents is due to Providence (*Pronoia*). Strabo, in a passage probably deriving from Posidonius, takes care to emphasize that if nature alone were at work, no continents would ever have existed:

The work of Nature is this, that all things converge to one thing, the centre of the whole, and form a sphere around this; and the densest and most central thing is the earth, and the thing that is less so and next in order after it is the water . . .

Therefore, Providence must be involved too:

But since water surrounds the earth, and man is not an aquatic animal, but a land animal that needs air and requires much light, Providence has made numerous elevations and hollows on the earth, so that the whole, or the most, of the water is received in the hollows, hiding the earth beneath it, and the earth projects in the elevations . . .[21]

As it stands, this account leaves however unexplained the changes taking place on the surface of the earth, changes which the Stoics were the first to underscore. If, in fact, erosion — i.e. nature — planes down rocks and mountains, may it not eventually nullify the work of Providence? Moreover, can the tenet that the continents are due to Providence be reconciled with the numerous observations — pertaining, e.g., to fossils — indicating that sea and land have occasionally changed their respective places? The Stoics sought to resolve the difficulty by the idea of continued *reciprocal* changes between sea and land, earth and water:

We must take it for granted, first, that the earth is not always so constant that it is always of this or that size, adding nothing to itself nor subtracting anything, and, secondly, that the water is not, and, thirdly, that neither of the two keeps the same fixed place, especially since the reciprocal change of one into the other is most natural and very near at hand; and also that much of the earth changes into water, and many of the waters become dry land . . . Why, then, is it marvellous if some parts of the earth which are at present inhabited were covered with sea in earlier times, and if what are now seas were inhabited in earlier times?[22]

Thus the Stoics were confronted with a problem not unlike the one with which they themselves confronted the Peripatetics: they disposed of the problem of the initial formation of the continents by attributing it to Providence; yet in order to account for the changes and the continued existence of the continents they too had to show how water can become land. Thus, an account of the efficient casue of mountains became a desideratum for both Peripatetic and Stoic philosophies.

The two Stoic geological arguments we have considered were eagerly seized upon by Arab upholders of a creationist cosmogony. The Ikhwân al-Safâ', for example (joined by many others such as al-Bîrûnî and al-Ghazâlî),[23] reproduced the Stoic argument for Providence: they argued that were it not for the mountains, water would have covered the entire surface of the earth; the world would have consisted of four concentric spheres of the elements, and man could not exist. The continents, therefore, were created by God, and they testify of His kindness.[24] Yet the Ikhwân recognized that both the land and the sea undergo perpetual changes. They maintained that these changes are cyclical and attributed them to the 36 000 years long revolution of the fixed stars, which modifies the conditions of heat and cold in the different parts of the world. The Ikhwân now had to show how changes in temperature transform land into sea and vice versa. The first half was relatively easy: when mountains are heated, they argued, their moisture evaporates and the remaining dry substance crumbles. The second half of the demonstration, however, remained a problem: the Ikhwân maintained that the rivers convey sand to the sea, at the bottom of which it is 'baked' into the hills and mountains. Yet the precise nature of this 'baking' was not specified, nor was it explained why the mountains should rise over the seas in which they were formed. The de-deficiency of the Ikhwân's account thus again highlights the difficulty of providing an adequate account of the formation of mountains.

Some more 'orthodox' authors adopted the position that no changes whatsoever in the respective positions of sea and land have taken place since the creation. Thus, the Arab author of the book known in Latin as *De elementis* argued on a double front. Against the Ikhwân al-Safâ' he correctly pointed out that the supposed 36 000 years cycle is by far too short: on their hypothesis the sea would have had to shift by one degree in a century, yet 'history teaches us that a great number of cities have been located for many centuries at the same distance from the sea.'[25] He then directs his criticism against those who, on Aristotelian premises, tried to account for the existence of the continents on purely physical grounds. 'Some philosophers,' the author of *De elementis* reports,

> claim that when the earth was formed it was perfectly round, without valleys or mountains. Its shape was then precisely spherical, like that of the heavenly bodies. Those valleys and mountains that we see on the surface of the earth are due to no other cause than the action of the waters. The waters hollowed out the less compact parts of the soil, and so the mountains were formed. These less compact regions, once hollowed out, became the places of the seas.

Aristotelian philosophers, we thus learn, sought to turn the argument from erosion in their favor. Yet is was an easy task to refute this argument: on the mentioned Aristotelian premises, the sublunary world initially consisted of four concentric spheres of the elements. Therefore:

> Suppose that at the beginning the earth were a body perfectly spherical and smooth, without any valley or mountain. The terrestrial mass was then necessarily covered up entirely with a uniform layer of the mass of waters. Then, however, the rain falling from the upper regions of the air fell on the layer of water covering the earth.[26]

Evidently, then, this rain could produce no erosion, much less lead up to the formation of mountains.

Some Aristotelians, lastly, ignored the orogenic problem altogether by treating the cosmological issue on a conveniently abstract level. This is notably the case of Ibn Rushd. In the section of his *Epitome of [Aristotle's] De generatione et corruptione* corresponding to *De gen. et corr.* II, 10, Ibn Rushd reviews the range of sublunar phenomena in which the heavenly bodies play a part. The stars, Ibn Rushd, argues, are eternal and so are the processes of generation and corruption depending on them, notably the very fact that there is life on earth. Now a condition for the continued, eternal, existence of the living species is the existence of dry places. 'It is evident,' Ibn Rushd says, 'that it is the celestial bodies that will continue to preserve this [kind of dry] place in species. Otherwise water would prevail over it, for the natural thing for earth *qua* heavy is that it be submerged in all its parts under water, since it has already been proved that that is its appropriate limit. Thus it is apparent that this function of the stars and especially of the sun is an essential one.'[27] Ibn Rushd contents himself with the assurance that sea and dry land will continue 'in species' — he does not seem to be interested in the down-to-earth, concrete, geological processes some of his predecessors had considered.[28]

Everyone, in sum, except those who simply maintained that the world has subsisted changeless since creation, needed a theory of the formation of mountains: the Aristotelian *falâsifah* and their opponents who denied the eternity of the world agreed that the changes in the respective places of sea and land were reciprocal. While erosion and subsequent overflooding easily accounted for the transformation of land into sea, the converse change was generally postulated without its efficient cause being indicated.

III. IBN SÎNÂ: TOWARD A THEORY OF THE FORMATION OF STONES AND MOUNTAINS

A philosopher and a physician, Ibn Sînâ (980—1037) touched both upon the summits of metaphysical speculation and the most tangible particular natural phenomena. We will now see how Ibn Sînâ, drawing on contemporary chemical ideas, framed an innovative theory of the formation of stones and mountains in order to secure the Aristotelian tenet of the eternity of the world. We may observe in passing that within the context of Arab philosophy this tenet changed somewhat its significance. Indeed, Aristotelians writing in Arabic, foremost among them al-Fârâbî and Ibn Sînâ, incorporated into their metaphysics significant Neoplatonic elements, notably the idea that all that exists proceeds from the first Being through emanation. Thus, as is well known, Ibn Sînâ held that there is one Necessary Existent, God, who is the necessitating cause of all existents in the world. Since God is eternal, this stance implied that the entire universe — immaterial (intellects) and material (celestial and sublunar matter) — is necessarily eternal too. The Neoplatonically-colored Peripatetics thus went beyond Aristotle in making the coming-to-be of the world and its eternal existence into a necessary corollary of their metaphysics.

In the *Dânesh-namé*, his encyclopedic work written in Persian, Ibn Sînâ clearly indicates that the existence of land poses a serious problem to his general metaphysical scheme.[29] Following Aristotle, Ibn Sînâ defines the *place* of a body as the inner limit of the body containing it,[30] and moves on to elaborate a conception of natural place which slightly deviates from that of the Stagirite: the natural place of fire is the inner surface of the firmament; those of air, water and earth are the inner surfaces of the spheres of fire, air, and water, respectively. In fact, Ibn Sînâ explains, since all simple bodies necessarily have a single natural movement, it follows that each of them has only one natural place. A second consequence is that the spatial arrangement of the elements is spherical: it is impossible, Ibn Sînâ argues, that one and the same nature (form) should produce different figures. This reasoning implies that the entire sublunar world necessarily — and therefore *eternally* — forms one body constituted of four contiguous concentric spheres. This account, founded as it is on general metaphysical considerations involving necessity, raises with particular acuity the question why earth and water visibly do not form eternal static and concentric spheres. In answer, Ibn Sînâ says that 'the cause for the water not covering the entire surface of the earth is that water becomes earth, and vice versa. Where earth becomes something else than itself a cavity ensues; and where something else other than earth becomes earth, there is an accretion.' In the *Dânesh-namé* Ibn Sînâ does not elaborate these general assertions: he does not specify how something that is not earth can become earth. This missing link is supplied by Ibn Sînâ's geology, for which we must look elsewhere, namely in Ibn Sînâ's comprehensive *Kitâb al-Shifâ'*.

We have a succinct summary of Ibn Sînâ's views from the pen of Shmuel b. Yehudah Ibn Tibbon, the well-known 13th century Jewish translator and philosopher. Shmuel Ibn Tibbon is best known for his translation into Hebrew of Maimonides' *The Guide of the Perplexed*, but he was also interested in subjects directly relevant to us here, as can be seen from the fact that he translated and commented on Aristotle's *Meteorologica*.[31] His own work, *Ma'amar Yiqqawu ha-Mayim* (i.e., 'Let the Water be Gathered'; *cf. Genesis* 1:9), written not long after 1221,[32] takes as its point of departure precisely the problem we are concerned with: one of his erudite friends, the author tells us, has asked him to find out what the philosophers say of the fact that the element water does not surround the earth and does not cover it entirely.[33] After a brief survey of the opinions of 'the majority of the [Aristotelian] philosophers,' including Ibn Rushd, Ibn Tibbon gives a short account of what Ibn Sînâ — 'also a follower of Aristotle' — says of the problem 'in his great book called *Kitâb al-Shifâ'*, in its part devoted to natural science, in the section on meteorology.' Ibn Tibbon's expostion is so lucid and insightful (in addition to being faithful[34]) that it is worth translating *in extenso*:

> Ibn Sînâ affirms the following: From what we have said previously it is clear that the nature of water and [that] of earth imply that the water should be above the earth, the earth in the midst of the water, so that the water would surround the earth from all sides. But existence [i.e., reality; *mezi'ût*] is not so. Rather, existence is according to what is necessitated by the global order.[35] Indeed, since it is the nature of the elements that parts of them change into one another, it is impossible for the earth to persevere in its natural state [*ha-ᶜinyan ha-tivᶜi la*], for it is in the nature of the earth that parts of it change into water or other elements; similarly, the other elements also change into earth. Now whatever earth chagnes into another element will be subtracted from the global body of the earth. Necessarily, therefore, there will be a defectiveness in the sphericity and the depth [of the earth], for earth is dry, and cannot aggregate so as to regain its natural [i.e., spherical] figure [*tekhûnah*; lit.: property]; rather, it will preserve its acquired figure, which is not natural to it. Again, whatever other elements will be transformed into earth and come down on it, will doubtless become an accretion and an eminence added upon it: indeed, unlike solidified [i.e., frozen] water, which, when poured [sic!] onto other [i.e., liquid] water, forms with it a single spherical body, it will not spread on [the globe of the earth]. It thus follows necessarily that on the globe of the earth crags, depths (?), and hills are formed.
>
> The action of the stars is involved in making necessary this change [of earth into water — and thus of dry land into sea — and vice versa], in accordance with their position at the zenith of the thing which is changing and in accordance with their motion. This holds in particular of the planets which at times move northward, at times southward. . . . It would seem that these are major causes [*sibot gedolot*] in bringing about an increase of water at one place [through] a displacement of water toward it, and a diminution of water at [another] place [through] a displacement of water away from it. Over a long span of time, each of these two [displacements] increases, until it achieves a notable effect in making the water pour to low places and in exposing the hills.
>
> Other causes help along, however, for it is impossible [that the phenomenon in question comes about] unless clay is formed from water and earth, and unless the sun and the stars have an influence on this clay, causing it to become stone when it is exposed, thus forming the mountains. This being so, it is impossible but that there be dry land and sea. So far what has been said by the above-mentioned savant.[36]

The context of Ibn Sînâ's geology is thus defined by the very same ideas we have encountered in the *Dânesh-namé*. In addition, we find here the notion — so often invoked by the astrologers as a knock-out argument, but shared by the natural philosophers too — that the displacement from one place to another of masses of water on the surface of the globe is caused by the celestial bodies (notably by the moon). This sets the stage for the really crucial question: how is clay formed from earth and water, and how are stones and mountains formed from clay? For an answer to these queries we have to turn to Ibn Sînâ himself, namely to Treatise (*fann*) V of the Physics of his *Kitâb al-Shifâ'* (composed about 1022), where Ibn Sînâ expounds his petrological and orogenic theory. In 1927 Holmyard and Mandeville showed that the well-known medieval tractate appearing under the titles of *De mineralibus* and *De congelatione et conglutinatione lapidum*, is a translation of sections from this part of the *Shifâ'*.[37]

To establish the 'conditions of the formation of mountains,' Ibn Sînâ says, one has to examine successively the conditions of the formation of stones, rocks and, lastly, mountains: petrology is to provide the foundation for the theory of orogeny. How, then, are stones formed? According to Aristotelian mineralogy, the underlying matter of stones is earth if they are opaque, and water if they are transparent. Ibn Sînâ accordingly distinguishes two processes[38]: stones are formed either through the hardening (*tafkhîr, conglutinatio*) of clay, or through the 'congelation' (*jumûd*) of water. We can quickly dispose of the later: congelation, the transformation of liquid into solid, is brought about, Ibn Sînâ holds, by a 'petrifying virtue,' instanced by the alchemists' Virgin Milk, which is 'compounded of two waters which coagulate into a hard solid.' Yet Ibn Sînâ is prudent on the matter and holds that, for the most part, stones and mountains are formed from earth.

How, then, does earth become stone? Here we must understand the terms of the problem as it faced Ibn Sînâ. On Aristotelian theory, 'moist' is 'that which, being readily adaptable in shape, is not determinable by any limit of its own'; 'dry,' on the other hand, is 'that which is readily determinable by its own limit, but not readily adaptable in shape.'[39] Stones, then, are obviously dry; they must be formed through desiccation. But desiccation of what? One would expect the answer to be: of a mixture of earth and water. But this poses a serious difficulty. According to Aristotle, if drying completely eliminates the moisture from a clump of earth, nothing solid can possibly result: 'earth,' Aristotle says, 'has no power of cohesion without the moist. On the contrary, *the moist is what holds it together*; for it would fall to pieces if the moist were eliminated from it completely.'[40] Now this postulate is precisely Ibn Sînâ's theoretical starting point: 'pure earth does not petrify,' (this is the well-known formula: *terra pura lapis non fit*), he echoes Aristotle, 'because the predominance of dryness over the earth endows it not with coherence but rather with crumbliness.'[41] This means that stones have to be dry while still containing moisture, an idea which would have seemed nothing short of a straightforward *contradictio in adjecto* to most of Aristotle's Greek commentators. In Ibn Sînâ's time, however, this was no longer the case. The idea of a *non-evaporable moisture* was readily available in Arab alchemy and chemistry: this was the notion of an *unctuous moisture*, and it is on this theoretical notion that Ibn Sînâ draws to account for the formation of stones through desiccation.

Beginning with the *observation* that

Often clay dries and is changed first into something intermediate between stone and clay, *viz.* a soft stone, and afterwards is changed into a stone [proper],

Ibn Sînâ goes on to present the *theoretical* contention that

The clay which most readily lends itself to this is that which is unctuous (*lazij*), for if it is not unctuous, it usually crumbles before it petrifies.[42]

The key notion here, on which the entire theory depends, is *unctuous*. What does it mean and why does unctuous clay not crumble upon desiccation? The question has more to it than meets the eye. We will now in fact briefly see that the concept of an unctuous moisture was a well entrenched one within the tradition of Arab natural philosophy. Far from simply using a convenient metaphor to avoid confronting the problem head on, Ibn Sînâ in fact sought to ground his petrology on a received theory within another part of natural philosophy.

IV. CHEMICAL PREMISES: AQUEOUS VS. UNCTUOUS MOISTURE
(AN HISTORICAL APERÇU)

On close inspection, we find that in his biological writings, Aristotle himself occasionally distinguishes two kinds of moisture, one *aqueous*, the other *unctuous* (or fat, greasy; *liparos*),[43] of which the latter is the less liable to decay. For instance, plants live longer than animals 'because they have an oiliness and a viscosity which makes them retain their moisture in a form not easily dried up' (notably, we infer from the context, through cold).[44] Similarly, 'Water animals have a shorter life than terrestrial creatures, not strictly because they are humid, but because they are watery, and watery moisture is easily destroyed, since it is cold and easily congealed' (i.e., dried up by cold).[45] Again, differences in the thickness of hair among animals are explained thus: 'if the moisture [of the skin] be watery it dries up quickly and the hairs do not gain in size, but if it be greasy the opposite happens for the greasy is not easily dried up.'[46] *Unctuous moisture, we may conclude, is resistant to desiccation by either heat or cold*; indeed, olive oil is dried up by neither heat nor cold.[47] And in a word: '*fat things are not liable to decay*'; indeed, 'a fat substance is incorruptible.'[48]

In the ps.-Aristotelian *Problems*, the notion is already used as a matter of course in a 'chemical' account:

For when the dough is kneaded and the lightest flour and the stickiest moisture are left, the bread, when it has been exposed to the fire, becomes glutinous and does not dry up; for that which is sticky cannot be separated.[49]

Aristotle's Greek commentators did not, as far as I am aware, make much of fats and of their perdurable properties. Indeed, the problem of cohesion gained in prominence only after a crucial development had taken place, namely when Aristotelian natural philosophy had to face — and to account for — a vast array of empirical findings, related to the disintegration of inanimate substances, which accumulated within alchemy, independently of the developments of philosophical reflection within the Schools. These findings emerged from the systematical use of distillation (a procedure unknown to Aristotle), first by Greek, but then, and mainly, by Arab alchemists. Arab natural philosophy sought to integrate the new facts (and indeed the alchemists' theoretical notions too[50]) within the theoretical framework it took over from Aristotle.

Arab alchemists distilled practically every mineral and animal matter.[51] The typical result of a fractional distillation was the following: first, a soft fire would raise a vapor which, upon condensation, became a clear liquid: this liquid was referred to as 'water.' Then, a second, stronger, heating would raise a further, colored and unctuous liquid, referred to as 'oil' (*duhn*). At the end of the process, a dry residue was left behind at the bottom of the alembic.[52] This kind of procedure and its typical results are described by the Renaissance chemist Conrad Gesner (1516–1565). Although his account is from a later period, Gesner describes the very same facts and is worth quoting for his clarity:

Of a plant or any other substance ordeined to be distilled, what parte of it is most meet to be extenuated and fyret (that is the purest parte, the lightest, the thinnest, the *moistest* and the most superficial parte . . .) being first of all fyret by the force of the heat, is lifted up; next suche other partes as in purenes cum nie to the first, and last such a *moisture* of the thinges as is more crosse *that held together the earthy partes*, a certain *fatness and oiliness*, by a stronger force of the fyre is separated, and taken up hooly; which once clean drawn forthe, the body remaineth dissolved and brought to ashes.[53]

Distillation thus *isolated* the two distinct kinds of moisture: the aqueous, which evaporates easily and which condenses to water; and the unctuous, which evaporates only with difficulty and upon the disappearance of which the body disintegrates. Only the unctuous moisture, then, brings about cohesion.

In the corpus of writings ascribed to Jâbir ibn Ḥayyân (for convenience, I will use the name 'Jâbir' to designate its authors), the number of fractions obtained was increased from three to four when it was observed that fractional distillation yields also an aeriform, volatile and inflammable substance (which we know today to be sal ammoniac[54]). This discovery allowed Jâbir to establish a one to one correspondence between the four fractions obtained in distillation and the four Aristotelian elements. Indeed, fractional distillation of organic matter became of capital importance to Jâbir because he took it to be a method by which the nature and composition of any given substance could be determined.[55] The first distilled, volatile and inflammable fraction, was naturally taken to derive from fire; the second was considered to be the element water; the third, oily, fraction was associated with the element air; lastly, the solid residue which remained behind was identified with the element earth.[56]

It is the notion of 'oil' that is of interest to us here. To Jâbir, just as to the other alchemists, oil was the second and last liquid to rise in distillation: its evaporation left behind only a powdery residue. This means that among the four components of a substance, the one responsible for cohesion is oil. Now, Jâbir believed that each of the four Aristotelian qualities could be isolated through distillation. Moisture, in particular, which — in conformity with Aristotle — Jâbir construed as the principle of liquidity and cohesion, he affirmed to be obtained through the distillation of oil: moisture, he claimed, is isolated 'when oil is distilled until a very glutinous and elastic substance is obtained ... This substance never solidifies.'[57] For Jâbir, then, oil, or unctuous moisture, is the principle of permanent liquidity and hence of cohesion. Whence the idea that it is the '*unctuous quality that brings about combination.*'[58] Indeed, the standard account of calcination, as given for instance by the famous physician and physicist al-Râzî, is this: calcination is the 'destruction of bodies' (i.e. metals) through 'the burning of the sulfurs and oils they contain, [resulting in] their reduction to white lime whose parts cannot further be divided.'[59]

Ibn Sînâ himself, let us now note, draws on the concept of unctuous moisture in his great medical work, the *Canon*, namely precisely when he seeks to explain how stones are formed within the human body. The formation of calculi (*ḥaṣâh*), Ibn Sînâ says, involves matter — which is affected — and an active, efficient, cause. The latter, to be sure, is (vital) heat, or, more precisely, *imbalanced*, i.e., excessively strong, heat. As for the material cause, Ibn Sînâ says that 'the matter [of calculi] is a *thick unctuous moisture* [*ruṭûbah lazijah ghalîzah*],' which has its origin in thick nutriment: milk and cheese, agglutinative (*lazij*) bread, indigestible fruits giving rise to unctuous moisture, etc.'[60] Ibn Sînâ, we see, was familiar with the chemical notion of 'unctuous moisture' and drew on it in order to account for cohesion: heat — which usually eliminates moisture — can give rise to hard bodies when acting on matter containing unctuous, that is non-evaporable, moisture.

By the tenth century, we may conclude, 'unctuous moisture' has become a well-entrenched theoretical concept, accounting for the cohesion of living and non-living matter. This was an indispensable concept if one wished — and many natural philosophers in fact did wish — to account for cohesion of substances within a largely Aristotelian framework, in which cohesion was imputed, almost by definition, to 'moisture.' The Ikhwân al-Ṣafâ', Albertus Magnus and many other medieval philosophers drew on it. As late as the second half of the seventeenth century the notion is still alive and well: even corpuscularians use it, e.g., to explain electrical attraction as brought about by tiny elastic — because unctuous — threads issuing from the rubbed electrical body.[61] It is on this well-entrenched theoretical notion that Ibn Sînâ founds his petrology and his geology.

V. CONCLUSION: METAPHYSICAL INTENTIONS, CHEMICAL FOUNDATIONS

Back to Ibn Sînâ's geology. We are now in a postion to understand why Ibn Sînâ insisted that stones and mountains are formed from specifically unctuous clay. This is not an alleged observational statement, nor a mere metaphor. Rather, Ibn Sînâ was drawing on an entire chemical *theory*, a theory that allowed him to explain how clay — a mixture of earth and water — can petrify. On this theory, the action of heat on the unctuous clay evaporates the aqueous moisture, but the unctuous, non-evaporable one, remains behind. This unctuous moisture continues to inhere in the clay and thus endows it with cohesion, preventing it from crumbling upon desiccation.

Once this petrology was established, the account of orogeny followed easily. This process is construed by Ibn Sînâ as follows. At first, water or clay petrify, presumably (Ibn Sînâ does not specify this) at low places, where they are carried by rivers as silt. Then mountains can be formed by one or the other of two causes: (i) The essential cause of mountains is a 'wind' (*rîh*) which arises in the bowels of the earth; this is the same 'wind' which is also the cause of earthquakes, i.e., Aristotle's dry exhalation.[62] This 'wind,' Ibn Sînâ says, 'raises a part of the ground and a height is suddenly formed.' (ii) Mountains can also be formed by an accidental cause, namely when the petrified mass remains where it was formed and 'certain parts of the ground are hollowed out while others are not, by the erosive action of winds and floods which carry away one part of the earth but not another.'[63]

Ibn Sînâ was now in a position to construe the surface of the earth as eternally subject to the actions of two counterbalancing forces. One is destructive and easily observable: 'at present time,' Ibn Sînâ says, 'most mountains are in the stage of decay and disintegration.' Yet theory establishes that an opposite, generative, geological process, one which is less accessible to observation, is at work too: some mountains 'by God's wills ... increase through the petrification of waters upon them, or through the floods which bring them a large quantity of clay that petrifies on them.'[64]

Ibn Sînâ thus succeeded where Theophrastus had failed. Contrary to Theophrastus, he could explain how a *cohering* mass of earthy matter can be formed, to be then raised by a 'volcanic' action. (Theophrastus speaks of 'fire,' Ibn Sînâ of 'wind,' but lurking behind both notions is in fact Aristotle's dry exhalation.) He thus gave substance to Theophrastus' metaphor in which the earth was likened to a tree which throws out its leaves again after having shed them: generation and corruption of the surface of the earth can go on eternally without leading up to a static final state. Peripatetic philosophers were now, at long last, in a position to rebuke the argument from erosion: the postulate of the eternity of the world was 'saved.'

The gist of Ibn Sînâ's petrological account — his invoking specifically unctuous moisture — has been recognized and correctly appreciated by Albertus Magnus.[65] 'It is perfectly clear,' Albertus says, that earth alone 'does not cohere into solid stone.' The reason is that 'the cause of coherence and mixing is moisture, which is so subtle that it makes every part of the earth flow into every other part.' Hence, 'if the moisture were not soaked all through the earthy parts, holding them fast, but evaporated when the stone solidified, then there would be left only loose, earthy dust.' Thus, one sometimes finds, compressed within a stone, earth which appears to be solid, but which finally falls into dust. 'And the cause of this is simply that its moisture, which was not unctuous or viscous enough, evaporated when the [surrounding] stone solidified.' The conclusion is that the moisture is necessarily non-evaporable, i.e., unctuous: 'There must be something viscous and sticky so that its parts join with the earthy parts like the links of a chain.'[66] It is, in sum, 'the viscous and unctuous moisture which gives coherence to the material of stone.' Ibn Sînâ was right, Albertus Magnus concludes, to insisit that stones are formed from unctuous clay.

The metaphysical intentions of Ibn Sînâ's chemically-founded theory did not escape notice either. They have been very clearly perceived and pointed out by Shmuel Ibn Tibbon, who perceptively compares them to the Aristotelian *opinio communis*, as represented by Ibn Rushd. Since Shmuel Ibn Tibbon thereby highlights the weighty metaphysical stakes lurking behind Ibn Sînâ's geological endeavor, his *exposé* is again worth translating *in extenso*:

It is manifest that his [Ibn Sînâ's] opion is that the exposed [i.e., dry] land is something which is generated [*nithawa*] after it had not been, an opinion which differs from that of the other philosophers following the opinion of Aristotle. He also dissents from them concerning the preservation of that [kind of dry] place. This disagreement follows upon [Ibn Sînâ's] above-mentioned view concerning generation. For the other philosophers say: the preservation [of dry land] is necessary — *qua* species, [dry land] can neither change nor be annihilated; for it is impossible that the entire earth be covered with water, with the result that all beings living only on land be annihilated. Thus in his commentaries, Ibn Rushd made clear his view that the existence of exposed land is necessary, not possible. To prove this he says: if it were possible that the entire earth be submerged [in water], then this possibility would [already] have been actualized; for time is eternal according to their [the philosophers'] opinion. Now the actualization [of that possibility] would entail the annihilation of the plants and animals. But, [Ibn Rushd continues,] there are many animal species in which generation and existence require the support of an individual of the same species: for instance, man acquires existence only through another man — from a father and the mother. The same holds of many animal and plant species. [Ibn Rushd] therefore says that [if the above hypothesis were true, then] these species would not [anymore] exist today: the [proposition stating] their existence would be false. But they do exist, as we know from sense experience. Consequently, their annihilation, their non-existence, is of the class of impossible things, not of the class of possible ones. Hence, what exists today has existed ever since and will exist eternally, infinitely. Ibn Rushd's view is indeed that the existence [i.e. generation] of man not from man is of the class of impossible things; this is for him a first postulate, one that is verified conjointly by sense experience and by the intellect that had been created in man. Hence it is by supposing this [the generation of man not from man] to be impossible, that he demonstrates the impossiblity of the earth being submerged in water.

[As against this,] Ibn Sînâ has written in the above-mentioned book that it is not impossible that a flood will occur and cover all the inhabited earth, or a part of it, and annihilate all or some animals, and that afterwards [these animals] come to be [again] from the *mixis* of the elements, helped along either by the stars only, or by one of the separate intellects [as well]. To explain his view: according to him, it is not impossible that, say, the species of man be annihilated and that subsequently, during the eternal time (as they believe), a *mixis* will come to be in the earth, which is suitable to receive the human form. For in his view, man's generation from man is not necessary, but only what is most frequent; it is the most appropriate and the easiest [mode of generation], just as most frequently a mouse is generated from a mouse, a frog from a frog, although occasionally a mouse comes to be from earth, a frog from rain water. The same holds of other species too. Also the generation of man from earth is possible, according to his opinion: the difference between the mouse, the frog and others which are born and give birth, and man is only a quantitative difference: this kind of [spontaneous] generation is very rare in man, indeed most infrequent, whereas in the mouse and the frog it is not all that rare, although it still is infrequent. This is what Ibn Sînâ's words amount to.[67]

Shmuel Ibn Tibbon thus makes clear that although in devising his novel geology, Ibn Sînâ responded to a need Peripatetic philosophers had felt at least since Theophrastus, he in fact pursued intentions that went far beyond what the other Aristotelians had in mind. For most philosophers assumed that *qua* species, dry land and sea had always existed, that, like all other species, they were eternal: this they interpreted as the work of a particularizer, and constructed upon it an argument from design.[68] Their problem was thus only to account for the *preservation* of the primeval balance between the elements, to show how, erosion notwithstanding, dry land and sea both persevere 'in species.' By contrast, Ibn Sînâ does away with the idea of a particularizer altogether: take a world consisting of the four concentric spheres of the elements and let time pass; through the sole interplay of natural forces dry land and sea will ultimately emerge, as indeed will all other species, including man. The actually existing species and structure of the surface of the earth are the work of nature, as will be also those which will, in due course, emerge subsequent upon a flood completely immersing the dry land.[69] Ibn Sînâ thus in fact succeeded in rebuking the argument adduced against the Aristotelians by the author of *De elementis*.[70] Through his novel, chemically-founded, geological theory, Ibn Sînâ the natural philosopher buttressed an essential doctrine of Ibn Sînâ the metaphysician: it allowed him, somewhat like Laplace some eight centuries later, to dispense with the hypothesis postulating a mindful, well-*intentioned* Creator, from whom issued the order 'Let the water be gathered!'

Writing in the third decade of the fourteenth century, the Hebrew poet Immanuel of Rome was of the opinion that for this geology Ibn Sînâ deserved nothing short of eternal hell. Following in Dante's footsteps, Immanuel describes his voyage through Hades, and reports having sighted there, among others, Aristotle, Galen, al-Fârâbî, Plato and Hippocrates; Ibn Sînâ is the last on this list of illustrious freethinkers:[71]

CNRS
Paris, France

NOTES

1. I. Kant, *Kritik der reinen Vernunft*, A 424—33; B 452—61.
2. For a fine collection of sources, *cf. La Naissance du monde* (= *Sources oriéntales*, 1) (Paris: Seuil, 1959).
3. Unfortunately (and surprisingly) there is no comprehensive and detailed study of this debate in its various contexts. For the earlier period, *cf.* J. Baudry, *Le Problème de l'origine et de l'éternité du monde dans la philosophie grecque de Platon à l'ère chrétienne* (Paris: Les Belles Lettres, 1931). A good study, albeit of limited scope, is W. Wieland, 'Die Ewigkeit der Welt (Der Streit zwischen

Johannes Philoponus und Simplicius),' in: *Die Gegenwart der Griechen im neueren Denken. Festschrift für H.-G. Gadamer zum 60. Geburtstag* (Tübingen: J. C. B. Mohr, 1960), 291—316. For the medieval period I know only of Ernst Behler, *Die Ewigkeit der Welt. Problemgeschichtliche Untersuchungen zu den Kontroversen um Weltanfang und Weltuntergang im Mittelalter. Erster Teil: Die Problemstellung in der arabischen und jüdischen Philosophie des Mittelalters* (Paderborn: F. Schöningh, 1965), which, however, is almost entirely derivative and, moreover, compiles its information mostly from outdated sources; Behler says some words on the context, and the stakes, of the debate in medieval philosophy on pp. 12—19. H. A. Davidson, *Proofs for Eternity, Creation and the Existence of God in Medieval Islamic and Jewish Philosophy* (New York/Oxford: Oxford University Press, 1987) provides an excellent overview of the arguments and of their history but does not seek to situate the debate in relation to other issues.

4. The subject of the following paragraph has been treated very extensively by P. Duhem in the chapters on 'L'équilibre de la terre et des mers' and on 'Les petits mouvements de la terre,' in *Le Système du monde*, Vol. IX (Paris: Hermann, 1958), pp. 79—323. In the few pages that follow my aim is not to add to Duhem's wealth of detail, but only to highlight how the geological problem is related to the metaphysical and physical contexts.

5. *Cf. De gen. et corr.* 2.10, 337a8; *Tim.* 58 A; and Friedrich Solmsen, *Aristotle's System of the Physical World* (Ithaca: Cornell University Press, 1960), p. 384 (n. 18), p. 394.

6. *Cf.* Solmsen, *Aristotle's System*, p. 407ff.

7. *Ibid.*, p. 420ff.

8. *Cf.*, notably, *Physics* 8.9 and Baudry, *Le Problème*, pp. 170—73.

9. *Cf. De gen. et corr.* 2.10—11; F. Solmsen, *Aristotle's System*, pp. 379—89; A. L. Peck, 'Appendix A' to his *Aristotle, Generation of Animals* (Cambridge, Mass.: Harvard University Press [The Loeb Classical Library], 1942), pp. 567—76.

10. *Cf.* Solmsen, *Aristotle's System*, p. 420ff.; and idem, 'Aristotle and Presocratic Cosmogony,' *Harvard Studies in Classical Philology, 63* (1958): pp. 265—82 (reprinted in his *Kleine Schriften* [Hildesheim: Olms, 1968], *I*, pp. 356—73), at p. 273ff.; I. Düring, *Aristoteles* (Heidelberg: C. Winter, 1966), pp. 386—7.

11. *Meteor.* 1.14 and 2.3.

12. *Cf.* Duhem, *Système du monde, IX*, p. 91ff.

13. Solmsen, *Aristotle's System*, pp. 431ff.; *cf.* also W. K. C. Guthrie, *A History of Greek Philosophy*, Vol. 1: *The Earlier Presocratics and the Pythagoreans* (Cambridge: Cambridge University Press, 1962), pp. 387—90. On the eternity of the human species see the remarkable study by K. Oehler, 'Ein Mensch zeugt einen Menschen. Über den Missbrauch der Sprachanalyse in der Aristotelesforschung,' reprinted in his *Antike Philosophie und byzantinisches Mittelalter* (München: C. H. Beck, 1969), pp. 95—145.

14. *Meteor.* 1.14, 351a19ff.; *cf.* Solmsen, *Aristotle's System*, p. 436f.

15. *Cf.* Thomas S. Hall, 'Life, Death and the Radical Moisture,' *Clio Medica, 6* (1971): pp. 3—23 at p. 6; Gad Freudenthal, 'The Theory of the Opposites and an Ordered Universe: Physics and Metaphysics in Anaximander,' *Phronesis 31* (1986): pp. 197—228, at p. 221.

16. *Cf.*, e.g., M. Pohlenz, *Die Stoa* (Göttingen: Vadenhoek & Ruprecht, 1948), *I*, pp. 77ff.; M. Lapidge, 'Stoic Cosmology,' in: J. M. Rist (ed.), *The Stoics* (Berkeley: University of California Press, 1978), pp. 161—85, at pp. 180ff.; D. E. Hahm, *The Origins of Stoic Cosmology* (Columbus, Ohio: Ohio University Press, 1977), pp. 185ff.
17. Philo, *De aeternitate mundi* 118—119, quoted after *Works*, vol. 9, trans. by F. H. Colson (Cambridge, Mass.: Harvard University Press [The Loeb Classical Library], 1941). The thesis that the arguments preserved by Philo are Theophrastus', who was already defending the eternity thesis against Zenon was the subject of some discussion. *Cf.* O. Regenbogen, 'Theophrastos,' in: Pauly-Wissowa, *Realencyclopädie*, Suppl. VII (Stuttgart, 1940), cols. 1539—40; Pohlenz, *Die Stoa, II*, 44; Baudry, *Le Problème*, pp. 219—20, pp. 236ff. The first to recognize the significance of these passages for the history of geology was Pierre Duhem; *cf.* his 'Léonard de Vinci et les origines de la géologie,' in his *Etudes sur Léonard de Vinci*, seconde série (Paris: Hermann, 1909), pp. 283—357, at pp. 286ff. The same passages are again discussed in *Système du monde, IX*, pp. 241ff.
18. The Stoic arguments are repeated by the Atomists, of whom, however, our evidence is scantier. *Cf.* Lucretius, *De rerum natura* 5.235ff. and Baudry, *Le Problème*, pp. 249ff.
19. Philo, *De aeternitate mundi*, 132—133. The passage seems to echo Aristotle, *Meteor.* 1.14, 351a27—29 referred to above. For the way this argument fits into Theophrastus' metaphysics *cf.* P. Steinmetz, *Die Physik des Theophrast* (Bad Homburg: Gehlen, 1964), p. 167. On the general context of the argument and on related arguments *cf.* H. A. Wolfson, 'Patristic Arguments Against the Eternity of the World,' in his *Studies in the History of Philosophy and Religion* (Cambridge, Mass.: Harvard University Press, 1973), *I*, pp. 182—206, on pp. 187ff.
20. Philo, *De aeternitate mundi*, 135—137.
21. Strabo, *The Geography*, 17.1.36, quoted after: *The Geography of Strabo*, with an English translation by H. L. Jones (Cambridge, Mass.: Harvard University Press [The Loeb Classical Library], 1932), vol. 8. On the attribution to Posidonius, *cf.* K. Reinhardt, *Poseidonius* (München: C. H. Beck, 1921), pp. 88ff. and *idem*, 'Poseidonius', in Pauly-Wissowa, *Realencyclopädie*, Vol. 43, pp. 665—6.
22. Strabo, *Geography*, 17.1.36.
23. For what follows, *cf.* Ikhwân al-Ṣafâ *Rasâ'il* (Beirut, 1957), **II**, pp. 91f.; F. Dieterici, *Die Naturanschauung und Philososphie der Araber im zehnten Jahrhundert. Aus den Schriften der Lauteren Brüder* (Berlin, 1861), pp. 99ff. For al-Bîrûnî, *cf.* his *The Determination of the Coordinates of Positions for the Correction of Distances Between Cities*, translated by Jamil Ali (Beirut: The American University of Beirut, 1967), pp. 23—5. For al-Ghazâlî, *cf.* Duhem, *Système du monde, IX*, p. 105. The argument was of course often rehearsed by Scholastic natural philosophers; *cf.*, e.g., Duhem, *Système du monde, IX*, pp. 126, 129, 131, 133.
24. Moreover, since the existence of land does not follow by natural *necessity*, its existence at one place rather than at another indicates that it has resulted from the voluntary action of a *particularizer* (*mukhaṣṣiṣ*); cf. H. A. Wolfson, *The Philosophy of the Kalam* (Cambridge, Mass.: Harvard University Press, 1976), p. 441. Quite evidently, the idea of God's providence, the doctrine of particularization, and the

argument from design are intimately related (*cf.* Maimonides, *The Guide of the Perplexed*, I, 74, 'The Fifth Method'); they are all discussed in Davidson, *Proofs for Eternity, Creation and the Existence of God* (n. 3, *supra*).

25. Quoted after Duhem, *Etudes sur Léonard de Vinci*, II, pp. 300f. *Cf.* also *Système du monde*, II, pp. 226—8 and IX, p. 256. The Arab original of *De elementis* (dating from the middle of the ninth century) seems to be lost and its author escapes identification to this day; cf. Charles B. Schmitt and Dilwyn Knox, *Pseudo-Aristotles Latinus. A Guide to the Works Falsely Attributed to Aristotle Before 1500* (Warburg Institute Surveys and Texts, Vol. XII) (London: The Warburg Institute, The University of London, 1985), p. 20.
26. *Ibid.*, p. 309; also in *Système du monde*, IX, pp. 256—7.
27. *Averroes on Aristotle's 'De generatione et corruptione'. Middle Commentary and Epitome*, translated by Samuel Kurland (Cambridge, Mass.: The Medieval Academy of America, 1958), p. 135.
28. Nor does Ibn Rushd describe here in any detail how the heavenly bodies are to 'preserve' the 'species' of dry places. He is a little more elaborate in his *Epitome of [Aristotle's] Meteorologica*. There he argues that the existence of dry land cannot be due to the sun's heat, for the southern hemisphere is warmer than the northern one and yet more submerged in water. Ibn Rushd believes that the drying effect is brought about by the rays of the sun immixed with those of the fixed stars, which are more numerous in the northern hemisphere. Levi ben Gerson (Gersonides, 1288—1344) severely criticizes Ibn Rushd's account, arguing that the existence of dry land is due to Providence and thus confirms the thesis of creation; *cf.* his *Milḥamot ha-Shem* 6.1.13 (Riva di Trento, 1560, fol. 57^b) and Ch. Touati, *La Pensée philosophique et théologique de Gersonide* (Paris: Minuit, 1973), pp. 185—7 (*cf.* n. 52 for references to some further authors who discussed the problem), as well as Davidson, *Proofs for Eternity, Creation and the Existence of God*, p. 231 (with note 108). (I have consulted Ibn Rushd's text as incorporated in Levi ben Gerson's Supercommentary on it in the manuscript of the Staatsbibliothek preussischer Kulturbesitz Berlin (West), Ms. Orient Fol. 1055, fol. 120^b.) As has been noted by M. Joel [*Lewi ben Gerson (Gersonides) als Religionsphilosoph* (Breslau, 1862), p. 77], a position very similar to Gersonides' had been advocated by his contemporary, the astronomer Isaac Israeli, in his *Sefer Yesod ᶜOlam* 2.2 [(Berlin, 1846—1848), I, pp. 17^bff.], written in 1310 (*ibid.*, pp. I, 1^b, II, 31^a). Levi's argument is rebuked by R. Ḥasdai Crescas (*Sefer Or ha-Shem* 3.1.3—4), but embraced by Crescas's student Joseph Albo (*Sefer ha-ᶜIqqarim*, 4.8), as well as by Shimeᶜon ben Zemah Duran (*Magen Avot* [Livorno, 1785], p. 9^a) and by the fifteenth-century philosopher Abraham Shalom (*cf.* H. A. Davidson, *The Philosophy of Abraham Shalom* [Berkeley and Los Angeles: University of California Press, 1964), pp. 62, 74). To be sure, the entire issue is closely bound up with the controversial question whether or not the world, though created, is destructible; *cf.* on this Seymour Feldman 'The End of the Universe in Medieval Jewish Philosophy,' *Association for Jewish Studies Review*, 11 (1986), pp. 53—77. Dante seems to have followed Ibn Rushd and to have further developed his account; *cf.* E. O. von Lippmann, *Beiträge zur Geschichte der Naturwissenschaften und der Technik*, Vol. 2 (Weinheim: Verlag Chemie, 1953), pp. 169—70.

29. For what follows, *cf.* Avicenne, *Le Livre de science*, translated by M. Achena and H. Massé (Paris: Les Belles Lettres, 1958), *II*, pp. 27, 31—3, 45f.
30. *Cf.* Aristotle, *Physics* 4.4, 212a5f., 20.
31. *Cf.* M. Steinschneider, *Die Hebraeischen Übersetzungen des Mittelalters* (Berlin, 1893), pp. 132—5.
32. Steinschneider (*ibid.*, p. 200, n. 676) concludes from internal evidence that it was written after 1221; Shmuel Ibn Tibbon was already dead in 1232 (*ibid.*, p. 132, n. 179).
33. Shmuel Ibn Tibbon, *Ma'amar Yiqqawu ha-Mayim*, ed. by M. L. Bisliches (Pressburg, 1837), p. 2.
34. Ibn Tibbon's *exposé* has been compared with Ibn Sînâ's own text in Georges Vajda, *Recherches sur la philosophie et la kabbale dans la pensée juive du Moyen Age* (Paris: Mouton, 1962), pp. 14—15. This chapter of Vajda's book has appeared in English as: 'An Analysis of the *Ma'amar Yiqqawu ha-Mayim* by Samuel b. Judah Ibn Tibbon,' *Journal of Jewish Studies, 10* (1959): pp. 137—49.
35. Hebrew: *kefî ha-mezi'ût ha-mehuyav le-seder ha-kôl*; in Arabic the last two words read: *linizâm al-kull. cf.* G. Vajda, *Recherches*, p. 14, where they are translated by 'l'ordre du Tout.'
36. Shmuel Ibn Tibbon, *Ma'amar Yiqqawu ha-Mayim*, pp. 7—8, I have somewhat corrected the printed text according to the manuscript Cod. Heb. 33 of the Bayerische Staatsbibliothek München (using the microfilm no. 1163 of the Institute of Microfilmed Hebrew Manuscripts in the Jewish National and University Library, Jerusalem), fol. 75^{v}f. This passage became known to a very wide public because it was quoted *in extenso*, without yet indicating its source, in the popular encyclopedia of R. Gershom b. Shlomo, *Sefer Sha‛ar ha-Shamayim* 2.2 (Rödelheim, 1801), pp. 6^v f.; (Warsaw, 1875), p. 13.
37. For their edition of the Arabic and Latin texts, accompanied by an English translation, *cf.* E. J. Holmyard and D. C. Mandeville (eds. and trans.), *Avicennae De congelatione et conglutinatione lapidum, Being Sections of the Kitâb al-Shifâ'* (Paris: Geuthner, 1927). That Ibn Sînâ is the author of the *De mineralibus* had been convincingly argued by Duhem, albeit without using the Arabic text, in 1909; *cf. Etudes sur Léonard de Vinci, II*, p. 302ff.
38. For what follows, *cf. ibid.*, pp. 71ff. (Arabic text), pp. 18ff. (English translation).
39. Aristotle, *De gen. et corr.* 2.2, 239b32f. All quotations from Aristotle are given after *The Complete Works of Aristotle. The Revised Oxford Translation*, edited by Jonathan Barnes (Princeton: Princeton University Press, 1984).
40. *De gen. et corr.* 2.8, 335a2f.,
41. Holmyard and Mandeville, *op. cit.*, p. 71 (Arabic), p. 18 (English).
42. *Ibid.*, p. 72 (Arabic), p. 19 (English). Holmyard and Mandeville render *lazij* by 'agglutinative'; I have preferred 'unctuous' so as to conform to medieval Latin usage.
43. The term and the notion go back at least to the ps.-Hippocratic treatise *On Fleshes* (fifth century B.C.), but tracing and dealing with this must remain outside the scope of the present overview.
44. *De long. et brev. vit.* 6, 467a6ff.
45. *Ibid.*, 5, 466b33ff.; *cf.* also 466b22f. and *De sensu* 2, 438a20f. for similar accounts.

46. *Generation of Animals* 5.3, 782b2ff.
47. *Meteor.* 4.7, 383b34.
48. *De long. et brev. vit.* 5, 466a23; *History of Animals* 3.19, 521a1. As has already become clear, the notion of unctuous moisture seeks to provide a physical or 'chemical' account of the cohesion of individual substances compounded of the four sublunary elements. But in Aristotelian philosophy the cohesion of substances is also subsumed under the notion of form, which, where living substances (plants and animals) are concerned, is their (vegetative) soul. It is therefore clear that the present *aperçu* artificially isolates one aspect of a general and fundamental problem in Aristotelian philosophy of nature, that of cohesion — or better: transtemporal stability [*cf.* M. Furth, 'Transtemporal Stability in Aristotelian Substances,' *Journal of Philosophy*, 75 (1978), pp. 624—46] — of material sublunary substances. Elsewhere I intend to deal with the entire problem in greater detail and try to show how the notion of unctuous moisture relates to the Aristotelian notion of soul, through that of vital heat.
49. *Problems* 21.12, 928a26ff.; *cf.* also 21.6.
50. The alchemists themselves accounted for their observations with their own distinctive sulfur-mercury theory, which the natural philosophers incorporated into their own system. Although the question of how this alchemical synthesis came about is highly germane to our topic and is crucial for an adequate understanding of the explanatory import of the notion of unctuous moisture, I will have to disregard it in the present paper. I have given a brief sketch of the historical developments involved in my paper 'Die elektrische Anziehung im 17. Jahrhundert zwischen korpuskularer und alchemischer Deutung,' in Chr. Meinel (ed.), *Die Alchemie in der europäischen Kultur- und Wissenschaftsgeschichte (Wolfenbütteler Forschungen* Band 32) (Wiesbaden: Otto Harrassowitz, 1986), pp. 315—326, and intend to come back to the topic in greater detail in the future.
51. R. J. Forbes, *Short History of the Art of Distillation* (Leiden: Brill, 1948).
52. *Cf.*, e.g., the 'Syriac and Arabic Alchemical Treatise,' in M. Berthelot, *La Chimie au Moyen Age* (Paris, 1893; reprinted Osnabrück/Amsterdam, 1967), *II*, pp. 184—5; J. Ruska, *Al-Râzî's Buch Geheimnis der Geheimnisse* (Berlin: J. Springer [*Quellen und Studien zur Geschichte der Naturwissenschaften und der Medizin, VII*], 1937), pp. 78ff.; 204ff.; 207ff.
53. Conrad Gesner, *The Treasure of Evonymus* (London, 1559), p. 2 (my italics). Similar descriptions can repeatedly be gleaned in al-Râzî's writings. *Cf.* Ruska, *Al-Râzî's Buch Geheimnis der Geheimnisse*, pp. 78ff., 204, 211, 216, 217, 218, 225.
54. J. Ruska, 'Der Salmiak in der Geschichte der Alchemie,' *Zeitschrift für angewandte Chemie, 41* (1928): pp. 1321—4.
55. For this and for what follows, *cf.* P. Kraus, *Jâbir ibn Ḥayyân: Contribution à l'histoire des idées scientifiques dans l'Islam*. Vol. II: *Jâbir et la science grecque* (= *Mémoires présentés à l'Institut d'Egypte*, Vol. 45) (Cairo, 1945 [reprinted: Paris: Les Belles Lettres, 1986]), pp. 5ff.; 41f.
56. *Cf.* also F. Rex, *Zur Theorie der Naturprozesse in der früharabischen Wissenschaft. Das 'Kitâb al-Ikhrâǧ', übersetzt und erklärt. Ein Beitrag zum alchemistischen Weltbild der Ǧâbir-Schriften* [= *Collection des travaux de l'Académie internationale d'histoire des sciences*, n° 22] (Wiesbaden: F. Steiner, 1975), p. 42.
57. Kraus, *op. cit.*, p. 10.

58. H. E. Stapleton, R. F. Azo and Hidâyat Husain, 'Chemistry in Iraq and Persia in the Tenth Century A. D.,' *Memoirs of the Asiatic Society of Bengal*, 8 (1927): pp. 317—418, at p. 338f. [Added in proof: In antiquity, the natural philosophers most concerned with the problem of cohesion were the Stoics, who accounted for it in terms of their fundamental concept of *pneuma*. I have recently suggested that this concept may have gone into the Arab alchemists' notion of unctuous moisture: *cf.* my "The Problem of Cohesion Between Alchemy and Natural Philosophy: From Unctuous Moisture to Phlogiston," in Z.R.W.M. von Martels (ed.), *Alchemy Revisited. Proceedings of the International Conference on the History of Alchemy at the University of Groningen, 17—19 April 1989* (Collection de travaux de l'Académie internationale d'histoire des sciences, vol. 33) (Leiden: Brill, 1991), pp. 107—16.]
59. Ruska, *Al-Razi's Buch*, p. 126 (*cf.* n. 52 *supra*); *cf.* also p. 74.
60. Ibn Sînâ, *Kitâb al-Qânûn fi-l-Ṭibb*, Book VI, Treatise 18, Article 2, Chapter 6 (Rome, 1593), p. 434.
61. For more details *cf.* my paper referred to *supra*, note 50.
62. *Cf.* Aristotle, *Meteorologica* 2.8. In this chapter Aristotle uses the terms *pneuma* and *anemos* to designate the dry exhalation, the cause of earthquakes; *cf.* H. D. P. Lee, *Aristotle, Meteorologica* (Cambridge, Mass.: Harvard University Press [The Loeb Classical Library], 1952), p. 203. In the Arabic translation of the *Meteorologica* both terms are rendered as *rîḥ*; *cf.* C. Petraitis, *The Arabic Version of Aristotle's Meteorology* (= *Pensée arabe et musulmane*, tome 39) (Beirut: Dar el-Machreq, 1967), Arabic pagination, pp. 73ff., p. 138 *sub rîḥ*.
63. Holmyard and Mandeville, *op. cit.*, p. 77 (Arabic), p. 27 (English).
64. *Ibid.*, p. 78f. (Arabic), p. 29 (English).
65. For what follows, *cf.* Albertus Magnus, *Book of Minerals*, trans. by D. Wyckoff (Oxford: Clarendon, 1967), pp. 12—13. The first to emphasize the extent to which Albertus Magnus is dependent on Ibn Sînâ (whom he mentions by name) was P. Duhem; *cf.* his *Etudes sur Léonard de Vinci, II*, pp. 302ff., included almost *verbatim* in his *Système du monde, IX*, pp. 257ff. In these two works Duhem gives much information on the influence of Ibn Sînâ on the Latin Middle Ages.
66. As D. Wyckoff has noted, the account of 'viscous' in terms of parts hanging together like the links of a chain, goes back to Aristotle, *Meteorologica* 4.9, 387a11—19.
67. Shmuel Ibn Tibbon, *Ma'amar Yiqqawu ha-Mayim*, p. 8 (*cf.* n. 33 *supra*), corrected after the MS (*cf.* n. 36), fol. 76ʳf. G. Vajda, *Recherches* (*cf.* n. 34 *supra*), pp. 16—7, translates some passages and gives a summary of the rest. On Ibn Tibbon's views *cf.* also S. Rosenberg, "Remarks on the History of the Idea of a Restaurative Redemption in Medieval Jewish Philosophy" (in Hebrew), in *Ha-Raᶜyon ha-Meshiḥi be-Yisrael* (Jerusalem: The Israel Academy of Sciences, 1982), pp. 37—86, on pp. 50 ff. Ibn Sînâ's theory discussed here, in particular his 'outrageous' idea (*cf.* below in the text) that even man can come to be through spontaneous generation, so that creation is not a necessary condition for his existence, of course hinges on his own notion of the agent intellect, the giver of forms. The details of this theory are authoritatively exposed in H. A. Davidson, 'Alfarabi and Avicenna on the Active Intellect,' *Viator*, 3 (1972): pp. 109—78, particularly pp. 156—9. Ibn

Rushd's view, or rather changing views, on this subject is studied in H. A. Davidson, 'Averroes on the Active Intellect as a Cause of Existence,' *Viator, 18* (1987): pp. 191—225.
68. *Cf.* note 28 *supra*.
69. In other words, Ibn Sînâ invokes efficient rather than final causes; *cf.* Sh. Pines, 'What was Original in Arabic Science?,' in A. C. Crombie (ed.), *Scientific Change* (London, 1963), pp. 181—205, on p. 189 (= *The Collected Works of Shlomo Pines*, vol. II: *Studies in Arabic Versions of Greek Texts and in Medieval Science* [Jerusalem: Magnes and Leiden: Brill, 1986], pp. 329—353, p. 337).
70. *Cf.* text at notes 25 and 26 *supra*.
71. In all likelihood Immanuel derived his knowledge of Ibn Sînâ's geology from Shmuel Ibn Tibbon's resumé. *Cf. Mahbarot 'Immanuel ha-Romi* ed. by Dov Yarden (Jerusalem: Mosad Bialik, 1957), 28.90—98. Here is my translation, which has no pretensions whatsoever to do justice to the poetical qualities of the original:

Present there is Ibn Sînâ,
He has been put to ridicule and to laughter,
For he said that the generation of man not from man is possible at times,
And for he said that the generation of mountains is along the natural way.
Would that he had remained dumb!
For he followed the belief in the eternity of the world.

BERNARD R. GOLDSTEIN

LEVI BEN GERSON: ON ASTRONOMY AND PHYSICAL EXPERIMENTS*

INTRODUCTION

Levi ben Gerson (1288—1344), sometimes called Gersonides or Leo de Balneolis, is well known as a philosopher, biblical exegete, mathematician, and astronomer. He lived in Orange and occasionally visited Avignon where his brother was physician to the Pope. The family name was de Balneolis, but there is no evidence that he himself was born or ever lived in Bagnols (*cf.* Shatzmiller, 1972, 1974). Levi does not cite any contemporaries and little is known of his life. He is mentioned by a few Hebrew writers of the fourteenth century: the astronomer Immanuel ben Jacob Bonfils of Tarascon (fl. ca. 1360), who may have been his pupil; the philosopher Judah Cohen (fl. ca. 1320—1350) who called Levi 'the lion of the group' (*ha-ari shebahavura*: Renan, 1893, p. 654); and the physician and historian Isaac de Lattes (fl. ca. 1372: Renan, 1893, pp. 682, 689—90; *cf.* Touati, 1973, pp. 541—59).

In the fifteenth century, we find marginal comments in Hebrew by Mordecai Finza of Mantua on Levi's great astronomical work (MS Naples, heb. III. F. 9) as well as references in other astronomical texts by Finzi (cf. MS Bodleian [Neub.] 2052). There is also a Hebrew commentary on Levi's astronomical tables by Moses Farissol Botarel (fl. ca. 1465: MS Bodleian [Neub.] 2022, and now another copy in the possession of Mr. M. Meer, New York).

Levi's treatise on astronomy was translated into Latin in 1342, i.e., in his lifetime, and one version was dedicated to the Pope in Avignon (Renan, 1893, p. 621). The influence of the translation is difficult to assess because few references to it have been found: the most significant of them is a long citation of a chapter on the Jacob Staff and the experiment for finding the center of vision in the eye that appears in Commandino's commentary on Archimedes's *Sandreckoner* (Venice, 1558: *cf.* Roche, 1981, pp. 5, 8). It has recently been noted that the Latin copy used by Commandino was probably the manuscript that was in the library of Fulvio Orsini, now part of the Vatican Library (MS Vat. lat. 3380: *cf.* Rose, 1975, p. 189). A copy of a shorter version of Levi's *Astronomy* belonged to Bernhard Walther (d. 1504) who pre-

S. Unguru (ed.), Physics, Cosmology and Astronomy, 1300—1700, pp. 75—82.
© *1991 Kluwer Academic Publishers. Printed in the Netherlands.*

sumably acquired it from his teacher, Regiomontanus (d. 1476: *cf.* Rose, 1975, p. 107; Petz, 1888, p. 260).

In Europe of the sixteenth and seventeenth centuries Levi's best known contribution to astronomy was the Jacob Staff or *radius astronomicus*, and a recent paper has described the diffusion of Levi's ideas (Roche, 1981; *cf.* Goldstein, 1977). It was also shown that some of Levi's results were still being rediscovered in the mid-seventeenth century despite their earlier publication in European texts (for the case of the eccentricity of the eye, see Roche, 1981, pp. 5, 27). Another recent paper has dealt with a problem associated with pinhole images (i.e., the camera obscura) ultimately solved by Kepler (Straker, 1981): I shall argue here that, unbeknownst to Kepler, Levi already had the solution long before him. Kepler was aware of Levi's work from Commandino's commentary on Archimedes's *Sandreckoner* (Straker, 1971, p. 218), but he did not have a copy of Levi's *Astronomy* as we learn from a letter he wrote in 1629 (Kepler, 1959, p. 389 [ed. Caspar]; trans. in Straker, 1971, p. 220).

Levi ben Gerson's main work in astronomy forms Book V, Part 1, of the *Wars of the Lord*, his magnum opus on religious philosophy. However, the astronomical part is preserved in a different set of manuscripts, and was omitted in the printed editions (1569, 1866). Recently a translation and commentary on the section concerning creation has appeared (*Wars*, VI: 2, 1—8) in which Levi alludes to some of his astronomical views (Staub, 1982). (For a general survey of the literature on Levi, see Kellner, 1979.) The original Hebrew version of Levi's *Astronomy* consists of 136 chapters (about 250 folios in manuscript) and is preserved in 4 manuscripts (see Goldstein, 1974a, pp. 74—77): I have prepared an edition with translation and commentary of the first 20 chapters (Goldstein 1985). In addition to discussions of astronomical instruments, Levi introduced new models for lunar motion and new tables based on them, observations that he made of the planets and of lunar and solar eclipses, a new theory of cosmic distances that led him to propose a universe vastly larger than that of Ptolemy and his medieval successors, a refinement of the theory of precession, new values for the solar parameters based on new observations, etc. (*cf.* Goldstein, 1974a, 1975, 1977, 1979, 1980, 1985, 1986, 1988). Moreover, Levi emphasized his belief that astronomy must be investigated by a scholar thoroughly familiar both with natural philosophy and mathematics, and that his predecessors had, for the most part, neglected one or the other (Levi, *Astronomy*, chap. 1 [Goldstein, 1985, pp. 22–23]). The branch of physics that enters his discussions explicitly is optics, and his new lunar models seem to be based implicitly on an optical analogy involving reflection (*cf.* Goldstein, 1974a, pp. 53–74; 1974b). By way of contrast, in the seventeenth century it was mechanics that ultimately was found to serve as the foundation of the analysis of the motion of celestial bodies. Indeed, it was only from the end of the sixteenth century and on that the sublunary/superlunary distinction was dropped and that physics began to be based on the interaction of particles colliding with, or acting upon, one another (*cf.* Barker and Goldstein, 1984).

ASTRONOMY AND PHYSICAL EXPERIMENTS

The Jacob Staff that Levi invented was widely used for astronomical observations, and to use it correctly it was essential to compensate for the eccentricity of the eye, i.e., the distance from the end of the staff nearest the eye to the center of vision. Levi devised an experiment in which there were two cross-pieces of different sizes placed on the same staff such that the cross-piece closer to the eye exactly covered the one farther away. By a simple geometric argument he then could find the distance sought, and his result was 1/20 span or about 1 cm: 250 years later Harriot in England obtained the same result (see Goldstein 1985, pp. 51–54; *cf.* Roche, 1981, p. 6). Levi adds that 'we carried out this procedure many times and in as many ways as possible, and we found the [center of vision] at the center of the eye, i.e., at the middle of the crystalline lens' (Levi, *Astronomy*, chap. 6: 47 [Goldstein, 1985, p. 54]), and later remarks: 'when you follow these instructions, it will happen that the distance from the center of vision inside the head of the observer to the surface of the plate adjacent to the eye is 1/20 span, for most people, as we determined by experiment (*baḥannu*) with much diligence and effort' (chap. 7: 5 [Goldstein, 1985, p. 55]). Thus we have an example of a physical experiment described in an astronomical context which served two purposes: one qualitative, and the other quantitative. Qualitatively, Levi demonstrated that the center of vision lies inside the eye and not on its surface, and quantitatively, he found the precise location of the center of vision in order to graduate his instrument correctly (i.e., the zero point for measuring the distance of the cross-piece from the eye is not the end of the staff but the center of vision).

Levi began his discussion of the camera obscura, or pinhole camera, in chapter 5 of his *Astronomy* and in it he assumes the rectilinear propagation of light. The difficulty in accounting for the shape of pinhole images was already noted in antiquity in a work ascribed to Aristotle (*Problemata*, Book XV, chap. 6, 911bl): 'Why is it that when the sun passes through quadrilaterals, as for example wickerwork, it does not produce a figure rectangular in shape but circular?' (*cf.* Lindberg, 1968, p. 158). Levi's interest in this device was related to its usefulness for observing eclipses and, as we shall see, he even tried to apply it to other types of astronomical observations. He was clearly aware that the size of the image depends on the angular size of the luminary as well as on the size of the aperture, and he seems to have been the first astronomer in the West to have realized that for quantitative measurements the size of the aperture must be taken into account (see Levi's *Astronomy*, chaps. 5, 9 [Goldstein, 1985, pp. 48—50, 69—73]; *cf.* Straker, 1971, p. 200). Levi does not cite any sources and it cannot be determined if he knew of Ibn al-Haytham's *On the Shape of Eclipses*, a treatise never translated into Latin or Hebrew (*cf.* Sabra, 1972, pp. 195—96; Lindberg, 1968, pp. 155—56; for the German translation of this text, see Wiedemann, 1970, Vol. **2**, pp. 87—101). Chapter 5 of Levi's *Astronomy* has been treated extensively based largely on the published Latin version (Curtze, 1901; Carlebach 1910, pp. 30—34; Straker, 1971, pp. 197—219; Lindberg, 1970, pp. 303—8). In this chapter Levi explained that the shape of the image is not the same as that of the aperture: 'From the figure (*temuna*: Lat. *demonstratione*) it is clear that when a ray passes through a polygonal window, the image will not be polygonal because it expands in every direction at each corner [*lit.*: angle] by the amount of the angular radius of the luminary. A corner comes to be like a quadrant whose center is the point of the corner, and this matter is visually perceived for the rays that come from the Sun and the Moon through polygonal windows' (chap. 5: 15—16 [Goldstein, 1985, pp. 48f]; *cf.* Lindberg, 1970, pp. 304—6). Note that his result is again quantitative as well as qualitative because he introduces the angular radius of the luminary to account for the rounding of the corners in the image. This principle is applied in some later chapters (e.g., chap. 13 [Goldstein, 1985, pp. 86—92]) where other astronomical instruments are described.

In addition to Levi's discussion in chapter 5, we find further remarks on the camera obscura in chapter 9 (Goldstein, 1985, pp. 69—73). For example, we learn that Levi tried to use the camera obscura to measure the sizes of Venus and Jupiter (chap. 9: 12). For present purposes the crucial section of this chapter concerns an instrument that combines a staff with a camera obscura (chap. 9: 13—17). An experiment that appears to be idealized is then described (chap. 9: 18—24). I consider it an idealized experiment only because the numbers are not those one would expect from measurement: the breadth of the hole is 2, the breadth of the image is 3, and the ratio of the excess of the breadths is 1 in relation to the length of the staff of 100 (chap. 9: 19). The most significant step is found in the geometrical demonstration in the course of which we are told that the rays from point B on one side of the luminary pass points D and G (the two extremities of the aperture) such that their extensions to the image (DT and GE) are parallel (chap. 9: 24; see Fig. 1). It then follows that the size of the aperture must be

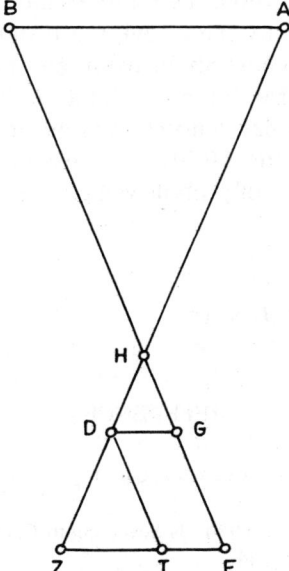

Fig. 1. This figure appears in the Hebrew manuscript of Levi's *Astronomy*, chapter 9, to illustrate the formation of the pinhole image EZ of luminary AB whose rays pass through the aperture GD.

subtracted from the size of the image. This is essentially the same argument and the same figure presented by Straker (1981, p. 275) to support his claim that at the end of the sixteenth century Tycho Brahe used this method to determine the apparent solar diameter (based in part on reports in Kepler's *Paralipomena*). Tycho's problem has been paraphrased as follows: 'Since the image of the sun in the dioptral camera is enlarged by an amount equal to the width of the aperture, then in the image of a partial eclipse of the sun, the lunar shadow (the image of the moon) will be diminished in size by just that amount. Consequently, the apparent diameter of the moon on the screen during an eclipse will be smaller than either good lunar theory would predict or a direct observation of the eclipse in the sky would reveal' (Straker, 1981, p. 278). To resolve this difficulty Kepler established a sound theory for the formation of pinhole images and published his original contributions in his *ad Vitellionem Paralipomena* (1604). In this book Kepler cites Levi in connection with the center of vision based on Commandino's report, and does not mention Levi with respect to pinhole images. Though Kepler certainly treated the problem of pinhole images in greater generality than Levi, the results are quite similar, and in both cases the effort came about from investigations of astronomical observations. It may also be noted that in his reports of his own observations of partial solar eclipses Levi never raises the problem that troubled Tycho (Goldstein, 1979). The reason, I contend, is that Levi understood the formation of pinhole images and did not see the need to comment on it there.

University of Pittsburgh
Pittsburgh, Pennsylvania, U.S.A.

REFERENCES

[*] This study was supported in part by a research grant from the National Endowment for the Humanities, U.S.A.

Barker, P., and Goldstein, B. R. 1984. 'Is Seventeenth Century Physics Indebted to the Stoics?,' *Centaurus* 27: 148—64.

Carlebach, J. 1910. *Lewi ben Gerson als Mathematiker*. Berlin: L. Lamm.

Curtze, M. 1901. 'Die Dunkelkammer,' *Himmel und Erde* 13: 225—236.

Goldstein, B. R. 1974a. *The Astronomical Tables of Levi ben Gerson*. Hamden, CT: Archon Books.

Goldstein, B. R. 1974b. 'Levi ben Gerson's Preliminary Lunar Model,' *Centaurus 18*: 275—88.
Goldstein, B. R. 1975. 'Levi ben Gerson's Analysis of Precession,' *Journal for the History of Astronomy 6*: 31—41.
Goldstein, B. R. 1977. 'Levi ben Gerson: On Instrumental Errors and the Transversal Scale,' *Journal for the History of Astronomy 8*: 102—12.
Goldstein, B. R. 1979. 'Medieval Observations of Solar and Lunar Eclipses,' *Archives Internationales d'Histoire des Sciences 29*: 101—56.
Goldstein, B. R. 1980. 'The Status of Models in Ancient and Medieval Astronomy,' *Centaurus 24*: 132—47.
Goldstein, B. R. 1985. *The Astronomy of Levi ben Gerson (1288—1344)*. New York, Berlin: Springer-Verlag.
Goldstein, B. R. 1986. 'Levi ben Gerson's Theory of Planetary Distances,' *Centaurus 29*: 272—313.
Goldstein, B. R. 1988. 'A New Set of Fourteenth Century Planetary Observations,' *Proceedings of the American Philosophical Society 132*: 371—99.
Kellner, M. M. 1979. 'R. Levi Ben Gerson: A Bibliographical Essay,' *Studies in Bibliography and Booklore 12*: 13—23.
Kepler, J. 1959. *Gesammelte Werke: Briefe 1620—1630*. Vol. **18**. M. Caspar (ed.), Munich: C. H. Beck'sche Verlagsbuchhandlung.
Lindberg, D. C. 1968. 'The Theory of Pinhole Images from Antiquity to the Thirteenth Century,' *Archive for History of Exact Sciences 5*: 154—76.
Lindberg, D. C. 1970. 'The Theory of Pinhole Images in the Fourteenth Century,' *Archive for History of Exact Sciences 6*: 299—325.
Petz, H. 1888. 'Urkundliche Nachrichten über den literarischen Nachlass Regiomontans und B. Walters 1478—1522,' *Mitteilungen des Vereins für Geschichte der Stadt Nürnberg 7*: 237—62.
Renan, E. 1893. *Les écrivains juifs français du XIVe siècle*. Paris: Imprimerie Nationale.
Roche, J. J. 1981. 'The Radius Astronomicus in England,' *Annals of Science 38*: 1—32.
Rose, P. L. 1975. *The Italian Renaissance of Mathematics*. Geneva: Librairie Droz.
Sabra, A. I. 1972. 'Ibn al-Haytham,' in *Dictionary of Scientific Biography 6*: 189—210. New York: Scribners.
Shatzmiller, J. 1972. 'Gersonides and the Jewish Community of Orange in his Day,' in B. Oded *et al.* (eds.), *Studies in the History of the Jewish People and the Land of Israel*, Vol. **2**, pp. 111—26. Haifa: University of Haifa [in Hebrew].
Shatzmiller, J. 1974. 'Further Information about Gersonides and the Orange Jewish Community in his Day,' in B. Oded *et al.* (eds.), *Studies in the History of the Jewish People and the Land of Israel*, Vol. **3**, pp. 139—43. Haifa: University of Haifa [in Hebrew].
Staub, J. J. 1982, *The Creation of the World According to Gersonides*. Chico, CA: Scholars Press.
Straker, S. 1971. *Kepler's Optics: A Study in the Foundations of 17th Century Natural Philosophy* (unpublished dissertation). Bloomington: Indiana University.
Straker, S. 1981. 'Kepler, Tycho, and the 'Optical Part of Astronomy': the Genesis of Kepler's Theory of Pinhole Images,' *Archive for History of Exact Sciences 24*: 267—93.

Touati, C. 1973. *La pensée philosophique et théologique de Gersonide*. Paris: Les Editions de minuit.

Wiedemann, E. 1970. 'Ueber die Camera obscura bei Ibn al Haitam,' in *Aufsätze zur arabischen Wissenschaftsgeschichte*, 2 Vols., **2**: 87—101. Hildesheim: Georg Olms Verlag.

Y. TZVI LANGERMANN

THE ASTRONOMY OF RABBI MOSES ISSERLES

I. INTRODUCTION

Rabbi Moses Isserles of Cracow (1525 or 1530—1572) was one of the major figures of Jewish thought in the sixteenth century.[1] Isserles is best known as a legist and, indeed, many of his rulings remain in force to this day for a large number of Jews. Our study of his astronomy is based on two of his writings: *Torat ha-Olah*[2] (hereafter *T.O.*), a work of religious thought in which philosophy, physics, and astronomy are all invoked, and a commentary to the Hebrew translation of Georg Peurbach's *Theoricae Novae Planetarum*.[3] The first of these works has drawn some attention from scholars, though not from historians of science. The second has never been published and, to the best of my knowledge, the manuscripts have never been the subject of modern investigation.

In this study we limit ourselves to a presentation of the actual thoughts and opinions of Isserles for, in our own opinion, the first step in intellectual history must always be the elucidation of the views themselves, however anxious we may be to address ourselves to the seemingly more significant questions of context, transmission and reformulation, historical patterns, and so forth. Nevertheless, even though we shall not go into these matters in the present study, we feel an obligation to, at least, sketch out in more general terms in what way Isserles' views may fit into the history of thought.

The first and most important point to be made is that Isserles' thought is of interest primarily for Jewish intellectual history. The relationships of trends and developments in Jewish thought to those taking place among non-Jewish neighbours have varied greatly from time to time and from place to place. Moreover, it has always been difficult for the historians who study this problem to detach themselves from their own beliefs as to what this relationship ought to have been (or, perhaps more accurately, what this ought to be). We shall therefore do our best to confine ourselves to a few observations of direct relevance to the investigation of Isserles' thought.

It appears to us that the Jewish thinkers who flourished in the Mediterranean basin and the Islamic Near East during the medieval period were, in their scientific and philosophical activity, very much part and parcel of the intellectual developments which were going on in the ambient cultures. In general, Jews and non-Jews had access to the same literature, sometimes met for study or discussion, and were concerned with more or less the same problems of science and philosophy, even though the different religious traditions may have led to different emphases or different twists to the same problem. More importantly for us is the observation that those Jews who made known the results of their research probably had some sense of advancing the frontiers of *human* knowledge (if I may be forgiven the cliché), that is to say, of participating in and contributing to an intellectual endeavour which transcends the interests of their particular community. I believe this to be true even though these scholars often published their works primarily with a Jewish audience in mind. For all of these reasons, the philosophical writings of Saʿadya Gaon, Isaac Israeli (the elder), and Maimonides, the medical writings of the last two, and the astronomical investigations of Isaac Israeli (the younger), Jacob ben Machir, and Levi ben Gerson — to give just a few examples — have an importance which is by no means limited to Jewish intellectual history and, indeed, their writings have been the object of great interest on the part of historians of philosophy and science in general.

This is not at all the case with Isserles. To the best of my knowledge, Isserles was limited to Hebrew source material; neither he nor any of his coreligionists in Cracow (which was at that time experiencing a cultural blossoming) had any meaningful scientific or philosophical exchanges with their Christian neighbours, and, perhaps most importantly, one gets the impression that Isserles did not regard his work as having any wider, not specifically Jewish, import.

We shall not attempt any explanation of just why the cases of the medieval Mediterranean and Renaissance Poland are so different. Obviously the different historical and cultural contexts, as well as internal developments within the Jewish world, must be taken into account. But the conclusion is clear regarding the present study: Isserles' thought must be viewed as a phenomenon which is rather separate — probably consciously so — from the intellectual activity of his time and place. The historian of thought must be particularly aware that he is dealing with the speculations of a member of a minority who was, most likely, uninterested in the implications of his thought for anything beyond his own tradition.

Yet I venture to suggest (and I certainly hope) that Isserles' thought will still be of some interest to those who deal in the history of science, the history of philosophy, and Polish cultural history. (I refer primarily to those who are not specifically interested in the intellectual efforts of minorities as such). Isserles' reasoning is not wholly out of place in these contexts, and for good reason. After all, his speculations are based on Aristotelian natural philosophy and Ptolemaic astronomy (both taken with a good dosage of medieval elaboration), and both of these still served as the fundaments of thought in his time and place. As a matter of fact, some specific parallels may be drawn between Isserles' activities and those of his Christian neighbours. Both the Jewish and Christian communities of Cracow were served by physicians who trained at Padua and who brought back to Poland, one assumes, something of the Italian Renaissance.[4] Isserles corresponded with Meir Katzenellenbogen, the rabbi of Padua.[5] Moreover, Isserles, like his Polish predecessor Wojcieh (Albertus) of Brudzewo (1445—1497), chose to comment on Peurbach's *Theoricae*, then considered to be the standard astronomical textbook.[6]

However, these remain strictly parallel developments. To the best of my knowledge, we still have no solid evidence of any intellectual contacts between Jews and Christians in sixteenth century Cracow.[7] In fact, it seems to me that we are dealing here with a peculiar juncture in Jewish intellectual history. For there was indeed a strong rationalist current in Jewish life in sixteenth century Poland.[8] Moreover, the Hebrew sources available to Isserles gave him a scholarly base not radically different from that of his Christian neighbours, and the still-living tradition of Jewish philosophical writing furnished him with an approach toward speculative inquiry which may also not have been terribly at odds with that of his non-Jewish contemporaries. As a result, his investigations may not seem too far out of line — in form if not in content — with those of his Gentile contemporaries. Yet I somehow get the feeling that there is something illusory about these similarities: Isserles seems to me to belong near the beginning of a chapter in Jewish intellectual history wherein speculative thought was directed increasingly inwards.

II. ASTRONOMY, PHYSICS, AND COSMOLOGY IN *TORAT HA-OLAH*

In *T.O.* Isserles entertains certain views regarding the motions of the heavenly bodies which are quite contrary to the accepted medieval cosmology. As we noted earlier, Isserles appears to rely for his knowledge of astronomy exclusively upon works written or translated into Hebrew. In particular, he does not know of the work of Copernicus. Nevertheless, from his Hebrew sources he absorbed two fundamental ideas, which may have provided the thrust for his cosmological speculations. First of all, from Maimonides, and others, he learned that the physical structure of the heavens was still an unsolved problem and, therefore, theories other than those of Ptolemy and Aristotle could and, perhaps, should be considered.[9] Second, actual alternative theories were also available to him, for example, the theory of al-Biṭrūjī as condensed by Isaac Israeli (the younger) in the latter's *Yesod Olam* 11,9.[10] Thus the Hebrew sources both encouraged cosmological speculation and offered substantive hints concerning alternative models. We shall now examine how Isserles handles this material.

Isserles states early on in *T.O.* (1, 23b) that there are three viable hypotheses which explain the motions of the heavens. Two of these are related to the discussion in the Talmud of a disagreement on this matter between Jewish and non-Jewish scholars.[11] The Gentile view is that the stars are fixed while the orbs revolve, and this Isserles identifies with 'the *Almagest*, which has been followed by all astronomers since that book [i.e. the *Almagest*] appeared.' The view of the Jewish sages was that the stars themselves move, and the orbs are fixed. The Talmud states that on this matter the Jews conceded to the Gentile scholars, but Isserles claims that this 'concession' was offered only because the hardships of exile prevented the Jewish scholars from fully developing their own theory. The third theory is that of the 'man [whose theory] shook [the world]' (*ha-'ish ha-marʿish*), i.e. al-Biṭrūjī, as described by Isaac Israeli.

Here as elsewhere in *T.O.* (e.g. III, 9a, 35b), Isserles insists upon the hypothetical nature of such theories. Some rules of medieval cosmology are nonetheless taken to be unquestionably true, as one would expect. The cosmos is surrounded by an outermost orb, and the orbs 'are spheres, round on all sides.' (I, 21a).[12] The orbs are physically contiguous, since there is no vacuum (I, 27b). Maimonides' count of eighteen orbs is accepted as correct (I, 48b).

The most striking departure from the medieval view is found in *T.O.* III, chapter 49 (pp. 62a ff.). Isserles is here arguing for *creatio ex nihilo*, and in countering the Aristotelian claim that the heavenly motions are eternal and unchanging, he makes two bold statements: that the moon has an intrinsic, and not merely apparent, motion which is not uniform and circular; and that it is quite possible, and perhaps desirous, to assert that the heavenly bodies undergo generation and corruption just like the sublunar bodies.

Let us examine these points in more detail. Here, as elsewhere in *T.O.*, Isserles is looking for philosophical or mystical symbolism in the Temple rites. In our specific case, Isserles wishes to show that the special sacrifices offered at the beginning of the lunar month symbolize the rebuttal of one of the claims of the philosophers who deny *creatio ex nihilo*, 'i.e. that the motion of the orbs has no causes of contrariety which would bring about corruption ... and if it does not corrupt, it does not come to be.' This assertion, Isserles answers, is refuted by the motions of the moon; hence the celebration of the new moon. To elaborate:

For on account of the moon we see the contrariety which exists in the movements of the orbs and their motions, some of which move from east to west, and some *vice versa*, some quickly, some slowly, some to the south and some to the north. Contrariety is found in their motions in the *lavlav* (spiral?) and *ʿiqqul* (wobbly?) motions, as it is clear to the astronomers, and as he (*sic*) wrote in the book known as *The Form of the Heaven and the Earth*, and the other astronomical works written on this, and there is no contrariety greater than this.

This problem, Isserles continues, compelled astronomers to assume many orbs for the moon. Isserles then brings several more examples of contrariety in the motion of the moon and the other planets as well.

In order to understand Isserles' argument, we recall that in the medieval conception, each planet is assigned one earth centered sphere or spherical shell. Within that sphere one distinguishes a number of orbs, not necessarily centered on the earth, whose combined motions are to account for the planet's observed path through the heavens.[13] Isserles here wishes to make two points. First of all, there are some individual orbs whose motion is characterized by contrariety (*hippukh*); that is to say, they violate the basic cosmological principle that to each body, only one uniform motion may be assigned. Examples of such contrariety are not limited to the moon. Isserles refers to 'the motion which the orb has in latitude as is shown in the theory (*tekhunah*) of Venus and Mercury in the books on astronomy, and it is the variation in the orbit (*nelizat ha-maʿagal*) of the moon of which Maimonides spoke ... as the commentator explained in his remarks.' Moreover, Isserles continues, the contrariety of the specific motions, one to another, of the different orbs nested within one particular sphere is also called intrinsic (*bi-ᶜaṣmo*) contrariety. Just like the different organs of the human body with their opposing properties and unequal powers bring about degeneration, so also the different orbs found within the planetary spheres bring about decay through their various motions.

We must offer brief explanations of three specific examples of contrariety given by Isserles which are themselves not without interest for the history of astronomy. The first and most intriguing is the reference to the motions called *lavlav* and *ᶜiqqul*. Both of these terms appear in the Hebrew translation of Peurbach used by Isserles. *ᶜIqqul* (*reflexio*) is used to describe the motion of the epicyclic apsidal diameter, one of the components of the motion in latitude of the inferior planets.[14] *Lavlav* (*trepidatio*) appears in two different contexts: to describe the well-known theory of the motion of the eighth sphere proposed by Thābit bin Qurrah[15]; and to describe the motion of the additional orb introduced by Peurbach in an attempt to give some physical account of the complicated motions in latitude of the inferior planets.[16]

Isserles, however, mentions the latitudinal motions of Venus and Mercury separately; here he is referring here to the *lunar* model. Moreover, his prime source for the motions of *lavlav* and *ᶜiqqul* is not Peurbach's tract (whose title, *Theoricae*, is written out in Hebrew characters at the beginning of Isserles commentary to that text), but a work entitled *The Form of the Heaven and the Earth*. I have as yet not been able to identify this treatise which, by its title, could be some sort of recension of Aristotle's *De Caelo* or, alternatively, an exemplar of the *Imago Mundi* literature. One important medieval figure who made much of these concepts was Ibn Rushd (Averroes), particularly in his long commentary to Aristotle's *Metaphysics*. In the Hebrew translation of that text, *ᶜiqqul* is used to translate *muḥādhah* and thus refers to the lunar *prosneusis*, a prime target for those wishing to criticize Ptolemaic theory on physical principles.[17] The Arabic counterpart of *lavlav* is *lawlab*, but the exact meaning of this term requires further study.[18]

The second piece of evidence for contrariety in the planetary motions presented by Isserles is the motion in latitude of the inferior planets. Here, as we have already seen, he has touched upon another sore point of the theory of the physical structure of the heavens. According to the theory expounded in the *Almagest*, the motion in latitude of Venus and Mercury has three components: an oscillation of the plane of the eccentric inclined orb about the ecliptic; and circular (in fact, eccentric) motions of the extremities of the epicyclic apsidal diameter and the diameter orthogonal to it. As far as I know, no ancient or medieval cosmological scheme provided a satisfactory physical account for these motions.[19]

The last item mentioned by Isserles is the 'variation in the orbit'[20] of the moon, mentioned by Maimonides in his *Laws of the Sanctification of the New Moon*. This particular correction long vexed both medieval and modern commentators. The correct explanation, namely that this 'variation' is a mathematical correction and not an independent motion of the moon, was given by Baneth and accepted by Neugebauer.[21] (I have recently found that this correction is none other than the *mamarr*, a known feature of medieval Islamic astronomy).[22]

Isserles here, however, cites specifically the explanation given by 'the commentator,' i.e., Obadiah ben David. According to Obadiah (commentary to *Sanctification* XVII, 9), this variation is a discovery of the later astronomers and is, in fact, an additional anomaly which the moon exhibits in its motion:

Only a few of the later astronomers sensed or understood it. However, the earlier ones, such as Ptolemy and his colleagues, did not mention this variation, nor did any of them note that the moon varies (*naloz*) in its orbit (*maʿagalo*) to the north or south. For the ancients hold that the belt of the orb of the orbit (!! *hagorat galgal hamaʿagal*) is in the plane of the circle of the eccentric orb and that the circle of the eccentric orb is in the plane of the inclined orb, and they did not note that the moon moves (*sovev*) on another circle, different from the circle of its orbit; rather it always moves on it and does not veer from it north or south. Therefore, according to the ancients, the moon has no variation whatsoever in its orbit. In the opinion of the later astronomers, who sensed this variation and found it after intense investigation by means of instruments, the circle of the orbit is always inclined from the inclined circle in the direction of the ecliptic...

Obadiah goes on to give the parameters of this supposed inclination, which match the corrections given by Maimonides. Now it is beyond the scope of this paper (and as yet an unsolved problem) to investigate just what discovery of the later astronomers is involved here. The point of interest for us is the fact that Obadiah regards the 'variation of the orbit' not as a mathematical correction to allow the calculation of the lunar visibility but as an actual, additional anomaly in the lunar motion. As such it would require for its production an additional body according to the medieval conception of one body, one motion. No such body is found in the standard cosmology, Obadiah's 'orb of the orbit" notwithstanding. Hence Isserles' remarks are in order.

This whole line of attack focusing on contrariety is said to derive from *Guide for the Perplexed* II, 19, where Maimonides explains that 'the variations in motion and other changes in the heavenly hosts demonstrate to us that the world is created' (*T.O.*, III 626). Maimonides does make much in that chapter of the differences in motion between the various orbs, but, it seems to me, his intentions are different from those of Isserles. Maimonides' goal is to show that the particular motions of the various orbs are not as they are of necessity and, therefore, their existence is due solely to 'the purpose of One who purposed,'[23] that is to say, God. It does not seem to be the case, however, that Maimonides wishes to claim that the dissimilar motions of the orbs imply that the heavens are subject to corruption.

Isserles, however, wants to go much further than Maimonides. He quite boldly proposes that the motion of the stars can and, perhaps, ought to be explained in terms of shedding forms and accepting new forms, that is to say, that the sublunar physics be applied to the heavens as well (*T.O.*, III, 63a):[24]

I say further that it is possible that there be there [in the heavens] casting-off of forms and taking-on of forms, as in the lower [regions] . . .

For proof Isserles refers to precession. The earliest astronomers could detect no motion of the eighth orb, whereas their successors detected a slow motion. Isserles is apparently not interested in the simple explanation that, precisely because precession is so slow, it can be detected only by reference to the observations of preceding generations, and hence the earliest astronomers could not have noticed it. Instead he maintains that both the earliest astronomers, who could detect no such motion, and the later astronomers, who did, were correct:

It is possible that the claim of the ancients, that it [the eighth orb] has no motion at all, is true. Rather, the place where the star was at first cast off its form, which is the form of the star, at that time, and lost its form, and another place on the orb became fit to accept the form of the star, and that place became crystalline (*sapiri*) and luminous and became a star.

As a justification of the view of the ancients, this explanation may be hard for us to swallow. Perhaps Isserles here regards local motion of the *body* of the star as a continuous process which, however slow, would nonetheless have been detected by the ancients. On the other hand, the migration of the stellar forms would be a discrete process, and the ancients simply never had a chance to observe the stellar forms 'leap' from one place to another.[25] In any event, the point is that this same process can be used to explain all of the heavenly motions, and Isserles knows it:

> Truly we could in this manner account for all matters astronomical. However, this would be in accord with the view of the sages who claimed that the orb is fixed and the stars revolve, and they already said that the Gentile scholars prevailed regarding that. Nevertheless, with regard to the eighth [orb], concerning which the ancients agreed that it has no motion, one ought to account for it by this.

In other words, the explanation just proposed for the stellar motions is to be identified with the ancient theory of the Jewish sages of fixed orbs and moving stars. Since the Jewish sages officially gave in on this point, this view ought not to be upheld. Considering, however, Isserles' previous remarks that this concession was forced by historical circumstances and is thus not a true reflection of the scientific worth of the theory, it seems reasonable to assume that Isserles took this alternative explanation quite seriously. In any event, he proposes that it be adopted for the motion of the eighth orb, concerning which the ancients put forth no theory of their own.

One should also point out that this explanation is in line with an important facet of Isserles' intellectual personality, namely the desire to compromise. Here the compromise is between the ancient and later astronomers. In another issue of cosmological import, Isserles again suggests a compromise (*T.O.*, II, 54b):

> It is known that a great controversy broke out between the investigative scholars regarding the orbs and their motions, whether they move because they have an intellecting (*maskelet*) soul, and they themselves are alive and intelligent, which is the opinion of the author of the *Guide* and his followers, or whether their motion is purely natural, which is the opinion of the author of the *ʿAqedah* [R. Isaac Aramah (c. 1420– 1494), Spain] and his group ... It is possible that the opinions of all the scholars are true and that some of the orbs are alive and intelligent, whereas the motion of others is natural.

This point is not pursued any further.

III. THE COMMENTARY TO PEURBACH'S *THEORICAE*

We have seen that in his philosophical work, Isserles is quite open to cosmological speculation. In his commentary to Peurbach, on the other hand, he adheres strictly to the medieval view, limiting himself to elaboration upon the terse text of the *Theoricae*. More than this: in the very places where Peurbach's text may hint at an alternative notion, Isserles chooses to reiterate in his commentary the accepted medieval viewpoint.

Several avenues to non-Ptolemaic ideas may by suggested. The first is found in Peurbach's exposition of the theory of Mercury. Ordinarily, the center of a planet's epicycle describes a circle which is the deferent orb of the planet. However, Mercury's deferent is not fixed and, hence, Peurbach suggests, it may be more appropriate to regard the path traced by the center of Mercury's epicycle as an oval. Here Isserles is quick to point out that the oval deferent spoken of by Peurbach does not exist as such. At any given instant, the center of Mercury's epicycle describes a circle about the deferent center. However, if one were to superimpose the circles produced at the apogee and perigee of the turning orb, the resulting figure would resemble an oval. In Isserles' words (f. 137a):

... for the other epicyclic centers always produce one circle, which is the circle which the center of the orb traces (*ha-roshem*!). However, concerning this [Mercury] since its center [Mercury's deferent] moves, therefore its circles also produce elongated (*arukim*) circles. Even though each circle is by itself spherical (*kaduri*), nevertheless if we imagine the circumference of the circle which it makes when the center of the turning [orb] is in its apogee [extended?] towards the circumference of the circle which it makes when the center is in perigee, we see that it moves with an elongated motion. See the figure. You will see that the two circles, which are ... (word missing) on this, that when they are joined together, they are elongated and round at the top and bottom, like an egg.

I am not sure to which figure Isserles is referring. There are two unlabeled figures at the end of the manuscript (f. 161a), and the one on the left looks appropriate for Mercury, though the oval deferent is not shown. However, in one of the figures accompanying the Latin text,[26] the oval is shown exactly as described by Isserles, that is, as the superposition of the deferent circles produced at the apogee and perigee of the turning orb.

The actual path of the center of the epicycle is not a circle, and the various circles which are superimposed to form the oval are never fully traced by the epicyclic center. Yet Isserles prefers his explanation, however forced, in order to safeguard the principle that the heavenly motions be described by circles alone.

Another opportunity for departure from the inherited medieval posture is presented by the description of the latitudinal motions of Venus and Mercury. These were specifically listed by Isserles in *T.O.* as a problematic feature of astronomy. Moreover, we have noted that the additional orb introduced by Peurbach in order to give a physical explanation of this motion is itself given a motion which is *lavlav*, and *lavlav* motion as well is included by Isserles among the troublesome aspects of contemporary theory. However, in his commentary to the passage wherein Peurbach introduces his additional orb, Isserles dryly indicates that, with this extra orb, the physical problems of the motion in latitude have been solved (f. 149a):

For it is impossible that one orb have two motions, therefore it is impossible that one orb have two motions, [one in] longitude and [one in] latitude, and therefore, they were forced to add another orb which compels (*ha-makhri'ah*) its motion in latitude the way the diurnal orb compels all of them [i.e. all celestial motions?]. For it is impossible to do without it, in order that one orb not have two intrinsic motions, as was explained as a principle (*haqdamah*) in the *Guide* as well.

There is no hint here whatsoever of the oscillations and motions of epicyclic diameters, nor of the problematic nature of *lavlav* motion.

A third opportunity for deviating from standard medieval cosmology is offered, perhaps, by the question of the trepidation of the equinoxes, or the motion of the ninth orb, discussed at length by Peurbach. Here again Isserles in his commentary is at pains to restate his allegiance to one of the fundamental doctrines of medieval cosmology: each motion is to be explained by one body which is responsible for that motion alone. In his summation of the planetary theory which precedes the discussion of the motion of the equinoxes, Isserles writes (f. 151a):

For it has already been explained above that it is impossible for one body to possess two intrinsic motions, [rather] one motion alone, and the second comes about by means of another body which gives it the second motion. Therefore, since those epicycles have many motions, they found it necessary to posit those numerous orbs within those [other orbs] in order that each one cause one motion...

Then, at the very beginning of his commentary to the section on precession, Isserles states (f. 151a):

Know that these three motions which the eighth orb has are not intrinsic to it. Rather, it has one intrinsic motion, and the two [other motions] are given to it by two other orbs. The author of this book maintains that there are ten orbs. The tenth is called the prime mover and imparts to it one motion, which is the daily motion common to all of the orbs. The ninth gives it a second motion. The third motion is intrinsic to it.

We see here Isserles' concern that trepidation not bring about any physical problems. In *T.O.* Isserles wavers between trepidation and precession. Note, however, that he correctly regards this issue not as a new problem for astronomers, but as a medieval development. Thus, although the only time the *Theoricae* is mentioned in *T.O.*, it is in connection with the problem of trepidation, it is there cited in the same breath as Abraham bar Hiyya's eleventh-century treatise, *Surat ha-'Ares* (III, 4a):

All of this is clear to those who know the properties (*tekhunot*) of the eighth orb, just like they were explained by the author of the *Theoricae* and the author of *Surat ha-'Ares*...

IV. SUMMARY AND CONCLUSIONS

In summing up his own analysis of the astronomical discussions of Isserles and another sixteenth century Jewish figure, Judah Loeb (better known as Maharal) of Prague, Herbert Davidson[27] claims that these two scholars were engaged primarily in harmonizing disparate texts, rather than in addressing scientific problems. Of course, Davidson notes, medieval scholars had also been concerned with textual harmonization but, it seems, Isserles and his contemporary have moved away from the philosophic spirit of their predecessors, in so far as they neglected the basic scientific issues which underlied and motivated the discussions found in authoritative texts.

On the whole these generalizations seem acceptable. I hope that the analysis presented in this paper may help deepen their meaning. For Isserles was not interested in textual harmonization for its own sake, as a mere academic exercise. Rather, he was addressing, in the main, the serious doctrinal issue of *creatio ex nihilo*. Now in our weariness from following the argumentation over this issue during the preceding centuries, we may be inclined simply to dismiss Isserles' musings as old hat. We must remember, however, that in the sixteenth century Aristotle was still perceived as an authority whose denial of creation greatly bothered religious thinkers. Moreover, in addressing himself to this issue, Isserles displays both technical expertise and power of imagination. Nonetheless, I still share Davidson's impression: something of the philosophic spirit is missing. The question of the true nature of the heavenly motions does not seem to weigh as heavily upon Isserles as it does, say, upon Maimonides or Gersonides. He seems instead to be searching for some refutation of Aristotle; if such a refutation can also square well with Talmudic pronouncements, so much the better.

The great difference of approach manifest in Isserles' two major writings on the subject of astronomy also calls to mind certain aspects of medieval scholarship. In his commentary to the standard textbook, Isserles sticks to accepted doctrine, even in those places where that doctrine may be called into question. The commentary was probably written for pedagogic, rather than investigative, purposes. On the other hand, in his philosophical work he displays an openness to alternative cosmological schemes. That he is able to do so is due, in no small measure, to the richness of the Hebrew literature at his disposal. In his boldest assertion he proposes that the apparent motions of the stars be explained in terms of sublunar physics.

Institute of Microfilmed Hebrew Manuscripts
Jewish National and University Library
Jerusalem, Israel

NOTES

1. On Isserles see Shlomo Tal, 'Isserles, Moses ben Israel,' *Encyclopaedia Judaica*, IX (1972), pp. 1081–1085, and the literature cited there. There is a biography of Isserles in Hebrew: A. Siev, *Rabbi Moses Isserles (Ramo)* (1957), and a Hebrew monograph by Yonah Ben-Sasson, *The Philosophical System of R. Moses Isserles* (5731 = 1971/2). A recent and, in my opinion, insightful treatment of Isserles is to be found in H. Davidson, 'Medieval Jewish Philosophy in the Sixteenth Century,' in B. D. Cooperman (ed.), *Jewish Thought in the Sixteenth Century* (Cambridge U.S.A., 1983), pp. 106–145, especially pp. 132–136.
2. First printed at Prague, 1570. I used the edition of 5614 = 1814/5 printed at Koenigsberg.
3. I used Bodley Hebrew manuscript Michael 195 (Neubauer 1332), ff. 112a–161a which, in my opinion, is Isserles' autograph. (Compare the MS. with the specimen of Isserles' handwriting reproduced at the beginning of Siev's biography). Bodley Opp. 1673 (Neubauer 2033), ff. 149a–194b, is another copy of Isserles' commentary, with an introduction by one Chaim Lisker, otherwise unknown, and a few

remarks of Lisker interspersed in the commentary. Another copy, unfortunately inaccessible at the present time, is Moscow Ginzburg 1069. (This is possibly identical to Aschkenazi 1857 = Fi. 43?, listed by M. Steinschneider, *Die hebraeischen Uebersetzungen des Mittelalters und die Jüden als Dolmetschers* (1893), 640).

Peurbach's work exists in several Hebrew translations and, in addition to Isserles, M. Delacrut and Moses Almosnino wrote Hebrew commentaries on the text. See Steinschneider, *op. cit.*, pp. 639—641, 645—646.

On Peurbach see C. Doris Hellman and Noel M. Swerdlow, 'Peurbach, Georg,' *Dictionary of Scientific Biography XV* (1978), pp. 473—479. Peurbach's treatise is now available in English: E. J. Aiton, 'Peurbach's *Theoricae novae planetarum*: A Translation with Commentary,' *Osiris*, 2nd series, *3* (1987), pp. 5—43.

4. S. Dubnow, *History of the Jews of Russia and Poland* (trans. I. Friedlander), Vol. 1 (Philadelphia, 1916), pp. 131—132.
5. A. Siev, *op. cit.*, p. 30, claims — without, however, offering any evidence — that Isserles' devotion to science and philosophy was due in large measure to the influence of Katzenellenbogen.
6. Albertus de Brudzewo, *Comentaria utilissima in theoricis planetarum* (Milan, 1495). See also J. Dobryzycki, 'Nicolaus Copernicus — His Life and Works,' in B. Bieńkowska (ed.), *The Scientific World of Copernicus* (Dordrecht-Boston, 1973), pp. 13—37, at p. 15.
7. H. H. Ben-Sasson, writing in the *Encyclopaedia Judaica* (part one of 'Poland,' in v. 13 (Jerusalem, 1972), pp. 710—722) quotes (pp. 721—722) two Christian sources, the chronicler Maciej Miechowicz and the cardinal legate Lemendone, both of whom note the interest in astronomy and medicine which was shown by sixteenth century Lithuanian Jews. Dubnow, *loc. cit.*, notes that Jewish physicians served at the Polish court. However, there is still no hard evidence of direct intellectual exchanges unless, perhaps, one wishes to include religious confrontations. On the latter see Judah Rosenthal, 'Marcin Czechowic and Jacob of Bełżyce: Aryan-Jewish Encounters in Sixteenth Century Poland,' *Proceedings of the American Academy for Jewish Research, 34* (1966), pp. 77—95.
8. This point is emphasized and substantiated by both Dubnow and Ben-Sasson (see preceding note).
9. The incompatibility of Aristotelian physics and Ptolemaic astronomy was for medieval thinkers 'a *skandalon* of science.' (S. Pines, 'Translator's Introduction,' Maimonides, *The Guide for the Perplexed*, Vol. I (1963), lxiii). The views of Maimonides and some later Jewish thinkers are discussed in my paper, 'The 'True Perplexity': *Guide for the Perplexed* II, 24,' in J. L. Kraemer (ed.), *Perspectives on Maimonides* (Oxford, 1990), 193—208. See also A. I. Sabra, 'The Andalusian Revolt against Ptolemaic Astronomy,' in E. Mendelsohn (ed.), *Transformation and Tradition in the Sciences* (Cambridge Univ. Press, 1984), pp. 133—153.
10. It seems clear that Israeli has al-Biṭrūjī in mind. *Cf.* B. R. Goldstein, *Al-Biṭrūjī: On the Principles of Astronomy*, 2 vols. (New Haven: Yale Univ. Press, 1971), *I*, p. 43.
11. The very brief discussion in the Babylonian Talmud (Pesaḥim, 94b) clearly has roots in ancient debates over cosmology. *Cf.* Aristotle, *De Caelo*, II, 8. Medieval thinkers kept the issue alive. I looked at Ibn Rushd's middle commentary to *De Caelo* in Hebrew translation, Bodley 1375, 27b ff., and Levi ben Gerson's

supercommentary in Casanatensa 152, 33b. In my opinion this issue is worthy of a closer study.

12. We consistently use the following translations for the technical terms of medieval cosmology:

   galgal: orb (Arabic *falak*, Latin *orbis*)
   kadur: sphere (Arabic *kurah*, Latin *sphaera*)
   ʿiggul: circle (Arabic *dā'irah*, Latin *circulum*)

   Sphere always refers to a three-dimensional figure, and circle to a figure of two dimensions. Orb is rather ambiguous and can be used for both two and three dimensional figures. See our 'A Note on the Use of the Term *Orbis (Falak)* in Ibn al-Haytham's *Maqālah fī Hay'at al-ʿĀlam*,' *Archives Internationales d'histoire des Sciences, 32* (1982), pp. 112—113.

13. For a recent summation of medieval cosmology, see Edward Grant, 'Cosmology,' in David C. Lindberg (ed.), *Science in the Middle Ages* (Chicago, 1978), pp. 265—302.

14. This is the third motion listed by Peurbach (I used the edition of Vitebergae, 1553, 220b; it corresponds to the Hebrew translation in ms. Bodley 1332 (sec note 3), 148b): 'ex parte reflexionis diametri longitudinem mediarum respectu augis verae, quae reflexio appelatur.' In the Hebrew translation of al-Biṭrūjī, *ʿiqqul* (Arabic *inḥirāf*) is used to denote the slant of the lunar epicycle, a feature of the latitude theory. (Goldstein, *al-Biṭrūjī, I,* 149; *II,* 412). For clarification of these and other technical features of ancient and medieval astronomy, see Olaf Pedersen, *A Survey of the Almagest* (Odense, 1974). See also note 19 below.

15. Peurbach, *Theoricae,* 245a = Bodley 1332 156b: 'Thabit vero duplicem tantum octavae spaerae motuum inesse dixit. Unum ... alium vero proprium scilicet trepidationis ...' Al-Biṭrūjī as well used *lawlab* to describe the trepidation of the equinoxes (Goldstein, *I,* 23).

16. *Theoricae,* 221b = Bodley 1332, 149a: 'propter dictas autem deviationis orbibus preanumeratis alium mundo concentricum praedictas omnes includentem superaddi videtur oportere ad cuius motum trepidationis praedictae deviatio nec accidant.'

17. *Cf.* volume 3 of the long commentary, edited by M. Bouyges as volume VII of *Bibliotheca Arabica Scholasticorum* (1948), pp. 1657, 1659. The *prosneusis* is attacked by Ibn al-Haytham in his *Al-Shukūk ʿalā Baṭlamyūs* (ed. A. I. Sabra and N. Shehaby, 1971), pp. 15 ff.

18. In the Latin translation of Averroes, the term used is *gyratio* (Bouyges, p. 1669). In one place, Ibn Rushd also suggests that the *lawlab* motion may refer to the trepidation spoken of by the Spanish astronomers (p. 1675). Nevertheless, it seems that Ibn Rushd has in mind some sort of spiral motion with more general applications for astronomy and, as noted, the matter needs further study. See in particular Sabra (cited in note 9), note 7.

19. The latitude theory is presented in the *Almagest* XIII, 2, with an apology on the part of Ptolemy for its complicated nature. For clear presentations with figures of this theory, see O. Neugebauer, *A History of Ancient Mathematical Astronomy,* 3 volumes, vol. I (1975), pp. 216—17, and Noel Swerdlow, 'The Derivation and First Draft of Copernicus' Planetary Theory: A Translation of the Commentariolus with Commentary,' *Proceedings of the American Philosophical Society,* Vol. 117, no. 6

(1973), pp. 423—512, at pp. 494—496. For Peurbach's suggestion of an additional orb to account, at least partially, for latitudinal motion, see above and note 16.
20. 'Orbit' is used here to translate the technical term *ma'agal* (following Gandz, see following note) and should not be confused with the modern meaning of the term nor with the medieval term 'orb' (*galgal*).
21. *Cf.* O. Neugebauer, 'Astronomical Commentary,' pp. 140—141, in Maimonides, *Sanctification of the New Moon*, translated by S. Gandz (New Haven, 1956).
22. See Y. T. Langermann, *The Jews of Yemen and the Exact Sciences* (Jerusalem, 1987), English section, p. 5, note 10, and Y. T. Langermann, P. Kunitzsch, and K. A. F. Fischer, 'The Hebrew Astronomical Codex Ms. Sassoon 823,' *Jewish Quarterly Review, 78* (1988), pp. 253—292, note 47.
23. Maimonides, *Guide* (trans. Pines), v. 2, p. 303.
24. Earlier writers had also countered the Aristotelian position with an assertion that the heavens are corruptible. Among Jewish philosophers Sa'adya had taken this view; the argument may go back to John Philoponus. See H. A. Wolfson, *The Philosophy of the Kalam* (Cambridge U.S.A. and London, 1946) pp. 514—516; and H. Davidson, 'John Philoponus as a Source for Medieval Islamic and Jewish Proofs for Creation,' *Journal of The American Oriental Society, 89* (1969), pp. 357—391. It appears that Isserles arrived to his particular viewpoint on his own; but see the following note.
25. The concept of 'leap' (Arabic *tafrah*), i.e., that a body can move from A to B without passing through the intervening space, was known to the Muslim *mutakallimūn* (speculative religious thinkers). The notion is thought to have been introduced by al-Nazzām in the ninth century. See H. A. Wolfson, *op. cit.*, pp. 514—516, and M. Fakhry, *History of Islamic Philosophy* (1970), pp. 241—242. However, I cannot say whether Isserles had some source for this idea or whether he came upon it on his own. *Tafrah* is not among the doctrines of the *kalām* which are refuted at length by Maimonides, *Guide*, I, ch. 73.

Now a Hebrew text found at the Bodlean Library at Oxford (Neubauer 1331), which dates probably from the fourteenth or fifteenth centuries, addresses on f. 75b the question: 'Does the angel, in moving from one place to another, pass over the places in between?' In my judgement (I plan a fuller study) this text was written by a Jew who was deeply imbued with Latin scholasticism. However, I can say nothing as to how this text fits into any developments in European thought, Jewish or Christian. Scholars have yet to investigate the influence of *kalām* on European thought. (There are some useful remarks in Wolfson, *op. cit., passim*).
26. The figure appears in the edition of Nuremberg, 1472 or 1473, 9r, and is reproduced in Willy Hartner, 'The Mercury Horoscope of Marcantonio Michiel of Venice,' *Vistas in Astronomy, 1* (1955), pp. 84—138, at p. 130.
27. Davidson (see note 1), p. 138. Siev (see also note 1), whose biography is written in a traditional, laudatory style, notes repeatedly the eclectic and unoriginal nature of Isserles' thought (see especially p. 238). One should also take note of a social factor which may have motivated Isserles' eclecticism and efforts at harmonization, namely, an attempt on his part to tone down any possible conflicts between narrow-minded traditionalists, kabbalistic mystics, and those bent on philosophical speculation.

PART THREE

Medieval Cosmology, Natural Philosophy, and Optics

EDWARD GRANT

CELESTIAL INCORRUPTIBILITY IN MEDIEVAL COSMOLOGY 1200–1687

I. ARISTOTLE

Prior to the introduction of Aristotle's physical works into the Latin West during the late twelfth and early thirteenth centuries, the idea of celestial incorruptibility was probably a minority opinion. It was not uncommon for scholars in late antiquity and the early Middle Ages to assume that the heavens were composed of one or more of the four elements. Since the elements were thought of as changeable entities, those who held that the whole world, including the heavens, was composed of one or more of them were committed, implicitly or explicitly, to the idea of a changeable or corruptible heaven.[1] The introduction of Latin translations of Aristotle's works during the twelfth and thirteenth centuries radically altered this tradition. A vital ingredient of Aristotle's 'new' cosmology was the belief in celestial incorruptibility.

Aristotle distinguished two kinds of incorruptibility: one associated with eternality and *unchangeability*, the other linked with eternality and *changeability*. The cosmos, which for Aristotle, was ungenerated, eternal, and indestructible, contained within itself both kinds of incorruptibility, one associated with each of two radically different parts into which he assumed the world was divided, namely the sublunar and celestial regions. Taken as a whole, the four elements of the sublunar, or terrestrial, region were as indestructible as the fifth element, or ether, of the celestial region. The totality of terrestrial matter that was composed of the four elements was constant and eternal, without beginning or end. Although the four elements are eternal, incorruptible, and indestructible as a whole, they are always changing, one into the other. Parts of fire were incessantly transformed into earth, and vice versa, while parts of air were always being transformed into water, and vice versa. Things were otherwise in the celestial region, where no part of the heaven could be transformed, or was transformable, into anything else. Apart from changes of position that arose as a consequence of regular circular motion, scholastic authors were agreed that substantial changes could not occur in the heavens. Our concern will be with this seemingly absolute sense of celestial incorruptibility.

II. THE MEDIEVAL DEFENSE OF CELESTIAL INCORRUPTIBILITY

The medieval defense of celestial incorruptibility included theory, observation, and an intuitive sense, expressed over many centuries, that the heavens, associated with angels and spiritual substances, ought to be incomparably superior to the terrestrial region. The theory depended on Aristotle's conception of generation and corruption, a conception that distinguished radically between terrestrial matter and celestial ether. Generation and corruption, or change, occurred in all substances that consisted of matter possessed of a form or quality that was potentially replaceable by its contrary. While one form was actualized in matter, its contrary was said to be in privation, but potentially capable of replacing it. Moreover it was assumed that that potentiality would eventually have to be realized; otherwise the form would remain unactualized and nature would have produced it in vain. Thus while one of a pair of contrary forms was actualized in matter, the other was absent and in privation, since two contrary forms could not exist simultaneously in one and the same body. Generation and corruption, and therefore all change, involved the possession and expulsion of one or the other of a pair of contrary forms or qualities.

While the kind of change just described was assumed to occur incessantly in terrestrial bodies, it was denied for the ether, or fifth element, that filled the celestial region beyond the concave surface of the lunar sphere.[2] Whatever opinion one might have held about the nature of the celestial ether — whether it was matter and form, matter only, or form only[3] — all who accepted its existence, and very few denied it,[4] were agreed that it did not take on contrary forms and could not therefore suffer generation and corruption by natural means. It followed that celestial matter was naturally incorruptible. Arguments in defense of celestial incorruptibility relied heavily on the presence or absence of contrary forms. It was usually conceded, however, that God, who had created the world supernaturally, could, if He wished, destroy it supernaturally.[5]

John Buridan incorporated these and other ideas into his defense of celestial incorruptibility. The celestial region is incorruptible because it lacks matter, where it is understood that 'every naturally generable or corruptible thing has matter.' The absence of matter in the heaven is inferred from the assumption that the celestial ether cannot receive a form or quality contrary to any of those that it already possesses.[6] Substantial change was therefore impossible in the heavens.

Is it perhaps possible that the heaven could suffer change and corruptibility by increasing or decreasing its size? Following Averroes, Buridan envisioned increase, or augmentation, in three ways:[7] by nutrition, as with plants and animals; by rarefaction; and by adding things from outside, as in heaping up a pile of stones. The celestial region, however, could increase in none of these ways. Nutrition requires substantial generation in which the nutrient is converted into the substance of the thing that ingests it. But substantial change was impossible in the heavens. Rarefaction depends on the primary qualities of hot and cold, which do not exist in the heaven because they would cause generation and corruption. Nor, finally, can the heaven be augmented from outside. Such an augmentation would require substances that possess a celestial nature. Where could such substances exist before they attached themselves to the heavens? If they were located some distance away, rectilinear motion would be required to reach their goal. But the celestial substance lacks any natural capacity for rectilinear motion.[8]

Was celestial change not also implied by the universally held assumption that the celestial region was the cause of virtually all changes in the terrestrial, or sublunar, region? In the process of causing terrestrial changes, would not the heaven itself inevitably suffer some degree of change? How, for example, could the sun and the stars affect the earth by light and other influences without the assistance of the inferior celestial spheres? Was it not plausible to suppose that the dissemination of influences from heavenly bodies in the farthermost reaches of the heavens could only be multiplied with the aid of the planets and spheres in the lowest parts of the heavens? The celestial recipients of these influences must surely be altered in some sense. To cope with this fundamental problem, Buridan and others distinguished a proper from an improper alteration. The dissemination of light and other influences involved the multiplication and reception of qualities that did not involve contraries nor any form of resistance to their transmission. In a similar vein, John of Jandun denied that the apparent variability of the sun's intensity in winter and summer was an instance of celestial corruptibility. The sun's power was identical in summer and winter. During the summer, however, its rays produced greater heat because they were refracted more nearly at right angles than in the winter. These variations in refraction of the sun's rays occurred only in the sublunar region and were caused by alterations in the media. Like Buridan, Jandun justified his defense of incorruptibility by distinguishing a proper from an improper sense of 'alteration' (*alteratio*).[9] By such means, natural philosophers also denied the claims of astrologers that the power of a planet varied in different signs.[10]

John Buridan and John of Jandun give us a sense of the problems that were raised about celestial incorruptibility. Apart from the apparent changes that one might observe in the heavens, about which more will be said in the next paragraph, Buridan and Jandun, and most medieval scholastics, imagined situations that tested the claim for incorruptibility. In all such instances they would 'save the imaginary phenomena' by denying the existence of contraries in the heavens or by distinguishing an improper from a proper sense of 'alteration.' Thus, as we saw, light and other influences were said to alter nothing in the heavens and to meet no resistance from the ether they moved through.

Of equal importance with the arguments just described, and perhaps even more significant, was Aristotle's declaration that no changes in the celestial region had ever been observed or recorded.[11] Buridan reiterated this powerful defense of celestial incorruptibility when he asserted that 'from all the writings that we have from most ancient times, it does not appear that the heaven from that time to the present has been corrupted or worsened.'[12] If the heavens were corruptible, some obvious sign of that corruptibility should have been detectable over a long period of time. Celestial incorruptibility was further reinforced by the fact that everyone assigned God or gods to the celestial region; for why would an eternal God or gods be assigned to a corruptible place?

From these and similar arguments, we may rightly infer that scholastic authors from the thirteenth to fifteenth centuries had developed an elaborate defense of celestial incorruptibility, which acquired the status of a virtually self-evident truth. Aristotle had greatly facilitated the near unanimous acceptance of incorruptibility by denying celestial locations to shooting stars, comets, and similar phenomena, and placing them instead below the moon in the upper atmosphere of the terrestrial region.[13] Without challenges to the Aristotelian system, celestial incorruptibility was taken as an essential ingredient of world order. Only when challenges to that system materialized in the sixteenth century, did some scholastics, especially in the seventeenth century, conclude that absolute incorruptibility was, after all, dispensable. Although the Copernican system made the earth a planet and therefore as perfect as the other planets, or, conversely, made the planets as imperfect as the earth, it was not until Tycho Brahe located the New Star of 1572 and the Comet of 1577 in the celestial region that belief in celestial incorruptibility would begin seriously to waver.

Claims in favor of celestial corruptibility were hardly new. As we saw (above, n. 1), a number of Church Fathers had argued for, or at least implied, a mutable heaven. Plato, whose works had become available in the late fifteenth century, was regularly included among those who believed in celestial change. A number of prestigious Greek commentators of late antiquity, such as John Philoponus and Themistius, also argued for celestial corruptibility. The works of all these figures were readily available during the course of the sixteenth century. Indeed even before Tycho's astronomical contributions, Bernardino Telesio and Girolamo Cardano had both rejected an unchanging heaven.[14]

III. SCHOLASTIC INTERPRETATIONS OF CELESTIAL INCORRUPTIBILITY IN THE SIXTEENTH AND SEVENTEENTH CENTURIES

By the seventeenth century, the consequences of Tycho's assault on celestial incorruptibility began to tell. The Aristotelian dogma of an unchanging heaven was under serious attack.[15] Scholastic authors who wrote *questiones* on Aristotle's *De caelo* — usually within a *cursus philosophicus* ranging over Aristotle's physical and metaphysical works — frequently included a question on whether the heavens were incorruptible. What had been a rather routine question in the Middle Ages was now a problem that threatened a vital aspect of traditional Aristotelian cosmology.

1. *Arguments in the Medieval Mode: Is the Celestial Substance Incorruptible?*

Three opinions about celestial incorruptibility were distinguishable: (1) that the heaven is intrinsically corruptible; (2) that by its very nature it is incorruptible; and (3) that the first two opinions are equally probable because the arguments for each are plausible. In elaborating the third opinion, Bartholomew Amicus (1562—1649), a Jesuit theologian, presents illuminating insights into the attitudes of theologians and Church Fathers toward the problem of celestial incorruptibility. The doctors who considered these matters never condemned the other opinions on doctrinal grounds because they were agreed that each opinion was consistent with the faith and without danger of error. Unlike the problem of the centrality of the sun or the earth, Amicus and other seventeenth century scholastics viewed the incorruptibility, or corruptibility, of the heavens as a philosophical problem rather than one of doctrine or faith. Indeed the 'principles of Platonic philosophy' were considered the ultimate source of support for those who assumed the corruptibility of the heavens.[16] The absence of doctrinal conflict, and an apparent sense that celestial corruptibility would not of itself destroy Aristotelian cosmology, eventually led some scholastics, as we shall see, to accept the corruptibility of the heavens.

Most scholastics, however, remained faithful to the medieval tradition and defended celestial incorruptibility.[17] Their defense was a combination of old and new, the latter consisting largely of arguments devised to cope with the consequences of the well-attested astronomical phenomena that had been witnessed in the 1570s, namely the New Star of 1572 and the Comet of 1577. Despite such significant evidence of change, to which, as we shall see, scholastics reacted in a variety of ways, the Coimbra Jesuits, Bartholomew Amicus and Franciscus de Oviedo could still proclaim that the best evidence for celestial incorruptibility was the experience of the ages which revealed a remarkable constancy in the celestial region.[18]

Most seventeenth century scholastics who defended celestial incorruptibility based their arguments on the intrinsic nature of the heavens rather than on some external cause.[19] Sometime around 1648, Johannes Poncius (John Punch; 1599—1661), a Franciscan Professor of Philosophy and Theology, declared that the most common opinion amongst peripatetics and theologians was that the heaven was incorruptible by its internal nature and could be altered neither by any natural agent nor by God in his ordinary power.[20] In support of his position, Poncius invoked the most basic and common medieval defense of celestial incorruptibility: the absence of contrary qualities in the heavens. Without contrary qualities, the qualities possessed by celestial matter were all capable of preserving the heaven because change could only occur if they were destroyed by contrary forms or qualities. Sublunar substances, by contrast, are alterable because it is their inherent nature to require contrary qualities.[21]

Bartholomew Amicus invoked additional metaphysical arguments to demonstrate the intrinsic incorruptibility of the celestial region. One appealed to the perfection of the universe. If imperfect things are part of a perfect universe, how much more ought incorruptible things to be part of it. But the heaven is the only reasonable choice for incorruptibility because all other bodies are imperfect and corruptible.[22] Another argument took its departure from the well-ordered perpetual generation and corruption of sublunar things.[23] Because every effect requires a proportionate cause, an invariable, incorruptible body is required to preserve the perpetual changes that occur in the determinate order of sublunar activity. Since the effect is perpetual, so also must the cause be perpetual and indeed incorruptible. This argument depends on the assumption — which Amicus makes — that experience teaches us that the order of the universe (*ordo universi*) in inferior things is determinate and invariable and cannot arise from the nature of inferior things, which are themselves variable and corruptible. Hence it ought to arise from an eternal and invariable principle, which must be the heaven.[24] In support of this inference, Amicus presents confirming arguments. Because the nature of a thing is proportionate to its goal or end and the goal of the heaven is to conserve the world perpetually by its continuous influences, it follows that the heaven must be incorruptible to act perpetually as a constant cause. Moreover, if the heaven was corruptible and some of it was destroyed, the entire universe would lack proper regulation. But God created the heaven so that it might serve as a rule or model (*regula*) for all corruptible things and human actions.

2. *Arguments in the Medieval Mode: Are Celestial Accidents Incorruptible?*

The incorruptibility considered thus far was relevant only to the celestial substance. But scholastics were also concerned about the nature of celestial accidents, or qualities, that inhered in the celestial substance or were produced by it. Were such accidents also incorruptible? By distinguishing two kinds of accidental changes, Bartholomew

Amicus argued that real accidental changes cannot occur in the heavens. In a separate question devoted to this problem, Amicus, as Galileo did before him, divided accidents into 'corruptive' and 'perfective.'[25] A corruptive accidental change (*mutatio accidentalis corruptiva*) is not possible in the heavens because it required the presence of contrary qualities, like hot and cold, which produced substantial mutations, that is, caused the corruption of one substance and the generation of another, as in the transformation of a cold substance to a hot substance. Corruptive changes were only possible in sublunar bodies composed of the four elements.

By contrast, a 'perfective accidental change' (*mutatio accidentalis perfectiva*) can occur regularly in the heavens because it does not involve positive contrary qualities. With perfective accidents, changes occur from the privation, or absence, of a quality, to the appearance of a positive quality; or vice versa.[26] Such accidents appear in the transmission of both light and various celestial influences that of necessity pass through successive parts of the heaven because they cannot act at a distance. As examples of perfective accidents, Amicus mentions lunar eclipses and the continual generation and corruption of the sun's light as it is moved around the sky by its orb.[27]

3. Challenge to Incorruptibility: The New Celestial Discoveries of Brahe and Galileo

Whatever departures from medieval cosmology these arguments may represent, they were clearly in the medieval tradition and represented nothing really new. Departures from that tradition would eventually arise of necessity by way of reactions to the new cosmology that was being shaped as a consequence of the astronomical achievements of Tycho Brahe and Galileo, the former demonstrating the celestial nature of new stars and comets, the latter detecting the variable nature of sunspots. The inference of celestial corruptibility seemed to follow inevitably. However plausible the inference, it was resisted by many, if we are to believe statements by Amicus and Poncius, who, as we saw, claimed that most scholastics and theologians believed in celestial incorruptibility.[28] Based on appearances, scholastic theologians seem to have been reasonably united in their defense of the traditional sense of celestial incorruptibility. Actually their responses to the powerful threat to the traditional sense of incorruptibility varied considerably. Some would deny that the new stars, comets, and sunspots were really celestial phenomena; others would accept their celestial location but deny that any substantial change had really occurred; a few would even abandon celestial incorruptibility. Variation, rather than uniformity, characterized the scholastic response during the seventeenth century.

Scholastics were aware of at least three instances of celestial change that had been reported prior to the sixteenth century, one by Pliny, who mentions that Hipparchus had observed a new star that eventually disappeared, and two by St. Augustine, one involving changes in the magnitude and course of Venus and the other a diminution in size of the star Spica in the contellation Virgo.[29] But such claims were unverifiable. Only when Tycho Brahe failed to detect parallax in his observations of the New Star of 1572 did many astronomers and natural philosophers come to believe that a new celestial phenomenon had actually occurred in the heavens. The lack of parallax seemed to certify a location in the sphere of the fixed stars above the planets.[30] A brief, but spectacular, two year life span left a considerable impact as it first emerged in brightness and then gradually dimmed and disappeared. Here indeed was an apparent case of celestial generation and corruption.

On the common assumption that comets were generated below the moon,[31] Tycho had denied that the new star of 1572 could be a comet. Five years later, the Comet of 1577 permanently altered his opinion on the location of comets. From its parallax, Tycho judged the comet to be above the moon in the region of Venus. With this knowledge, Tycho repudiated two fundamental Aristotelian and medieval beliefs: the conception of comets as sublunar and the immutability of the heavens.[32] By the time many of our seventeenth century scholastic authors wrote, celestial incorruptibility had received yet another blow, when, in 1611, Galileo, Christopher Scheiner, and Johann Fabricius used the telescope to observe the variability of sunspots.[33] 'If blemishes could appear and disappear on the face of the sun itself,' Drake observes, 'the incorruptibility and inalterability of the heavenly bodies was destroyed.'[34] Galileo's telescopic observations of the moon, too, indicated a surface of great unevenness covered with mountains and valleys thus suggesting a corruptible body like the earth.

4. *The Scholastic Reaction: Bartholomew Amicus*

All of these discoveries were well known to scholastic authors of the seventeenth century and all were specifically mentioned by Bartholomew Amicus, who rejects them one by one.[35] Although he did not himself propose definitive explanations, Amicus presents four possible approaches that scholastic authors could, and did, employ as the basis for appropriate interpretations of the new discoveries. All denied the occurrence of substantial celestial changes.

The first preserved the Aristotelian position that all the seemingly new phenomena are actually sublunar. They were the physical effects of various external causes, such as an impure medium, the extreme distance of the objects, the falsity of the instruments, and so on.[36] The objective in this approach was to exploit the inevitable differences in observation, interpretation, and mode of argumentation amongst those who upheld the celestial location of the new discoveries. These differences were taken as a sign of uncertainty and weakness of argument. Indeed Amicus believes that all comets after the time of Aristotle were sublunar. In particular, the comet of 1618 was below the moon because to the naked eye and to those viewing through a telescope (*tubum opticum*) it appeared weaker, duller and more obscure than the light of the stars. Nor was its parallax sufficient to place it beyond the moon. Indeed those who place the comet above the moon do not agree on its altitude.[37]

The second approach[38] relies on God's supernatural power. One could readily concede that the newly observed phenomena are in the etherial heaven in both substance and accidents, placed there by a supernatural power. Many reasons might have motivated God to create these celestial displays: as the sign of a great effect; to terrify mortals; to demonstrate his dominion and power over his creatures, and so on. Because of their supernatural origin, the new celestial phenomena should not count as natural celestial alterations. Celestial incorruptibility remained inviolate.

The third approach was clearly the most important. It was a response against those who located some, or all, comets above the moon. To counter such claims, it was necessary to explain the emergence and disappearance of each comet and new star 'without [the assumption of] a new generation and corruption of celestial substance.'[39] Amicus reports a number of ways in which others had sought to attain this objective. One way was to assume that a new star was nothing more than a concentration of reflected light (*splendor*) from an aggregate of planets and stars located in a particular part of the heaven. A second explanation involved a union of small, ordinarily invisible stars (*stellae*), which upon coming together formed a sufficient mass to become visible. A third way involved certain stars in the firmament each of which is fixed in an epicycle that makes the star visible when the epicycle is turned toward us and invisible when it is turned away from us. To all of these possibilities, Amicus himself raised fatal objections.[40]

He was more receptive to a fourth possibility in which, once again, the idea was to produce a new star from something that preexisted in the sky and was therefore ungenerated. Because visible stars remained at fixed distances from each other, they could not unite in any way to form a brighter, new star. To side-step this difficulty, Amicus assumed numerous epicycles carrying denser, but invisible, parts of celestial ether. When perchance three of these epicycles were aligned in such a way that their denser etherial parts are clustered together, the sun, which shines on all the stars, would illuminate the etherial cluster causing the latter to appear as a new star. As the epicycles move away from each other, the new star will gradually fade away. Although Amicus found this the most probable of the alternatives described — it was apparently defended in a public debate on the comet of 1618 at the University of Ingoldstadt — he could not understand how the New Star of 1572 in Cassiopeia could be a first magnitude star during its first year and then jump to a third magnitude star in the second year. Why, if the epicyclic motions were continuous, did they not produce a star of the second magnitude before they generated a star of the third magnitude?

Amicus suggested yet other explanations for new stars and comets,[41] but like many scholastics, would not commit himself to any single explanation to account for the new celestial phenomena. He was content to present a number of possible explanations that denied corruptibility. This was a reasonable procedure since no causal explanations for the new phenomena were sufficiently compelling to make the case for celestial corruptibility.

5. *The Scholastic Reaction: Christopher Clavius and Raphael Aversa*

Where Bartholomew Amicus used contemporary responses to defend vigorously traditional cosmology, Christopher Clavius (1538—1612) and Raphael Aversa (1589—1671) sought, however modestly, to adapt to the new astronomical discoveries of the sixteenth and early seventeenth centuries.

Clavius was one of the first scholastics to support Tycho Brahe. If the New Star were no further away than the atmosphere, or air, it should have revealed different aspects. None had been observed. Nor indeed could it be in any of the regular planetary orbs because no astronomer had yet detected any motions that might indicate this. Clavius concluded that it had to be in the most remote parts of the celestial region, that is, in the firmament among the fixed stars.

But what could have caused the New Star? Clavius suggests that it was either created by God as a major portent, or, if it were a natural celestial event, then one would also have to concede that comets could also be created in the heavens, just as they could be created in air. But if this were true, perhaps 'the heaven is not a certain fifth element, but a mutable body, although less corruptible than inferior [terrestrial] bodies.'[42] Many philosophers of the ancient world believed this before and after Plato and Aristotle. Indeed some of the most eminent Church Fathers — Basil, Gregory of Nyssa, and Ambrose — taught the corruptibility of the heavens.

Just as we expect Clavius to opt for celestial corruptibility, he decides to withold his opinion (*meam enim sententiam in tanta re non interpono*)[43] and rest content to have demonstrated that the new star is in the starry firmament. Nor did he wish to attempt answers to other difficult questions associated with the new star: what it portended; why it vanished after two years, and so on. Only God knows the answers to such questions. Thus did one of the most influential astronomers of the sixteenth and early seventeenth centuries decide in favor of the real existence of the new star but avoid a decision on the perplexing question as to whether that new star signified a corruptible or incorruptible celestial region.

Raphael Aversa sharply distinguished between new stars and comets: the former were celestial; the latter, phenomena of the upper atmosphere just below the moon.[44] Thus Aversa agreed with Clavius that new stars were celestial phenomena. Acceptance of a celestial location for new stars was based on Aversa's high regard for astronomers like Tycho Brahe, who could find no parallax for the New Star of 1572. Unlike Clavius, however, Aversa chose one of the positive alternatives: the celestial region was incorruptible.

The assumption of new stars as celestial phenomena posed a major problem for Aversa and others who shared his opinions: how could the incorruptibility of the heavens be preserved while simultaneously accepting the real celestial location of new stars? After rejecting the arguments of those who located the New Star of 1572 in the upper atmosphere,[45] Aversa opposed similarly those who conceded that it was a celestial phenomenon, but denied that it was newly generated in the sky. The defenders of this opinion insisted that the new star was there all the time, but hidden from our view until somehow rendered visible. In this explanation, the traditional assumption was made that the heaven is not of equal density, but that stars are denser accumulations of the celestial ether. Some stars, however, were invisible because they lacked the requisite density to reflect sufficient light. But somehow, in the course of celestial rotations, other denser parts of the heaven are added to, or aligned with, a previously invisible star, rendering the whole aggregation visible. Conversely, when these parts are withdrawn from the new star it gradually disappears from sight and resumes its previously invisible state.[46] Among his refutations of this explanation,[47] Aversa observes that if it were true and all celestial motions are uniform, a given denser portion of the sky ought to be augmented and diminished in the same way in every one of its revolutions. Hence a new star ought to come into being and pass away in every revolution.

Not only did Aversa refute other physical explanations, but he firmly rejected the notion that the new star of 1572 was a miracle.[48] In the absence of any seemingly plausible natural explanation, the Coimbra Jesuits and others thought it reasonable to invoke a miracle. But Aversa insisted that the subsequent appearance of other new stars tended to weaken the case for a miracle. Moreover, when God performed celestial miracles in Sacred Scripture, the reason for it was made evident. But what was the divine purpose of the new star? No satisfactory response had been given and Aversa opted to omit God from his explanation.

He was, however, convinced that new stars were really new productions in the sky. How this might occur is described in a final opinion, with which Aversa seems to agree.[49] New stars are brought into being not by a substantial change but only by an accidental mutation. Aversa's explanation is unusual because it departs from certain ideas that were in vogue when he wrote. It was assumed by many that the formation and disappearance of new stars was caused by condensation and rarefaction of celestial matter. For those who assumed that the heaven was composed of fluid matter akin to air — as Tycho Brahe did and, by the 1620s, many others — a greater condensation of the fluid at one point in the heavens could produce a new star and its subsequent gradual rarefaction would cause it to disappear. Some supporters of solid celestial spheres offered a similar explanation but found it more difficult to explain how condensation and rarefaction could occur in solid spheres. Aversa, who believed in the solidity of the celestial spheres, was aware of the difficulty and sought to avoid it by abandoning altogether the idea that the heaven was distinguishable into rare and dense. It was better to conceive the heaven as displaying opacity (*opacitas*) and transparency (*diaphaneitas*) because those qualitative properties are more appropriate to the heaven than the quantitative properties, rare and dense.[50] Aversa makes his point by comparing crystal and wood. Although crystal and glass are very dense, they are not opaque but transparent whereas wood, which is more opaque than crystal or glass, is less dense than either. What makes the heaven transparent like air is its great density, or solidity. By analogy with crystal, celestial transparency is evidence of solidity rather than rarity.

But Aversa thinks it inappropriate to say that stars are more solid than the heaven or that the heaven is sometimes more solid than at other times. Changes in the heaven cannot occur by mutations of substance or quantity, but only by qualitative means. A qualitative change occurs when a diaphanous part of the heaven becomes somewhat opaque and can thus receive, and presumably reflect, sufficient light to become visible as a star (just as, presumably, the dense crystal is slightly opaque so that it reflects light). The new star will gradually disappear as its opacity diminishes and it ceases to reflect light.[51] But what could make a part of the heaven opaque and then diaphanous? Aversa admits that the cause of such effects — whether the result of a very powerful cause, or a particular alignment of stars — is unknown, although he reports, with considerable scepticism, that some astrologers have attributed this process to certain planetary aspects.

Despite the concession that new stars were unequivocally celestial phenomena, Aversa denied that substantial or quantitative changes could occur in the celestial region. By rejection of the essentially quantitative terms 'dense' and 'rare' and their replacement by the presumably qualitative terms 'diaphaneity' and 'opacity,' Aversa believed that he had somehow preserved celestial incorruptibility.

6. Repudiation of Celestial Incorruptibility

While there were certainly other arguments in support of celestial incorruptibility, those by Amicus and Aversa represent a good sampling of the kinds that scholastics believed most effective in the seventeenth century.[52] During the first three-quarters of that century, most scholastic authors continued to defend celestial incorruptibility. But not all.

In the third quarter of the seventeenth century, three who did not were Giovanni Battista Riccioli (1598—1671), Melchior Cornaeus (1598—1665), and George de Rhodes (1597—1661), all Jesuits. Of the three, Riccioli, is by far the most significant. The most famous Jesuit astronomer of his day, he published the *New Almagest* (*Almagestum novum*) in 1651,[53] where he included virtually all the major problems of Aristotelian cosmology and considered how the latter was affected by the new Copernican astronomy, as well as by the recent appearance of new stars, comets, and sunspots. Many of these problems were taken up in the ninth book titled 'On the Systems of the World.'[54] Here, in Section 1, Riccioli treated 'The Creation and Nature of Celestial Bodies' within which context — in chapters 5 and 6 — he discussed the problems relevant to our subject. The fifth chapter[55] is devoted to the nature of celestial matter — whether it is a simple or composite body and whether it is the same or different from that of elementary matter. The decisions on these questions are relevant to the sixth chapter[56] in which Riccioli actually considers 'whether the heaven is generable and corruptible.'

After following the usual scholastic procedure and presenting pro and contra arguments for the various relevant questions, Riccioli concluded not only that the heaven is composed of matter and form — a popular and not very controversial opinion in the sixteenth and seventeenth centuries — but that celestial matter is probably the same as the matter of the sublunar region,[57] a thesis defended in the Middle Ages by Aegidius Romanus (ca. 1245—1316) and William of Ockham (ca. 1285—ca. 1350).[58] Aegidius Romanus, however, like his medieval colleagues, believed unquestioningly in celestial incorruptibility. He was therefore confronted with a serious problem: if celestial and terrestrial matter are the same pure potencies capable of receiving forms, why do generation and corruption occur only in the sublunar region? For Aegidius the answer lay in the differences between the forms in the two regions. In the terrestrial region, every form has its contrary which will inevitably replace it and cause a generation and corruption; in the heaven contrary forms and privations do not exist. Although the matter in the heaven is the same as that in the sublunar region, the form in celestial matter cannot be displaced and change cannot occur. Celestial incorruptibility was thus preserved without serious challenge.

Riccioli, however, drew radically different consequences from his assumption of the essential identity of celestial and terrestrial matter. For him the conclusion that celestial and terrestrial matter were of the same kind, signified that the heaven must be composed of one or more of the elemental bodies: fire, air, and water.[59] Thus where Aegidius Romanus made prime matter the basis of the identity of celestial and terrestrial matter, Riccioli made it elemental matter, that is matter that had already been actualized beyond the level of prime matter. It was thus appropriate for Riccioli to identify the elements from which the heaven was composed.[60] He concluded that 'it is more probable that the heaven in which the fixed stars are is watery; the heaven in which the planets are is fiery.'[61] The probability of this interpretation derived not from reasoned arguments, but appeared to be more consonant with the manner in which Patristic authorities had interpreted Sacred Scripture. Some Church Fathers had believed that the waters above the heaven had congealed to form the sphere of the fixed stars, while others believed that the planets moved by themselves through a fluid planetary matter. Riccioli accepted both versions[62] and thus followed a trend that was well established by 1651, when he published the *Almagestum novum*.[63]

With the fixed stars embedded in a single solid sphere and the planets moving freely through a fluid medium, Riccioli had to identify the elements that constituted these heavens. He readily admitted that no genuine evidence or precise arguments could be offered in support of the claim that the heaven of the fixed stars was a congealed watery solid and the heaven of the planets was a fiery fluid. Patristic authorities were however at hand. Because some fathers had held that the heaven consisted of elementary water and others that it was composed of elementary fire,[64] it seemed a good compromise to identify the sphere of the fixed stars as the solid and watery sphere because the stars themselves remained fixed and unchanging and seemed to enclose the world and because the term 'firmamentum' was used to describe the starry sphere; and to view the heaven through which the planets moved as a fiery fluid since the paths of the planets varied.[65]

Although Riccioli's assumption of a fluid planetary heaven was not of itself a sufficient indication of a belief in celestial corruptibility,[66] his belief that the heaven actually consisted of two terrestrial elements was. In his chapter on the corruptibility or incorruptibility of the celestial region, which follows immediately after the chapter that identified celestial and terrestrial matter, Riccioli declares the corruptibility of the celestial region. On the basis of his assumption that the heaven of the fixed stars is most probably watery and that the heaven of the planets is fiery, he infers 'that from its very internal nature, the heaven has the capacity for generation and corruption.'[67]

Perhaps because of an awareness of how radically he had departed from traditional Aristotelian cosmology, Riccioli sought to salvage a remnant of celestial incorruptibility. Although, by its elemental nature, the heaven is intrinsically corruptible, it is not corruptible by any naturally created external agent. Thus, for Riccioli, the celestial region was 'accidentally incorruptible' (*per accidens esse incorruptibile*) because no natural, external agent could corrupt it. This immunity from corruption by natural agents was perhaps a consequence of the heaven's great distance from the terrestrial region, which was external to it; or perhaps attributable to the great mass of the celestial region; or because of the distinctive nature of the primary qualities that God placed in the heaven. Whatever the reason, Riccioli's concession to incorruptibility was of little consequence as is evident when he likens celestial incorruptibility to that incorruptibility which applies to the whole earth and to the totality of air, each of which is really incorruptible as a totality even if its parts suffer continual change. Despite its overall incorruptibility, the heaven was nevertheless capable of suffering corruption in its parts. In this the celestial region was just like the earth or air: it suffered change in its parts while the whole endured unchanged.

Only with regard to the Empyrean sphere did Riccioli accept the traditional opinion of incorruptibility. The Empyrean sphere was not, however, a visible sphere, although it was required for the perfection of the universe and for the incorruptibility and eternal well-being of our bodies.[68]

Melchior Cornaeus rejected celestial incorruptibility in a single paragraph.[69] Because Cornaeus believed in the identity of celestial and sublunar matter, and the latter is a principle of generation and corruption, he inferred corruption for the heavens.[70] Partial corruption in the heavens is also evident from the many new stars that have been reported from as far back as 125 B.C., including those of 1572, 1600, and 1604. Corruptibility is also implied by Scriptural predictions of a judgment day in which the heavens will be destroyed. And finally Cornaeus argues that the sun, the most beautiful part of the heavens, is corrupted almost daily by fires.

CONCLUSION

It is evident that in the seventeenth century, scholastic opinions about celestial incorruptibility changed rather dramatically from what they had been during the period between the Middle Ages and the end of the sixteenth century. Even if the majority of seventeenth century scholastics retained the traditional opinion — and this is by no means certain — scholastics like Riccioli, Cornaeus, and de Rhodes were prepared to abandon it and concede that substantial generations and corruptions could and did occur in the celestial region. In answering the charge that Aristotle had declared the heavens to be immutable and incorruptible, Cornaeus would even declare that

if Aristotle were alive today and could see the alterations and conflagrations that we now perceive in the sun, he would, without doubt, change his opinion and join us. Surely the same could be said about the planets, of which the Philosopher knew no more than seven. But in our time, through the works of the telescope, which was lacking to him, we know for an absolute certainty that there are more.[71]

Even some traditionalists like Aversa were prepared to break with Aristotle and allow that new stars and sunspots are celestial, rather than terrestrial, phenomena.

But why did scholastic Aristotelians yield on this seemingly important element in Aristotle's cosmology? On this, we can only speculate. Although many scholastics denied a celestial location to the new phenomena, others must have realized, as did Aversa, that astronomical data from the most respected astronomers of the day could not be ignored indefinitely. Thus the first breakthrough — to concede the celestial location of the new phenomena — was probably made rather painlessly because it was still feasible, and even easy, to insist that such phenomena represented only accidental, rather than substantial, changes.

The eventual transition to the concept of celestial corruptibility was probably aided in no small measure by a widespread belief in the sixteenth and seventeenth centuries that Plato, Sacred Scripture, and many Church Fathers were agreed that the heaven was constituted of one or more terrestrial elements and that the heaven was therefore capable of substantial change.[72] Indeed, as we saw, the eighth sphere of the fixed stars was often identified as the frozen or crystalline form of the Scriptural waters above the firmament. Other Church Fathers had followed the Platonic idea that the heavens were made of the fourth and purest element, fire. Although, as we saw earlier, such ideas were known during the late Middle Ages, Aristotle's conception of a fifth incorruptible element, or ether, had replaced Platonic and Patristic interpretations. To works containing ideas about the corruptibility of the heavens in the Middle Ages, others were added as they became available in the sixteenth century, as, for example, the works of Plato, John Philoponus and Saint Basil. In the developing anti-Aristotelianism of the sixteenth century, celestial corruptibility was an opinion that became more difficult to ignore than during the Middle Ages. The dramatic celestial discoveries of the late sixteenth and early seventeenth centuries provided the scientific basis for abandoning incorruptibility. Scholastics who found the celestial discoveries of Tycho Brahe and Galileo compelling could justify support for celestial corruptibility by direct appeal to Plato and, more significantly, to the Church Fathers. Or they may have found the astronomical arguments compelling only because of the corroborating statements of the Church Fathers.

What role did the gradual acceptance of a fluid, rather than solid, heaven play in the abandonment of celestial incorruptibility? Tycho Brahe's claim that the comet of 1577 was moving among the planets clearly implied the non-existence of solid planetary spheres.[73] For those who accepted comets as supralunar, a gradual but inexorable shift toward a fluid heaven began. But did a fluid heaven imply a corruptible heaven? At least one Jesuit scholastic, Antonio Rubio, in a work of 1615, believed that a fluid heaven would have to be corruptible (presumably because of divisibility) and therefore rejected it.[74] But others, for example, Bartholomew Amicus, Johannes Poncius and Franciscus de Oviedo, thought that the solidity or fluidity of the heavens was irrelevant to the issue of incorruptibility.[75] Indeed Oviedo believed that the heaven was both fluid and incorruptible.[76] For some scholastics, then, fluidity alone did not necessarily entail divisibility. The matter of the heaven might be such that it was only capable of receiving a single form; or celestial matter might be incorruptible by virtue of its form, a form that adhered to its matter so firmly that another could not be received.[77] A seventeenth century scholastic could therefore accept fluidity and incorruptibility. Although the shift from solidity to fluidity was a significant change from the medieval tradition, it was not crucial for the issue of celestial incorruptibility.

Indiana University
Bloomington, Indiana, U.S.A.

NOTES

*   I am grateful to the Program in History and Philosophy of Science of the National Science Foundation for its generous support of my research on medieval cosmology of which this article forms a part.
     Because this article considers medieval scholastic thought from the thirteenth to seventeenth centuries, I have chosen a termination date of 1687 for dramatic effect. As the date of publication of Isaac Newton's *Philosophiae Naturalis Principia Mathematica*, it marks the effective end of medieval cosmology.
1.   Two who were explicit in their assumption of celestial corruptibility were John Philoponus, the sixth century Greek Neoplatonic commentator on the works of Aristotle, and John Damascene. For selections on celestial corruptibility drawn from the works of Philoponus, see Walter Böhm, (ed. and tr.), *Johannes Philoponos Grammatikos von Alexandrien |6. Jh. n. Chr.|* (Munich: Verlag Ferdinand Schöningh, 1967), pp. 326—27, 329—31. Although the works of Philoponus, from which Böhm drew his selections, were unknown in the Middle Ages, some of the relevant passages had been quoted by his contemporary, Simplicius, who included them in his Greek commentary on Aristotle's *De caelo*, which was translated into Latin by William of Moerbeke in 1271. For a brief discussion, see S. Sambursky, *The Physical World of Late Antiquity* (New York: Basic Books, Inc., 1962), pp. 158—166, especially p. 164. John Damascene assumed the corruptibility of the heavens when he declared that 'it is evident that the sun, moon, and stars are composite, and by their very nature subject to corruption.' The passage is from Damascene's *De fide orthodoxa* as translated in *Saint John of Damascus, Writings*, (tr.) Frederic H. Chase, Jr. in *The Fathers of the Church, A New Translation* (New York: Fathers of the Church, Inc., 1958), bk. 2, p. 221. For much the same reason, St. Augustine probably believed in celestial corruptibility, and perhaps also St. Basil. But they were not explicit.
2.   Aristotle, *De caelo* 1.3.270a.13—24.
3.   On the different interpretations of celestial matter, see my article 'Celestial Matter: a Medieval and Galilean Cosmological Problem,' *Journal of Medieval and Renaissance Studies, 13* (1983), pp. 157—186.
4.   One of the few who denied that the whole heaven was composed of a fifth element, or ether, was Robert Grosseteste, who, in his *De generatione stellarum*, argued that

the stars were composed of the four elements and were therefore corruptible. The heaven was therefore at least partially, if not wholly, corruptible. See Ludwig Baur, (ed.), *Die philosophischen Werke des Robert Grosseteste, Bischofs von Lincoln, Beiträge zur Geschichte der Philosophie des Mittelalters*, Vol. **9** (Münster, 1912), p. 33, pp. 35—36. I am grateful to Dr. Peter Sobol for bringing this treatise to my attention.

5. In their defenses of celestial incorruptibility, both John Buridan and Galileo concede God's power to destroy the world. For Buridan, see *Iohannis Buridani Quaestiones super libris quattuor De caelo et mundo*, (ed.), Ernest A. Moody (Cambridge, Mass.: The Mediaeval Academy of America, 1942), bk. 1, question 10 ('Whether the heaven is generable, corruptible, augmentable, diminishable, and alterable'), p. 46; for Galileo, see *Galileo's Early Notebooks: The Physical Questions. A translation from the Latin, with Historical and Paleographical Commentary*, (tr.), William A. Wallace (Notre Dame, Ind.: University of Notre Dame Press, 1977), p. 101. William Ockham insisted that because God could destroy the heaven if He wished, the incorruptibility of the celestial region was not absolute. Indeed it was potentially corruptible. For the references and further discussion, see Grant, 'Celestial Matter,' pp. 171—172. The passage from Galileo cited above was intended as a rejection of Ockham's position, which is described by Galileo on p. 94, 6b (Ockham is not mentioned).
6. See Buridan, *Quaestiones super libris quattuor De caelo*, p. 46. By adopting a strict Aristotelian interpretation of matter, Buridan was led to deny the existence of matter in the heavens. Thus he stood in opposition to Thomas Aquinas, who assumed a celestial matter that was radically different from its terrestrial counterpart, and Aegidius Romanus (Giles of Rome), who argued that celestial and terrestrial matter were identical. Despite these radically different interpretations of the celestial ether, all believed in celestial incorruptibility. For an exposition and analysis of these different views, see Grant, 'Celestial Matter.'
7. *Ibid.*, pp. 46—47.
8. Both Albert of Saxony and Galileo adopted the same interpretation as Buridan. For Albert, see *his Questions on De caelo*, bk. 1, question 17 (*quaestio ultima*), in *Questiones et decisiones physicales insignium virorum: Alberti de Saxonia in octo libros Physicorum; tres libros De celo et mundo; duos lib. De generatione et corruptione; Thimonis in quatuor libros Meteororum; Buridani in tres lib. De anima; lib. De sensu et sensato . . . Aristotelis. Recognitae rursus et emendatae summa accuratione et iudicio Magistri Georgii Lokert Scotia quo sunt Tractatus proportionum additi* (Paris, 1518), fol. 102r, col. 1; for Galileo, see Wallace, (tr.), *Galileo's Early Notebooks*, p. 100.
9. *Ioannis de Ianduno in libros Aristotelis De coelo et mundo quae extant quaestiones subtilissimae: quibus nuper consulto adiecimus Averrois sermonem De substantia orbis cum eiusdem Ioannis commentario ac quaestionibus* . . . (Venice: apud Iuntas, 1552), bk. 1, question 17 ('Whether the heaven is alterable'), fols. 12r, col. 2 — 12v, col. 2.
10. Buridan flatly denied that internal alterations could occur within a planet, although he allowed that its zodiacal location might play a role in the planet's influence. As he explained it, 'a hot planet seems of greater power if it is in a hot sign than if it

were in a cold sign because the sign and the planet can [then] simultaneously influence heating and thus a great hotness arises here below. But if the hot planet is in a cold sign, the influence of the sign prevents the influence of the planet from acting because the sign acts in a way contrary [to the planet]; [under these circumstances] the planet appears to possess little power.' Buridan, *op. cit.*, p. 46 for the astrologers' claim and p. 48 for Buridan's reply.

11. *De caelo*, 1.3.270b.13—17.
12. *Op. cit.*, p. 46.
13. See Aristotle, *Meteorologica* 1.4.342a.30—33; also 1.3.341a.33—35.
14. See William H. Donahue, *The Dissolution of the Celestial Spheres 1595—1650* (Ph.D. dissertation, University of Cambridge, 1972; reprint New York, Arno Press, 1981), pp. 51, 52.
15. See Donahue, *ibid.* and numerous other references on the problem of celestial incorruptibility scattered throughout the dissertation.
16. Because of its importance, I quote the relevant part of the third opinion: 'Tertia opinio est dicentium utramque opinionem esse probabilem quia utriusque partis rationes solvi possunt. Unde etiam putant utramque opinionem salva fide ac sine erroris periculo, ac temeritatis ulliusque gravioris censure nota defendi posse, ut patet ex Doctoribus, qui de re tractant inter quos nemo alterius partis opinionem damnat ut censura dignam. Nam licet aliqui ex patribus dicant coelos natura sua esse corruptibiles, tamen non significant pertinere ad fidem, nec illam deducunt ex principiis fidei, sed ex principiis philosophiae Platonicae . . . ' Amicus, *In Aristotelis libros De caelo et mundo dilucida textus explicatio et disputationes in quibus illustrium scholarum Averrois, D. Thomae, Scoti, et Nominalium sententiae expenduntur earumque tuendarum probabiliores modi afferuntur* (Naples, 1626), p. 232, col. 2.
17. In their commentary on *De caelo*, the Coimbra Jesuits declared that 'by common consent, almost all the Peripatetic Schools' (*tota fere Peripatetica schola communi assensu*) supported the incorruptibility of the heavens (*Commentarii Collegii Conimbricensis Societatis Iesu: In quatuor libros De coelo Aristotelis Stagiritae* [Lyon, 1598; first published in Coimbra, 1592], p. 66; the work was actually written by Emmanuel de Goes, S. J. [1542—1597]). Similarly Franciscus de Oviedo declared that 'almost all philosophers, theologians, and interpreters of Scripture' (*fere omnes Philosophi, Theologi, atque Scripturae interpretes*) believed that the heavens are incorruptible (*Integer cursus philosophicus ad unum corpus redactus*, Vol. **1** [Lyon, 1640], p. 464).
18. Coimbra Jesuits, *De caelo*, p. 66; Amicus, *De caelo*, p. 232, col. 2; Oviedo, *Cursus philosophicus*, p. 464, col. 1, par. 19.
19. As examples of external causes, Amicus, *De caelo*, pp. 230, col. 2—231, col. 1, mentions God, who may preserve incorruptibility by his own power; or He may add a preservative quality to the heavens that makes them incorruptible, just 'as the bodies of the blessed are said to become incorruptible by the gift of incorruptibility.'
20. *Philosophiae ad mentem Scoti cursus integer primum quidem editus in collegio Romano Fratrum Minorum Hibernorum. Nunc vero demum ab ipso authore in conventu magno Parisiensi recognitus* . . . (Lyon, 1672), question 4 ('Whether the

heaven is corruptible'), p. 617, col. 1. Although the work was first published in three volumes between 1642 and 1645, the section on *De caelo* was added to the editions after 1648. Poncius, who was a Scotist, assisted Luke Wadding in the publication of the edition of the works of John Duns Scotus. See Charles H. Lohr, 'Renaissance Latin Aristotle Commentaries: Authors Pi-Sm,' *Renaissance Quarterly*, Vol. 33, no. 4 (winter 1980), p. 665.
21. Indeed even if celestial and sublunar matter were identical, Poncius insists that the heaven would be incorruptible and the terrestrial region corruptible, as we see in the first conclusion of question 4, where he declares (*ibid.*) that 'The heaven could be incorruptible from its very internal nature whether its matter were of the same species as that of sublunar matter or different.'
22. Amicus, *De caelo*, p. 235, cols. 1—2.
23. *Ibid.*, p. 235, col. 2.
24. *Ibid.*
25. 'Whether [the heaven] is corruptible as to accidents' in Amicus, *De caelo*, pp. 247, col. 2 — 249, col. 1. In support of this distinction, Amicus cites John of Jandun, whose ideas were briefly mentioned above.
26. In his *Questions on De caelo*, Galileo declared his agreement with Simplicius, Averroes, and St. Thomas that 'alteration is twofold: one corruptive and the other perfective. The first is between contraries and involves corruption, and this has no place in the heavens; the other involves no contrariety and is found even in spiritual things — for which reason it is found also in the heavens' (Wallace, tr., *Galileo's Early Notebooks*, pp. 100—101).
27. Amicus also classified the generation and corruption of new stars and sunspots as accidental perfective mutations because they are formed by the natural motions of bodies and epicycles. For further discussion, see below.
28. According to Johannes Hofer, who wrote on the question of celestial incorruptibility in the early eighteenth century, a great number of peripatetics continued to support celestial incorruptibility. By that time, however, Hofer could report that the authors who thought the heavens corruptible were far more numerous ('Corruptibiles autem caelos asservant auctores longe plures'). See *Promptuarium philosophicum complectens argumenta e nobilioribus philosophiae totius controversiis authore P. Joanne Baptista Hofer, S. J. in Alma Catholica et Electorali Universitate Ingoldstadiensi*. Editio tertia (Ingolstadt, 1732), p. 644. Indeed Hofer claimed (p. 646, col. 2) more than 80 supporters of celestial corruptibility comprised of classical authors, Church Fathers, astronomers and mathematicians.
29. For Pliny, see *Natural History*, bk. 2, ch. 26 and for Augustine's statement on Venus, see *City of God*, bk. 21, ch. 8. I have not located a reference to Spica, which is mentioned by Amicus (*De caelo*, p. 242, col. 1, where the reference to Augustine is given incorrectly as bk. 21, ch. 10). Although Raphael Aversa (*Philosophia metaphysicam physicamque complectens quaestionibus contexta in duos tomos distributa auctore P. Raphaele Aversa* [Rome: Apud Iacobum Mascardum, 1627], Vol. 2, pp. 83—84) makes no mention of Spica, he does include correct references to both Pliny and Augustine and adds a few more examples of alleged celestial corruptibility.
30. Aversa, *Philosophia*, p. 84.

31. See C. Doris Hellman, *The Comet of 1577: Its Place in the History of Astronomy* (New York: Columbia University Press, 1944), p. 121.
32. Hellman, *The Comet of 1577*, p. 130. Tycho wrote two treatises on the comet of 1577, one in Latin, the other in German. The former, *De mundi aetherei recentioribus phaenomenis*, was first published in 1588 but was begun immediately after the comet's disappearance (Hellman describes the volume on pp. 337—338). Its wider dissemination began only at the start of the seventeenth century. The German work, according to Hellman (p. 122), was 'probably written immediately after the disappearance of the comet in 1578 but first printed in 1922.' Also see Hellman's discussion in *Dictionary of Scientific Biography*, Vol. 2, pp. 406—408.
33. See Stillman Drake (tr.), *Discoveries and Opinions of Galileo* (Garden City, N.Y.: Doubleday & Co., 1957), p. 82.
34. Drake, *Discoveries and Opinions*, p. 83.
35. Amicus, *De caelo*, p. 242, for the five arguments indicating celestial corruptibility and p. 246 for his rebuttals.
36. Drake (*Discoveries and Opinions of Galileo*, p. 73) describes Galileo's attacks against those who sought to subvert his telescopic discoveries by challenging the validity of the instrument.
37. Amicus, *De caelo*, p. 243.
38. *Ibid.*, pp. 243, col. 2; 244, col. 2.
39. *Ibid.*, pp. 244, col. 2; 245, col. 2.
40. Among a number of arguments against the second explanation, Amicus insists that such a union of stars could not occur in the orb of the fixed stars because the latter can have no motion relative to one another. Although this reason would also have sufficed against the third explanation, Amicus denies the existence of stellar epicycles on grounds that their existence should cause new stars to occur more frequently.
41. *Ibid.*, p. 246, col. 1. In one explanation, Amicus mistakenly attributes to Christopher Clavius the opinion that the New Star of 1572 was not real but was rather a regular star that only appeared larger because of terrestrial exhalations that lie between us and the star, just as a coin placed in water appears greater because of refraction. Clavius, however, reported this as the opinion of others (see *Christophori Clavii Bambergensis ex Societate Iesu in Sphaeram Iohannis de Sacro Bosco Commentarius*, 4th ed. (Lyon, 1593), p. 208. His own opinion, as we shall see below, upheld Tycho Brahe's judgment that the New Star was really in the heaven.
42. Clavius, *ibid.*, p. 211.
43. *Ibid.*
44. The arguments pro and con, with an ultimate resolution in favor of the sublunar nature of comets, appear in Aversa's *Philosophia metaphysicam physicamque complectens quaestionibus contexta*, 2 vols. (Rome: Apud Iacobum Mascardum, 1625, 1627), Vol. 2, pp. 91—100.
45. For his arguments against locating new stars in the sublunar region, see Aversa, *Philosophia*, Vol. 2, pp. 85—86.
46. *Ibid.*, p. 86, col. 2. According to Aversa, Franciscus Vallesius, who upheld this interpretation, invoked Genesis 2.1 and Ecclestiastes 3.14 to show that God would

not create a new star. In the former, we learn that God completed the heaven and earth in the latter that all of His works are eternally preserved. It followed that God would not create a new star nor allow one to be generated.
47. *Ibid.*, pp. 86, col. 2—87, col. 2.
48. This is the seventh of eight opinions that Aversa describes (*ibid.*, p. 89, col. 1).
49. The eighth opinion in *ibid.*, pp. 89, col. 2—90, col. 1. Because he mustered serious objections to all but the eighth opinion, it appears that Aversa favored it over the others, although he failed to make his support explicit.
50. Sigismundus Serbellonus seems to agree with Aversa (see his *Philosophia Ticinensis*, 2 vols. [Milan: ex typographia Ludovici Montiae in Collegio S. Alexandrii PP. Barnabitarum, 1657—1663], Vol. 2, p. 35).
51. In their joint work, Bartholomeus Mastrius and Bonaventura Bellutus adopted the same opinion, and even mention Aversa and Amicus. They explain that 'with regard to the stars newly seen, we say that they have occurred from a certain accidental mutation made in the heaven from a certain concourse of stars unknown to us with respect to opacity and diaphaneity, so that the part that was previously transparent (*diaphana*) emerged opaque. Thus it could reflect light to us and be seen.' And, on the basis of parallax, they, like Aversa, also placed the new star at approximately the distance of the sun but definitely not among the fixed stars. See *Bartholomaei Mastrii de Meldula et Bonaventurae Belluti De Catana Ord. Minor Convent. Magistr. Philosophiae ad mentem cursus in integer.* Vol. 3: *continens disputationes ad mentem Scoti in Aristotelis Stagiritae libros De anima, De generatione et corruptione, De coelo, et Metheoris. Editio novissima a mendis expurgata* (Venice, 1727), p. 500, par. 114.
52. In the 1590s, Galileo himself had defended celestial incorruptibility with quite similar, but even more traditional arguments, arguments that were representative of the Jesuit theologians and natural philosophers at the Collegio Romano between 1570 and the early 1590s. See Wallace, (tr.), *Galileo's Early Notebooks*, qu. 4 ('Are the Heavens Incorruptible?'), pp. 93—102. Galileo's relations to the Jesuits at the Collegio Romano are described by William A. Wallace, *Prelude to Galileo, Essays on Medieval and Sixteenth-Century Sources of Galileo's Thought* (Dordrecht and Boston, 1981), pp. 281, 308, 309. For a list of the Jesuit authors on whom Galileo may have relied, see Wallace, (tr.), *Galileo's Early Notebooks*, pp. 12—21. Of the group, Christopher Clavius is the best known.
53. *Almagestum novum astronomiam veterem novamque complectens observationibus aliorum, et propriis novisque theorematibus, problematibus, ac tabulis promotam; in tres tomos distributam quorum argumentum sequens pagina explicabit* (Bologna: ex typographia Haeredis Vitorij Benatij, 1651). Only the first volume, in two parts, appeared.
54. Book 9 is in the second part (*pars secunda* or *pars posterior*) of the first volume.
55. *Ibid.*, pp. 232—236.
56. *Ibid.*, pp. 237—238.
57. Riccioli presents his opinion in the following conclusion: 'Licet non possit a nobis demonstrative atque evidenter sciri, quaenam sit caeli visibilis substantia et natura, probabilius tamen est illud constare ex materia eiusdem rationis cum elementari.' *Ibid.*, p. 235, col. 1

58. For a discussion of their arguments, see my article, 'Celestial Matter' (n. 3 *supra*).
59. By omitting any discussion of earth as a possible component of celestial matter, Riccioli indicates his rejection of it.
60. Riccioli, *Almagestum novum*, pars posterior, p. 235.
61. 'Probabilius est caelum in quo sunt stelle fixae aqueum; caelum autem in quo sunt planetae igneum esse' (*ibid.*, p. 236, col. 1).
62. In ch. 7 (*ibid.*, pp. 238—244), Riccioli took up the question 'Whether the heavens are solid or whether, indeed, some or all are fluid' ('An caeli solidi sint, an vero fluidi omnes vel aliqui'). At the end of the question, in a 'unica conclusio,' Riccioli declares that (p. 244, col. 1) 'although it is scarcely evident mathematically or physically, it is much more probable that the heaven of the fixed stars is solid, that of the planets fluid.'
63. The transition from the universally held belief that the heaven was comprised of a large number of solid spheres to a belief that the heaven was entirely fluid is described by Donahue, *Dissolution of the Celestial Spheres*. (n. 14 *supra*). 'By the end of the 1620's,' he declares (p. 188), 'the debate over the fluidity of the heavens was very nearly concluded. Although belief in solid spheres was not quite dead, it was, even in the universities, the opinion of a minority of authors.' Although Donahue mentions a number of scholastic authors, it is by no means obvious whether his judgment would apply to the majority of those who wrote in the course of the seventeenth century, or even after 1620.
64. *Almagestum novum*, p. 233, col. 2.
65. *Ibid.*, p. 236, cols. 1—2.
66. A number of scholastics had argued that 'whether the heaven was fluid or solid' had no relevance to its corruptibility or incorruptibility (see the conclusion, below).
67. 'Sequitur caelos hosce esse ab intrinseco et natura sua generationis et corruptionis capaces. . . .' *Almagestum novum*, pars posterior, p. 238, col. 1. When in this question on celestial incorruptibility, Riccioli described the basis for the belief in celestial corruptibility, he declared that 'the foundation of this opinion is threefold: namely, the authority of Sacred Scripture, the testimony of the Fathers, and the arguments derived from experience concerning spots and torches near the solar disk that were discovered by the telescope and from certain comets that have come into being and passed away above the moon. These changes are more naturally explained by generation and corruption than by other more violent means or by nonviolent miracles' (*ibid.*, p. 239, col. 2). Despite the empirical and scientific arguments for believing in celestial corruptibility, Riccioli chose to base his decision on the first two reasons.
68. George de Rhodes presented similar ideas in *R. P. Georgii de Rhodes Avenionensis, e Societate Iesu, Philosophia peripatetica ad veram Aristotelis mentem libris quatuor digesta et disputata. Pharus ad theologiam scholasticam nunc primum in lucem prodit* (Lyons: sumptibus Antonii Hugetan & Guillielmi Barbier, 1671), pp. 278—281. Since de Rhodes died in 1661 and his work was first published in 1671, the actual date of composition is unknown. For what it is worth, there is no mention of Riccioli in the sections that I read. De Rhodes argued for the fluidity of the entire heavens, including the sphere of the fixed stars. He specifically refuted the explanations of Vallesius (that new stars are not 'new' but are in the heaven all

the time and only seen when they become sufficiently dense) and Aversa (that new stars are produced by an accidental generation of opacity; *Philosophia*, p. 279, col. 1).
69. *Curriculum philosophiae peripateticae, uti hoc tempore in scholis decurri solet . . . auctore R. P. Melchiore Cornaeo, Soc. Iesu, SS. Theologiae doctore eiusdemque in alma universitate Herbipolensi professore ordinario* (Herbipolis, 1657), p. 489.
70. In the very next section, Cornaeus rejects the existence of a celestial ether, or fifth element, and suggests that fire is the most probable matter of the heavens (*ibid.*, pp. 490—491).
71. 'Si Aristoteles hodie viveret et quas modo nos in sole alterationes et conflagrationes deprehendimus, videret absque mutata sententia nobiscum faceret. Idem sane est de planetis quos Philosophus septenis plures non agnoscit. At nos hoc tempore opera telescopii (quo ille caruit) plures omnino esse certo scimus.' Cornaeus, *Curriculum philosophiae peripateticae*, p. 503.
72. Riccioli discusses all three corroborating sources and provides a lengthy list of biblical passages and Church Fathers (*Almagestum Novum*, pp. 237—238).
73. Because the parallax of the comets placed them below the fixed stars, one could continue to believe, as did Riccioli, that the fixed stars were embedded in a solid sphere (Donahue [*The Dissolution of the Celestial Spheres*, p. 117] holds that the sphere of the fixed stars was the last element of the old cosmos to go).
74. Donahue, *op. cit.*, p. 105.
75. Amicus, *De caelo*, p. 270, col. 2; Poncius, *Philosophiae ad mentem Scoti cursus integer*, p. 620, col. 1; Oviedo, *Integer cursus philosophicus*, p. 462, par. 2.
76. Oviedo, *ibid.*, p. 464, col. 1, par. 17.
77. Oviedo, *ibid.*, p. 462, col. 1, par. 2.

EDITH D. SYLLA

THE OXFORD CALCULATORS AND MATHEMATICAL PHYSICS: JOHN DUMBLETON'S *SUMMA LOGICAE ET PHILOSOPHIAE NATURALIS*, PARTS II AND III

What made the Oxford Calculators famous, or infamous, in the fourteenth and fifteenth centuries was their subtlety. In this, although their distinctiveness lay in their use of mathematics within physics, their reputation merged with that of English logicians such as William of Ockham, their work being put along with terminist logic into the general category of 'Anglican subtleties.' In recent papers I have tried to show that this linking of mathematical and logical work arose in part because of the importance of certain disputations at Oxford, in particular the disputations *de sophismatibus* and the disputations that constituted the 'determinations' that occurred at the time of the student becoming a bachelor of arts.[1] Within the context of these student disputations, the Calculators' facility with mathematics was valued because it provided students with complex, often counterintuitive, results which they might use to defeat their opponents.

In the present paper, however, I want to direct attention to the Calculators' use of mathematics in contexts more directly tied to natural philosophy or physics and less associated with sophismata or disputation. Of all the Calculators' works, William Heytesbury's *Rules for Solving Sophismata* may be taken as the best known example of the linking of their work to sophismata or disputations. When seen in the light of Heytesbury's *Rules*, Richard Swineshead's *Book of Calculations* appears to have many of the same disputational-sophismatical links. Others of the Calculators' works, however, have many fewer obvious links to solving sophismata. In this latter category I would include Thomas Bradwardine's *On the Ratios of Velocities in Motions*, Roger Swineshead's *On Natural Motions*, and John Dumbleton's *Summa of Logic and Natural Philosophy*.

In comparing the Oxford Calculators' work to that, say, of Copernicus or Galileo, it distorts the situation in some respects to concentrate on how the Calculators' work appears in a disputational-sophismatical context, because in that context the prime purpose was not to expound natural philosophy or physics as such, but rather to train students to think clearly and reason effectively. If we want to know how the

Calculators operated when they were dealing with physics or natural philosophy as such, we will do better to look at Bradwardine's *On Ratios*, Roger Swineshead's *On Natural Motions*, or Dumbleton's *Summa*.

In this paper, then, I concentrate on the last named work and in particular, because of length constraints, on Parts II and III of that work. I argue for two general theses relevant to the topic of our conference. First of all, when they are observed doing physics, the Calculators do not fit the description that 'many of the more adventurous notions of the high Middle Ages were never advanced as true, or even credible.'[2] On the whole, Dumbleton in his *Summa* is arguing forcefully in favor of physical propositions which he believes to be true. Only occasionally does he offer alternate solutions, as might be offered to a student preparing for an assigned role in disputations. In this respect, then, Copernicus's acceptance of the hypotheses of his system as realities does not entail an 'epistemological revolution' and is not a 'drastic departure from the scholastic tradition.'[3] Of course, in the medieval division of labor between physics and astronomy, the physicist was generally supposed to be discovering how things really are, even if the astronomer was not. I might add to that picture, however, the claim that even when physics was made mathematical like astronomy, it did not, at least in the case of a Bradwardine or a Dumbleton, give up its claims to be describing reality. If at Paris in the fourteenth century there was a tendency to downplay, or deny, the demonstrative character of natural science, in case science should conflict with theology — and I think this may have been less common than is sometimes asserted — then perhaps Oxford was different from Paris in having natural philosophers working more autonomously.

In an article on 'Aristotelianism and the Longevity of the Medieval World View,' Edward Grant has argued that the scholastic *questio* form:

produced an atomization of Aristotle's physical treatises into sequences of particular questions and problems which focussed attention on the independent question and thus severed its connections and associations with other related issues treated in the same treatise or elsewhere in the Aristotelian corpus. Not only were related topics left uninterpreted, but even single topics as, for example, the doctrine of place, were left in the form of specific questions that were never organized into a larger, coherent whole, which might have drawn attention to glaring inconsistencies and weaknesses.[4]

In a footnote,[5] Professor Grant refers to Dumbleton's *Summa* as a work to which this description might apply. I believe that it does not so apply and that Dumbleton's *Summa* offers a well-integrated as well as articulated system, not unlike what Thomas Aquinas does for theology in the *Summa Theologiae*. There are many cross references between questions, and I have not noticed any inconsistencies. This will be argued in detail below.

My second general thesis relevant to the theme of our conference has to do with the role of mathematics in the Calculators' work. It has often been pointed out that although the Calculators discussed, seemingly *ad infinitum*, how various physical variables should be measured, they rarely, if ever, made an actual physical measurement. Is this because they were inherently non- or even anti-empirical? How do the Calculators fit into a general transition between medieval qualitative physics and seventeenth-century quantitative physics? John North has argued that in many respects the science of kinematics in the Middle Ages was more highly advanced as applied in astronomy to celestial motions than as applied in physics to terrestrial motions.[6] Accepting this general thesis, I would add to it the observation that perhaps the Calculators' major problem in mathematical physics was the adequacy or inadequacy of the mathematics at hand to encompass the generally accepted phenomena. A typical complaint about a proposed mathematical description is that it is discontinuous where the data is continuous or that it is not general enough, i.e., that it can handle only certain cases and not others that are known to occur. Often the Calculators' own mathematics is adequate only for special cases, as when a value is constant, or linearly increasing or decreasing. One of Bradwardine's contributions was to make generally familiar the mathematics of the function he used to relate forces, resistances, and velocities. The most pressing problem for the Calculators' mathematical physics, then, may have been simply the need to find mathematical functions adequate for their purposes. As long as the mathematics they had was not even sufficient to account for the generally accepted phenomena, what would be the purpose of seeking out new facts? In their efforts to enlarge the repertoire of applicable mathematical functions, the Calculators were not very successful by seventeenth-century standards, but they were more successful than most of their contemporaries.

If the Calculators were attempting to devise a more adequate mathematical repertoire, a major constraint on their choice of descriptive functions was the assumption, clearly observable in Dumbleton's mathematical physics, that an adequate mathematics must arise, as if by abstraction, from the real physical entities or activities to be described. Dumbleton does not think, for instance, of approximating a continuous physical entity or process by a discontinuous mathematical function. Moreover, he even has objections to calculations that involve products lacking a clear physical interpretation. In the examples of Dumbleton's mathematical physics given below, the tight connections between his mathematics and his physics are quite clear, as well as his Aristotelian and Ockhamist attitudes towards the status of mathemtical entities. Given such attitudes, Dumbleton's mathematics, or quantification, plays an auxiliary role to his physics, or natural philosophy, which is quite different, moreover, from what occurs in modern physics, where the mathematical descriptions often seem to play the central role, with the physical interpretations of the mathematical entities in the auxiliary position.

THE STRUCTURE AND NATURE OF DUMBLETON'S *SUMMA*

Dumbleton's *Summa*, according to its preface, was planned to have ten parts, but only the first nine are present in the most complete extant manuscripts, and it is thought that the tenth part may never have been written. Part I of the *Summa* concerns selected logical topics more or less related to physics, and the other eight existing parts concern natural science. In the preface Dumbleton says that the *Summa* contains 'certain doubts from the five natural books, with the solution of logical and natural sophismata,'[7] but these are only a part of its contents. In Parts II and following the 'doubts' are fairly obvious, but no physical sophismata are identified as such so far as I have noticed. The first part of Part II, on the principles of nature, corresponds to Book 1 of Aristotle's *Physics*, but Part II then goes on to discuss the intension and remission of forms in considerable independence from Aristotle. The beginning of Part III of the *Summa*, on the relation of motion to the categories, corresponds loosely to the first part of Book 3 of Aristotle's *Physics*, but the majority of Part III, dealing with measures of velocity in local motion, augmentation, and alteration, corresponds, if to anything, to part of Book 7 of the *Physics*. The end of Part III of the *Summa*, dealing with the nature of motion and time, returns to topics of Books 3 and 4 of the *Physics*. A short outline of Parts II and III of the *Summa* may make this structure clearer:[8]

Dumbleton's Summa. Short Outline of Parts II and III

Part II

A. On first principles.
   1. Chapter 1. On the meaning of the terms 'principle,' 'cause,' and 'element.'
   2. Chapter 2. Aristotle uses the word 'principle' in different senses in different works.
   3. Chapter 3. Opinions of pre-Aristotelian thinkers on the principles.
   4. Chapters 4—6. The opinions of Plato and Aristotle on the principles. Aristotle's opinion is the most natural and in line with human sensations.
   5. Chapters 7—10. Clarifications, doubts, and replies concerning first principles.
   6. Chapters 11—19. Questions concerning matter and form.
B. Chapter 20. How something undergoes more and less.
   1. Whether substantial forms in the abstract undergo more and less. They do not. Chapters 21—24. How one thing is more perfect than another.
   2. Chapters 25—33. How qualities intend and remit. Discussion of various opinions, and choice of preferred opinion. Reply to objections.
   3. Chapters 33a—38. How difform qualities are intense and remiss. Denial of indivisibles.
   4. Chapter 39. Whether secondary qualities undergo intension and remission.
   5. Chapters 40—43. How the intensity of mixed qualities is determined.
   6. Chapter 44. Whether any substantial form intends and remits.

Part III

A. Chapters 1—4. On the cause of true motion. Why there is motion only in quantity, quality, and place.
B. How velocity in true successive motions is produced and caused.
   1. Local motion.
      a. Chapters 5—6. Opinions on local motion.

b. Chapters 7—9. How motion follows a ratio. Objections and replies.
   c. Chapter 10. A latitude corresponds to its mean degree.
   d. Chapters 11—15. Other conclusions.
  2. Chapters 16—18. On motion of alteration. Chapters 19—20. How mixed bodies are altered.
  3. Chapters 21—25. On motion of augmentation.
C. Chapters 26—44. On the nature of motion and time.
  1. Definitions of motion.
  2. Definitions of time.
  3. Solution of doubts.

Given the ordering of the topics in Parts II and III of the *Summa*, then, and the amount of space devoted to the different topics, it is misleading to think of the *Summa* as if it is a question commentary on Aristotle's five natural books. Dumbleton is really much more independent than that. If I had space to consider the later books here, the same view would be confirmed. Part IV considers the nature of the elements, the elements' qualities, density, rarity, rarefaction, powers, and so forth. Part V considers spiritual action or light. Part VI concerns the limits to powers and celestial movers. Part VII concerns the relation of primary and secondary causes to the generation of individuals. Part VIII concerns the generation of animals and the unity of the sensible and intellectual souls. Part IX concerns the five senses.

To confirm the independence of the *Summa* from a simple Aristotelian commentary format, it may be useful to look at Dumbleton's citations of Aristotle, Averroes, and other authors. The first nineteen chapters of Part II are, indeed, not unlike a commentary of sorts on Book 1 of Aristotle's *Physics*. I have made a rough tabulation of the citations in these chapters with the results shown in Table 1. Interestingly, Dumbleton often cites Averroes's comments first and then mentions, almost as an afterthought, that Aristotle says the same thing in the attached text. I have therefore combined the citations of the two authors. As can be seen by the table, Dumbleton is consistently talking about Book 1 of the *Physics*. This does not mean, however, that he discusses Book 1 in isolation from the rest of the *Physics*, or, indeed, from various other books of the Aristotelian corpus.

When Dumbleton turns to discuss the intension and remission of forms in Chapters 20—44 of Part II, the frequency of citations from

Book I of the *Physics* drops sharply[9] and there is little special link with the *Physics* as such. I have tabulated the citations in these chapters in Table 2. Though tabulating citations by chapter is only a very rough measure of Dumbleton's dependence on an Aristotelian text, Table 2 at least shows, I think, that in Chapters 20—44 Dumbleton's discussion is not tied specifically to any one text.

TABLE 1
Citations in Dumbleton's *Summa*, Part II, Chapters 1 through 19.

Chapter	Aristotle and/or Averroes	Other
1	Physics, Books 1, 6	
	De Caelo, Book 1	
	Metaphysics, Books 5, 12	
2	Posterior Analytics, Book 2	Liber de Regimine Principium
	Physics, Books 1, 2, 4, 8	
	De Caelo, Book 1	
	De Anima, Book 2	
	Metaphysics, Books 7, 12	
3	Physics, Book 1 (views of Parmenides, Melissus et al.)	
	Metaphysics, Books 4, 7, 13, 14	
4	Physics, Book 1	Plato, Timaeus
	De Anima, Book 2	
	De Sensibus [?]	
	Metaphysics, Book 5	
5	Physics, Book 1	
	De Generatione, Book 2	
6	Physics, Book 1	
	De Caelo, Book 1	
	Metaphysics	
7	Physics, Books 1, 2	
8	Physics, Books 1, 8	
	Metaphysics, Books 7, 12	
9	Physics, Book 1	
10	Physics, Books 1, 2	
	De Caelo, Book 1	
11	Physics, Book 1	
	Metaphysics, Book 7	
12	Physics, Books 1, 4	
13	Metaphysics, Books 5, 7	Lincoln (i.e. Robert Grosseteste)
14		

Table 1 (Continued)

Chapter	Aristotle and/or Averroes	Other
15	Metaphysics, Books 5, 7, 9, 12	
16	Physics, Books 1, 2, 5	
	De Anima, Books 2, 3	
	Metaphysics, Books 5, 7, 8, 9, 12	
	De Generatione, Book 1	
17	Categories	
	Physics, Book 1	
	De Caelo, Book 1	
	De Anima, Book 3	
	Metaphysics, Books 1, 7, 8, 9	
18	Categories	
	Physics, Books 1, 4, 5, 7	
	De Generatione, Book 1	
	De Anima, Book 2	
	Metaphysics, Books 5, 8, 12	
19	Categories, Posterior Analytics, Book 1	
	Topics, Physics, Book 4	
	De Anima, Book 1, Metaphysics, Book 8	

Citations taken without verification from MS Peterhouse 272.

TABLE 2
Citations in Dumbleton's *Summa*, Part II, Chapters 20 through 44.

Chapter	Aristotle and/or Averroes	Other
20	Categories	
	Physics, Books 2, 7	
	De Anima, Book 2	
	Metaphysics, Book 7	
21	Physics, Book 5	
22	Metaphysics, Book 8	
23		Porphyry
		Euclid, Elements, Book 5
		Campanus

Table 2 (Continued)

Chapter	Aristotle and/or Averroes	Other
24		
25	Physics, Book 2	
26	Physics, Book 6	
	De Generatione, Book 1	
27		
28	Physics, Book 5	
29	Categories	
	Physics, Book 1	
	De Generatione, Book 1	
	Meteora, Book 4	
	De Anima, Book 2	
	Metaphysics, Book 12	
30	Physics, Books 4, 5	Avicenna, Sufficientia
	Metaphysics, Book 2	
31	Physics, Books 1, 4	
	De Anima, Books 1, 2, 3	
32	Metaphysics, Book 9	
33		
33a	Metaphysics, Book 5	Auctor Sex Principiorum
34		
35		
36	Metaphysics. Books 3, 10	
37	De Caelo, Book 1	
38	De Caelo, Book 1	
	De Generatione, Book 1	
39	Physics, Books 5, 7	
	De Generatione, Book 1	
	De Anima, Book 2	
	De Sensu et Sensato	
	Metaphysics, Book 5	
40		
41		
42	Metaphysics, Book 5	
43		
44	Categories	Avicenna, Sufficientia
	Physics, Books 1, 5	Porphyry
	De Caelo, Books 1, 3, 4	
	Metaphysics, Books 8, 9	

Citations taken without verification from MS Peterhouse 272.

The citations in Book III of the *Summa* — see Table 3 — indicate an even greater independence from any one Aristotelian text for most of this part of the *Summa*. In the early chapters Dumbleton discusses the causes of motion — this accounts for the references to the *Categories, Physics,* and *Metaphysics* in Chapters 1—3. In Chapters 5—25 Dumbleton discusses the mathematics relating forces, resistances, and velocities in various kinds of motions, and here he has few references to Aristotle and Averroes and, instead, some to Euclid, Campanus, and the *auctores de ponderibus*. In Chapters 26—43, then, he discusses the nature of motion and time and, again, as in the beginning of Part II, his discussion could be considered a commentary of sorts on certain parts of the *Physics* and *Metaphysics*, with repeated references to the same sections.

TABLE 3
Citations in Dumbleton's *Summa*, Part III (entire).

Chapter	Aristotle and/or Averroes	Other
1	Physics, Books 1, 5	Avicenna, Sufficientia
	Metaphysics, Book 7	
2	Physics, Books 5, 7	
	Metaphysics, Book 11	
3	Categories	
	Physics, Books 1, 2, 5	
	De Anima, Book 2	
	Metaphysics, Book 12	
4		
5	Physics, Book 4	opinio Avempace
6		Euclid, Elements, Book 5
7	Physics, Books 4, 7	Campanus, Euclid, Book 5
8		Euclid, Elements, Book 5
9	Physics, Book 8	
10—13		
14	Physics, Book 1	
15		Auctores de Ponderibus
16—17		
18	Physics, Book 7	
19—21		
22	Physics, Book 7	
23	De Caelo, Book 4	

Table 3 (Continued)

Chapter	Aristotle and/or Averroes	Other
24—25		
26	Physics, Books 3, 5	
27	Physics, Book 3	
28	Physics, Books 4, 5, 7	
	De Caelo, Book 1	
	De Anima, Books 1, 2	
29		
30	Categories	
	Physics, Book 5	
	Metaphysics, Book 11	
31	Physics, Book 1	Boethius in Divisionibus
	Metaphysics, Book 10	
32	Physics, Book 4	
	Metaphysics, Book 12	
33	Physics, Book 4	
	Metaphysics, Book 12	
34		
35	Physics, Book 4	
	Metaphysics, Book 12	
36	Physics, Book 4	
37	Physics, Book 4	
	Metaphysics, Book 12	
38	Physics, Book 4	
39		
40	Physics, Book 4	
41	Physics, Book 4	
42	Metaphysics, Book 12	
43—44		

Citations taken without verification from MS Peterhouse 272.

Parts II and III of the *Summa*, then, constitute a general introduction to the principles of the physical world and to motion, taking a more mathematical slant in the second half of Part II and in the central section of Part III than might be expected in the usual medieval Aristotelian work. Although the commentary-like sections and the independent sections are somewhat dissimilar in their approaches, it is not an unintegrated work.[10]

This impression of integration is bolstered by a number of references to other parts of the *Summa*. In Part II, I have found fourteen references either back to Part I or ahead to later parts of the *Summa*, including references to Parts III, IV, VI, VII, VIII, and X. In Part III, I have noticed nine such references to other parts, including references to Parts II, IV, and X. Even if Dumbleton never wrote Part X, he evidently had clearly in mind what he planned to discuss there.[11]

Dumbleton's physics is not physics *secundum imaginationem*, but physics of the real world. Although I do not have space here to survey in detail Dumbleton's conclusions to show their real world connections, a final external measure of Dumbleton's commitment to the physical validity of the views he presents may be taken from his uses of variants of the word 'imaginatio,' the phrase 'secundum imaginationem' having been often taken by historians as a hallmark of fourteenth-century philosophers' detachment from the real world.[12] I have noted many uses of 'imaginatio' and related words in Parts II and III of the *Summa*. See Appendix for a concordance of these uses.

In a large number of cases, Dumbleton contrasts 'imaginary' to 'real.' In his discussion of alternate theories of the intension and remission of forms, for instance, one of Dumbleton's main points is that according to the theories to be rejected there will be no real latitude of quality or only an imaginary one, and if the latitude or distance to be gained in alteration is only imaginary, so will motion of alteration be only imaginary.[13] Since Dumbleton believes motion of alteration to be real, it must gain a real distance, namely the continuous latitude of quality that is contained between any degree of quality gained and zero degree. This subject is discussed in greater detail below.

Dumbleton often treats 'imagine' (*imaginare*) as a synonym for 'understand' (*intelligere*), and in collating manuscripts I find these words as variants for one another.[14] For something to exist only in imagination, then, is more or less the same thing as it existing only in the mind (*ratione*). If time, then, is supposed to exist only in human apprehension of motion, Dumbleton might well say that time is an imaginary quality. Interestingly, Dumbleton expounds Averroes as saying that motion in the formal sense is a real or imaginary quality of a moving body.[15] In other contexts, Dumbleton refers to motion as an imaginary quality, thus exhibiting his Ockhamist tendencies to minimize the number of entities or qualities supposed really to exist.[16] Dumbleton argues, for instance, in favor of measuring difform qualities by their maximum degree by saying:

This position posits that difform motion is as intense as any of its parts. Since the latitude of heat or of other qualities has the same relation to its subject and to qualities in the same species as the latitude of motion has to its subject, therefore a difform quality is as intense as any of its parts. . . . if imaginary qualities [i.e. motion] are as intense as any of their parts, far more [should this be true of] real qualities like heat and cold and other qualities.[17]

In the cases in which Dumbleton sets up imaginary situations, he is as likely to imagine a situation that could be naturally true as he is to imagine a naturally impossible situation.[18] Dumbleton does not, therefore, to any degree imagine a 'mathematical physics . . . which neither seeks nor claims to have application to the physical world.'[19]

Mathematical Physics in Dumbleton's Summa, Parts II and III

What is the nature of the mathematical physics that appears in Parts II and III of the *Summa*? First of all, it should be said that Dumbleton is an Aristotelian rather than a Platonist in his attitude toward the foundations of mathematics. In Part I of the *Summa*, for instance, he suggests that while the mathematician assumes that points, lines, and surfaces exist and goes on to prove theorems about them, it is perhaps only the physicist who can demonstrate the existence of points, lines, and surfaces on the grounds that bodies must have terminations.[20] Physics, therefore, is a more fundamental science than mathematics, since physics can prove the principles of mathematics. In Part II of the *Summa*, Dumbleton says:

Aristotle in the *Metaphysics*, Book 5, says that a line is long *per se* and implies that a body is long through a line separated in reality from the body, as a quality may be separated from its subject, because a line is a quantum *per se* (this is in the text of comment 19). And thus Aristotle speaks as if longitude and latitude were quanta in themselves and as if other things like bodies were quanta by means of these dimensions.

To these and others that might be alleged to the contrary, it should be said that Aristotle speaks in the *Metaphysics*, Book 5, defining words as they are used in diverse sciences, and therefore he is speaking as a geometer, or he defines the word as it is defined in geometry. He is not speaking as a natural philosopher, and things considered in geometry are not admitted unless for the sake of argument or information in other sciences. Similarly, Lincoln [?] is speaking understanding line and quantity as a geometer as if quantities were distinct from quantified things, which, simply speaking, should be denied.[21]

Mathematics, therefore, in Dumbleton's view, abstracts certain properties of physical entities and treats them as if they were separate, although they do not exist separately in reality. When, therefore, Dumbleton uses mathematics in Parts II and III, he does not so much impose prior mathematical structures upon physical entities, but rather tries to abstract mathematical measures from the physical situation. His mathematics is not mathematics as a separate science, but only mathematics as a tool for clarifying physical concepts and making them more exact and comprehensive.

This approach is clear in Dumbleton's discussion of the intension and remission of forms in Part II. One of the major points of this section is that in the most proper sense only qualities like hotness and coldness (and the bodies that have these qualities) undergo increase and decrease of degree. Substantial forms are not so variable.[22] This is the case, according to Dumbleton, because for intension and remission in a proper sense, the form must have within itself some latitude or formal dimension in which a body may participate to a greater or lesser degree. It is not enough, in Dumbleton's view, that it be possible to rank indivisible forms on some imaginary scale of greater and lesser. Dumbleton argues:

... there is no real or imaginary latitude to which all the degrees of natural things may be applied. ... Every specific essence is simply indivisible in its nature and from indivisibles a continuum can in no way be made. Therefore the perfections of things do not make a latitude nor are they properly comparable to a latitude. ... If species did make a latitude, then part of this latitude would correspond to each species, and thus the specific essence would be divisible, which has been disproved.[23]

Comparisons of degrees, Dumbleton says, may be made in the abstract or in the concrete. In the concrete we may say that a man is whiter than an ass, and this is a proper comparison, because the whitenesses of the man and ass are of the same species, and the greater whiteness is divisible into a part equal to the whiteness of the ass and a part by which it exceeds that whiteness. Since this is the case, it is possible to have a ratio between these two whitenesses, a ratio being defined by Euclid and Campanus as a comparison between two things which are of the same species.[24]

On the other hand, in the abstract, two things of different species might be compared to each other, as it might be said that a man is a greater substance than an ass. In this case, however, there would be no ratio involved in the comparison, because:

... in every ratio, whether arithmetic or geometric, the smaller always is related to the larger as an aliquot part which, taken some number of times (finite or infinite), produces a larger total. But an infinite number of perfections like the perfection of an ass are not more perfect than one such perfection. Therefore there is not a proper ratio in which a man is a greater entity than an ass is.[25]

In this argument, then, Dumbleton's point is that substantial forms are not properly comparable to each other because each is indivisible in its essence and indivisible things are not comparable with respect to quantity. Substantial forms must be of indivisible perfection because otherwise a subject might participate in them to a greater or lesser degree, and one might have a man with so small a part of the perfection of the essence of man that his perfection would equal that of a fly, which is impossible. On the other hand there is no external measure of substantial perfection (no Great Chain of Being that really exists physically) on which an individual substantial perfection might be located. In any case, for there to be a comparison of degree or size there must be some larger whole in that dimension of which parts may be taken.

The tight linking of the mathematical and physical in this section is obvious. If substantial forms were to be comparable in terms of more and less, there would have to be a real continuous physical dimension in which they could participate. The requirements that the physical dimension be continuous and that the substantial forms themselves not be indivisible with respect to perfection arise from mathematical considerations. But it is typical of Dumbleton's physics that he also requires this continuous physical dimension or latitude to have a basis in nature and not only to be imaginary.

These sorts of considerations again come into play when Dumbleton considers competing mathematical-physical theories of the intension and remission of those forms like hotness and coldness for which he did admit variation. Of the two main theories of the intension and remission of forms that Dumbleton rejects, the first says that qualities increase and decrease by the action of an agent without anything being acquired or lost, just as a body may be rarefied or condensed without gaining or losing matter from outside or just as a body might be moved in a vacuum and not traverse anything except in imagination, since in a vacuum there is no space.[26]

To this view Dumbleton replies that if it were true, there would be no real motion in quality. In every real continuous motion, he says, two really distinct things are necessary, namely a mobile and a divisible space. But according to this view there is no really divisible space in quality. Therefore in quality there could not be any real and continuous motion. Dumbleton's elaboration of this reply casts light on the role of *secundum imaginationem* arguments in his physics as I have noted above. He says:

> Let some quality, for instance the heat of fire, be intended uniformly up to the highest degree. According to this position no such distance is acquired by the heat because of the alteration, nor by the subject in which the heat inheres, unless according to imagination. This is the case because the position says that the heat remains simple from the beginning to end without anything acquired or lost. Therefore there is no true motion in this alteration unless it be imaginary, just as, although we might have imagined simple motion outside the heavens, nevertheless there would be no real motion there, since from our imagining (*ex hoc*) it does not follow that there is real motion outside the heavens, because things don't follow imaginations (*res non sequuntur ymaginationes*). Since, in the given alteration, there is nothing but an imagined space, it follows that there is no true motion, but imagined motion, just as if we were to understand, imagining, that point A moves in a straight line and acquires nothing in fact of a distance that is divisible in really distinct parts; then point A in no way is really moved unless according to imagination; so it is in the proposed case of uniform alteration.[27]

Thus, according to Dumbleton, for real motion to occur in quality, there has to be a real distance to be acquired (namely the latitude of form), which is really divisible into diverse parts. The comparison to rarefaction and condensation made by supporters of the rejected view is not apt because in rarefaction there is a real external place, acquisition of the parts of which can be used as a measure of rarefaction.[28] Dumbleton goes on:

> The cause of the deception of those who support the rejected opinion is that they imagine that a degree is more intense than another just because it is more or less distant from zero degree of the latitude of its species, even though the degree in itself is really indivisible intensively and contains in itself no latitude. This is the same as if one should imagine that a point terminating a foot length contained in itself more of quantity than a point terminating a half-foot length, and thus, by referring the points to the distances that they terminate [should say that] one contained more quantity than the other. A point, however, has no quantity in itself.[29]

Dumbleton's arguments against the second rejected opinion on alteration are similar. According to this second opinion, when a body becomes hotter it obtains in each instant of time a new more intense indivisible degree of heat and loses the degree it had previously.[30]

Again Dumbleton argues that in this case no real distance would be gained in alteration:

Let fire act on earth inducing heat uniformly in the whole body up to the highest degree in an hour. This latitude or this distance which is from zero degree up to the highest degree, by the acquisition of which the earth is supposed to be altered, would be composed of indivisibles. Therefore this latitude would be of no quantity. Or it follows that a continuum and distance in some category is composed of indivisibles, the opposite of which is proved by the Philosopher in the *Physics*, Book 6, and in *De Generatione*, Book 1.[31]

According to the view of alteration that Dumbleton accepts, on the other hand, each degree of quality possessed by a body contains in itself a real latitude of quality. A body increases in a given quality, then, by gaining new parts of that quality which, together with the old, constitute a higher degree. These parts of quality and the latitude they compose provide a real distance to be gained in a motion of alteration.[32]

It should be apparent, then, how close is the connection between mathematics and physics in this section of the *Summa*. Mathematical arguments — for instance that a continuum cannot be composed of indivisibles — are used to argue against the plausibility of certain physical theories, namely, in this case, the second rejected theory of alteration. A mathematical entity, e.g., the latitude of form in a given category or species, has to be abstracted from really existing entities, for instance from really divisible degrees of quality. Dumbleton does not postulate a mathematical entity, representative, say, of the Great Chain of Being, when there is no real entity (The Great Chain) from which it could be abstracted.

This use of mathematics in physics is a far cry from the standard picture of Ptolemaic astronomy in which any mathematical model using deferents, eccentrics, epicycles, equants, etc., might be equally acceptable so long as it predicted the observed positions of the planets. Dumbleton is not trying to make predictions, but rather to understand the physical structure of the world.

Having established divisible latitudes and degrees as the mathematical basis of the intension and remission of forms, Dumbleton faces a number of more complicated problems, for instance the problem of how the quality of a body with varying degrees of quality in different parts should be measured. His solution of this problem, namely that any such difformly qualified body should be considered as intensely qualified as its maximum degree, again shows the tight connection Dumbleton maintains between his mathematics and his physics. Suppose one were to say that a body is as hot as its average degree. To calculate this average degree, one would have to weight each degree the body has by the proportion or the quantity of the body it qualifies. But, Dumbleton seems to believe, this would imply that quantity is a cause of intensity, which it is not. He therefore opts for measuring difformly qualified bodies solely in terms of their maximum intensities. Dumbleton argues:

If intension and remission were acquired by something because of quantity or extension, with the latitude contained in it remaining equal, it would follow that quantity would produce intension of the quality and subject, which is a false conclusion. . . .
Nothing can be made more or less so, except with respect to a single qualitative distance. Since from quality and quantity a single distance cannot be produced, it follows that because of quality and quantity together nothing is intense.[33]
The precise and essential cause of intension in an intensible and remissible quality is the distance or latitude in which true motion in quality is produced. This is shown as follows: if the qualitative distance between the degrees that are the termini of motion of alteration is destroyed, nothing is more intense than another, neither a divisible thing nor an indivisible one. But if this latitude between degrees is posited, one thing is more intense than another. Therefore this qualitative distance between degrees properly and per se is the cause of intension in quality. . . .[34]
If it were said that the latitude of quality not related to quantity is of no intension per se, it would follow that there neither is, nor can there be, motion in quality, which is a false conclusion. And the inference holds, because then there would be no true distance in the category of quality.[35]
All qualities in the same species having the same latitude are equal to each other intensively.[36]

In this discussion, then, Dumbleton seems to imply that not only must mathematical entities, such as the latitude of quality, have a foundation in real physical entities, but also the steps in a mathematical calculation, such as the calculation one might make to determine an average degree, must have an acceptable physical interpretation. Since the product of intensity and quantity corresponds to no physical entity in Dumbleton's view,[37] he rejects the theory making use of this product.

Other examples of Dumbleton's mathematical physics may be taken from Part III of the *Summa*. Before Thomas Bradwardine's *On the Ratios of Velocities in Motions*, the major alternatives for relating mathematically forces, resistances, and velocities were the view ascribed to Aristotle, according to which velocity is proportional to the ratio of force divided by resistance, and the view ascribed to Avempace, according to which velocity is proportional to the difference of force and resistance.[38] Another possibility was to combine these two views, making velocity proportional to the ratio of force minus resistance to resistance. According to Bradwardine, however, velocity varies as the ratio of force to resistance, but this statement is interpreted treating the operation of 'compounding' ratios as analogous to adding numbers. In this interpretation, when, for instance, the ratio of 3 to 1 is compounded with, or 'added' to, itself, the result is the ratio of 9 to 1, so that the ratio of 9 to 1 is considered to be 'double' the ratio of 3 to 1. If a force equal to 3 moves a resistance equal to 1 with a velocity v, then, according to Bradwardine, a force equal to 9 will have a ratio to 1 that is 'double' the ratio of 3 to 1, and so it will move the resistance with a double velocity, or 2v.[39]

As a part of his argument for this view, Bradwardine asserted, implicitly, that this was the only way in which ratios could be treated. To express the view usually ascribed to Aristotle, Bradwardine did not say that velocity varies with the ratio of force to resistance, but rather that velocity varies as the force if the resistance is held constant and velocity varies inversely as the resistance if the force is held constant. In dividing the traditional Aristotelian view into two statements in this way, Bradwardine avoided presenting it as a view involving ratios together with an alternate interpretation of operations on ratios. Whether he sincerely thought that the variation of ratios can be understood in only one way or, rather, pretended to do so in order to support his own view more easily is unclear. In any case it is somewhat ironical that he also argued against his own two-proposition-version of the traditional Aristotelian view on the grounds that it does not deal with cases in which the force and resistance vary at the same time.[40]

In an earlier article I have described how Dumbleton used paired latitudes of motion and of the ratio of force to resistance to express Bradwardine's function.[41] Here I will concentrate, instead, on how Dumbleton argues against the two major previous theories concerning the relations of forces, resistances and velocities. It is notable that Dumbleton begins his discussion of this subject by saying not that he will discuss quantitative, or mathematical relations, or anything of the sort, but rather that he will discuss how velocity is produced and caused.[42] He begins with the opinion traditionally ascribed to Avempace, most of his contrary arguments involving the fact that

Avempace's theory predicts a finite velocity in a vacuum, where resistance is zero, whereas there was a long tradition of assuming, following Aristotle, that motion in a vacuum will be infinitely fast or impossible.

Dumbleton's arguments against the traditional Aristotelian view (here presented allowing only for variation in force with the resistance constant) are more interesting. The foundation for the Aristotelian view, Dumbleton says, is the following:

Let 3 act on 2 and another 3 on another 2, each according to the ratio 3:2. Each of these motions induces a certain latitude of heat or local motion. If, therefore, one makes a single agent from these two and applies it to one of the previous bodies acted on, since one agent does not impede the other, indeed it more likely assists, it follows that the two agents together will produce twice what one produced before, and consequently cause a motion twice as fast.[43]

Dumbleton's reply to this argument is to say that 'in every action the whole action is the action of the whole agent and all of its parts,'[44] so it does not necessarily follow that the parts of an agent can be supposed to produce what they would produce if they acted alone. In this case, since in Bradwardine's view, twice the velocity produced by the ratio 3:2 would be produced by double 3:2 or 9:4, which equals 4 1/2: 2, Dumbleton argues that with the combined agent of 6, each part aids the other, and more than double the original velocity is produced.[45] He has a number of further arguments to support the view that parts of an agent do not act independently.

This argument, that parts of an agent do not add their effects the way their powers might be added, helps explain why Dumbleton called this section a discussion of how velocity is produced and caused. In his arguments directly against the Aristotelian position this same sort of approach is again evident. Dumbleton tacitly assumes that if it is not possible to preserve a relation between the parts of a force and the velocities they produce, whether they act independently or together, then it is the ratios of force to resistance that preserve their identification with certain velocities produced, whether the ratios are combined (using the idea of compounding ratios) or not. Thus it is the ratios that produce, or cause, the velocities and not simply the forces, assuming the resistances constant.

So, Dumbleton argues, if the Aristotelian position asserts that 8 produces in 1 twice the velocity that 4 produces in 1, this must be related to the fact that 8:1 contains the ratio 4:1 one and half times [8:1 equals 4:1 to the three-halves power]. If this is the case, then a force A should move a resistance of 4 twice as fast as 3 moves 2 if the ratio A:4 equals the ratio 3:2 to the three-halves power. But in such a case A will be less than 9 (because the ratio 9:4 is equal to 3:2 squared, which is more than 3:2 to the three-halves power), whereas according to the standard Aristotelian position, it should take a force of 12 to move 4 twice as fast as 3 moves 2 (or 6 moves 4).[46]

In one sense in this argument Dumbleton is simply assuming Bradwardine's function in order to prove an alternative view false, which is hardly a satisfying proof. On the other hand, he can be seen as using the physical assumption that it is a ratio of force to resistance that produces a velocity. Thus he says in his fourth argument against the Aristotelian position:

If some latitude of motion is produced by three ratios of 2:1, if some fixed part [of the motion] corresponds to the part of the latitude of proportion from the ratio of 2:1 down to the ratio of equality, it follows that just as great a part of the same latitude of motion will correspond to just as much of the other latitude of proportion.[47]

This means that if three ratios of 2:1 or, in other words, if $(2:1)^3 = (8:1)$ produces a velocity of 3n, then the latitude of ratio from 1:1 to 2:1 produces the velocity n, the latitude from 2:1 to 4:1 produces the velocity n, and the latitude from 4:1 to 8:1 also produces n.

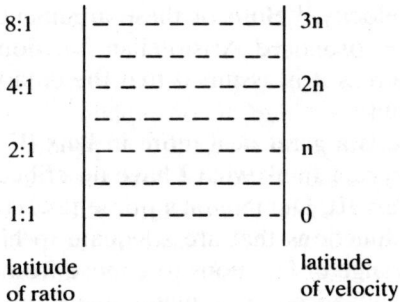

In his eighth argument, Dumbleton says:

It follows from the [Aristotelian] position that 9 acts just as quickly in 4 as 12 in 4, for 9 acts in 4 according to two ratios of 3:2. ... Therefore 9 acts twice as much in 4 as 6 acts in 4 because 6 acts [in 4] only according to one ratio of 3:2. This follows from the foundation of this position because one ratio compounded or joined (*coniuncta*) with another does not hinder the other as one agent hinders another joined to it....[48]

After a number of other arguments treating the mathematical compounding of ratios as if it involved a physical compounding of actions, Dumbleton finally says in his twelfth argument against the Aristotelian position:

If from a ratio 2:1 there arises a motion A, which cannot arise from a smaller ratio, it follows that a motion double A cannot be produced except by two ratios 2:1, that is by a ratio compounded of two double ratios ... because the effect is not doubled unless the cause is doubled.[49]

However strained this conflation of mathematical and physical relationships may seem to us, nevertheless I think it makes clear once again how closely tied together Dumbleton's mathematics and physics were.

Dumbleton's other arguments against the Aristotelian view are more nearly related to its mathematical adequacy. Assuming the Aristotelian view, there is no combination of force and resistance that can produce a velocity equal to or less than half the velocity produced by the ratio 2:1. This follows because halving the force 2 to get half the velocity results in a ratio 1:1, which is insufficient to produce velocity.[50] Looking at the same situation from another angle, it follows that a force equal to a resistance, if it increases only minutely, will produce a sudden leap in velocity.[51] Both of these arguments are telling because they show that the proposed Aristotelian function has a mathematical discontinuity, whereas it is assumed that the data to be explained have no such discontinuity.

Although there is a great deal more in Part III that I have not been able to mention here, I think what I have described is characteristic. As in Part II, so in Part III, Dumbleton's prime task is to find mathematical descriptions and functions that are adequate to his purposes. He does not have a wide range of functions to choose from, and he has to work hard for minor advances. His mathematics is further constrained, beyond his rather inadequate mathematical resources, by the extremely close tie he wants to preserve between the mathematical description and the underlying physical reality. In his own mind he surely was not doing mathematical physics *secundum imaginationem*, but trying, through mathematics, to make more exact and comprehensive his description of the Aristotelian, i.e. the physical, world.

CONCLUDING REMARKS

The novelty and impact of the Oxford Calculators' work lay in their intensive application of logical and mathematical techniques to the analysis of physical problems. With regard to physics they were, on the whole, loyal Aristotelians of Ockhamistic tendency, except that they made greater use of mathematics than either Aristotle or Ockham. When they were solving sophismata, for instance in William Heytesbury's *Rules for Solving Sophismata* and *Sophismata*, they were not primarily concerned with determining what was naturally the case, but when they were doing natural philosophy, for instance in all but the first part of John Dumbleton's *Summa Logicae et Philosophiae Naturalis*, physical reality *was* their concern.

The mathematical physics found in Dumbleton's *Summa* is an Aristotelian mathematical physics, in the sense that the mathematics depends on quantities abstracted from bodies, places, or qualities. Even mathematical operations are supposed to correspond to physical operations, mathematical additions to physical additions, etc. Dumbleton's *Summa* is a well-organized and connected synthesis and not a fragmented consideration of isolated questions. As he writes, Dumbleton seems to have the plan of his whole work in mind, so that he tries to be consistent from place to place, developing his ideas in a logical order, as variations on an Aristotelian theme.

To the possible objection that he does not systematically cover every part of the cosmos, I would respond, first, that he does cover many if not all the major points of Aristotelian natural philosophy, and, second, that except where he states otherwise, it can be assumed that he supports the major aspects of the Aristotelian world picture. The *Summa* is not, of course, a deductively self-contained system like Euclid's *Elements*, Newton's *Principia*, or even Bradwardine's *De proportionibus*. Contrary to the familiar picture of the fourteenth century as a time of intellectual disintegration and disarray, however, Dumbleton writes in a generally confident and assertive way, as if he felt himself to be in a period of stability and progress. He does not seem to feel the need for intellectual pacifism, to use the terminology of John North's paper elsewhere in this volume.

In assessing the nature of late scholastic natural philosophy, then, we perhaps need to pay more attention to the characteristics of a commentary tradition. The writing of any commentary apparently presupposes the value of the text being commented on. Furthermore most commentaries make use of previous commentaries on the same work. Novelty may be received differently in a commentary tradition than in traditions embodied in other sorts of works. A work like Dumbleton's *Summa*, although not an Aristotelian commentary as such, presupposes the existence of a commentary tradition and general familiarity with the outlines of the Aristotelian world view. I believe Dumbleton would have seen differences between his view and that of Aristotle as friendly amendments and improvements.

If Dumbleton's *Summa* had been as well known as Walter Burley's commentary of the *Physics* it might have contributed to an increased use of mathematical formulations within natural philosophy, for instance to a broader understanding of Thomas Bradwardine's ideas about the relations of forces, resistances and velocities. It would have given Italian scholastics a less sophismatical example of what Oxford calculatory work was like, perhaps leading to a tempering of later criticisms of Anglican subtleties. It might in this way have contributed to an undramatic, but longer lasting, change in natural philosophy. We could then study what the impact might have been by considering, for instance, Paul of Venice's *Summa of Natural Philosophy*, written by an Italian Augustinian who had studied at Oxford.

But instead of circulating widely on the Continent, Dumbleton's *Summa* languished mostly in English manuscripts, while the works of Bradwardine, Heytesbury, and Swineshead made a splash on the Continent and were the subject of printed editions and commentaries. Perhaps this was because the *Summa* was not as novel from a European perspective and because Dumbleton was not a member of any of the religious orders, but on this point I can at present only speculate.

Finally, then, in understanding the *fortuna* of the works of the Oxford Calculators, we should consider that the corpus that we have labelled 'works of the Oxford Calculators' contains works of rather diverse sorts written for rather diverse purposes, works, however, whose fates were intertwined. Although later medieval and modern commentators on the fourteenth century intellectual scene may have understood the Calculators' work to be highly hypothetical and imaginary, not to say counter-factual, such labels do not fit significant parts of their works. In particular, Dumbleton's mathematical physics is, if anything, not too hypothetical, but rather too closely tied to the physical situation as he understood it. If Dumbleton's *Summa* was not taken as the basis for an evolution in Aristotelian natural philosophy, it was because it was not the *Summa*, but other Calculatory works, whose nature determined the *fortuna* of the Calculatory tradition in later centuries.[52]

North Carolina State University
Raleigh, North Carolina, U.S.A.

NOTES

1. Edith Sylla, 'The Oxford Calculators,' in Norman Kretzmann, Anthony Kenny, and Jan Pinborg (eds.), *The Cambridge History of Later Medieval Philosophy* (Cambridge: Cambridge University Press, 1982), pp. 540—563.
2. A. Rupert Hall, 'On the Historical Singularity of the Scientific Revolution of the Seventeenth Century,' in J. H. Elliott and H. G. Koenigsberger (eds.), *The Diversity of History. Essays in honour of Sir Herbert Butterfield* (Ithaca, New York: Cornell University Press, 1970), p. 207.
3. *Ibid.* The latter phrase is quoted from Edward Grant, 'Late Medieval Thought, Copernicus, and the Scientific Revolution,' *Journal of the History of Ideas, 23* (1962), 197.
4. Edward Grant, 'Aristotelianism and the Longevity of the Medieval World View,' *History of Science, 16* (1978), p. 98.
5. *Ibid.*, note 16, p. 106.
6. John North, 'Kinematics — More Etherial than Elementary,' in Madeleine Pelner Cosman and Bruce Chandler (eds.), *Machaut's World: Science and Art in the Fourteenth Century, Annals of the New York Academy of Sciences*, Vol. **314** (New York: New York Academy of Sciences, 1978), pp. 89—102.
7. MS Cambridge, Peterhouse 272, f. 1ra. The following study is based for the most part on a transcription of Parts II—VI of the *Summa* as contained in this manuscript, with occasional variant readings from other manuscripts.
8. For a different and longer outline, but one which, I think, may overemphasize the role of the 'doubts,' see James A. Weisheipl, 'Developments in the Arts Curriculum at Oxford in the Early Fourteenth Century,' *Medieval Studies, 28* (1966), pp. 169—172.
9. The drop is sharper than might appear because a single mention in the table may refer to a brief mention or to discussion of the cited book repeatedly throughout the chapter.
10. In the preface to the *Summa* Dumbleton appears to say, if I have interpreted his somewhat obscure Latin correctly, that he has not only collected diverse bits or points into a *Summa*, but has in some way ordered and matured them: 'Plurimorum scribentium grati laboris dignique memoria participes ad mensuram mee facultatis doni ex logicali materia communi et physica quandam summam veluti spicarum

dispersarum manipulum quoquomodo maturatum et in compositum recolegi.' MS Cambridge, Peterhouse 272, f. 1 ra.
11. Dumbleton's references to other parts of the *Summa* in MS Peterhouse 272 are on ff. 11va, 11vb, 12vb, 13rb, 13va and vb, 15va, 16va, 19ra, rb, and va, 20rb, 26rb, 27va and vb, 28vb, 29ra and rb, 30rb and vb, and, finally, 31rb.
12. *Cf.* Edward Grant, 'Late Medieval Thought, Copernicus, and the Scientific Revolution,' (above note 3), p. 205 and note 32.
13. See Appendix, quotations from ff. 17va, 18va, 20rb, 29va.
14. For instance, when on 31 rb, line 3 of transcription, MS Peterhouse 272 says 'isto modo intelligendo,' MS Vatican Lat. 954 reads 'isto modo ymaginando.'
15. See Appendix, f. 29va.
16. See Appendix, quotations from ff. 18vb (first quotation), 19va, 29va, 31rb.
17. Latin in Appendix, quotation from f. 19va.
18. See Appendix, quotations from ff. 16vb, 17va, 18ra, 19rb, 26va, 28rb, 29vb, 33ra.
19. Curtis Wilson, *William Heytesbury. Medieval Logic and the Rise of Mathematical Physics* (Madison, Wisconsin: University of Wisconsin Press, 1956), p. 25, cited in Edward Grant, 'Later Medieval Thought . . .' (above, note 3), p. 205, n. 32.
20. MS Peterhouse 272, f. 8va. Later in the *Summa*, f. 20va, Dumbleton, in good Ockhamist fashion, denies that surfaces, points, and lines exist. *Cf.* also Appendix, quotation from f. 34ra, where Dumbleton says that points are imaginary.
21. MS Peterhouse 272, f. 14va. The reading 'Lincoln' is uncertain.
22. *Ibid.*, f. 16va: Tertio modo proprissime magis et minus suscipere dicitur quod natum est suscipere gradum intenciorem post remisciorem vel econtra. Et sic intensio et remissio solum competit qualitatibus in tertia specie qualitatum et earum subiectis propter ipsas competit intencio et remissio.
23. *Ibid.*, f. 16vb. *Cf.* Latin in Appendix.
24. *Ibid.*, f. 17rb.
25. *Ibid.*
26. MS Peterhouse 272, f. 17va. *Cf.* Latin in Appendix, 17va, first quotation.
27. *Ibid.*, *Cf.* partial Latin in Appendix, second quotation from this folio.
28. *Ibid.*, f. 17vb.
29. *Ibid.* See Latin text in Appendix.
30. *Ibid.*, f. 17va.
31. *Ibid.*, f. 17vb.
32. *Ibid.*, f. 18va ff.
33. *Ibid.*, f. 19va.
34. *Ibid.*, f. 20ra.
35. *Ibid.*, f. 20rb.
36. *Ibid.*, f. 20ra.
37. Nicole Oresme did assume the existence of a physical entity, the quantity of quality, corresponding to this product. *Cf.* Marshall Clagett, *Nicole Oresme and the Medieval Geometry of Qualities and Motions* (Madison, Wisconsin: University of Wisconsin Press, 1968), pp. 172–177, 404–407.
38. For a survey of this history see Marshall Clagett, *The Science of Mechanics in the Middle Ages* (Madison, Wisconsin: The University of Wisconsin Press, 1959), pp. 430–440.
39. See H. Lamar Crosby, Jr., (ed.), *Thomas of Bradwardine. His Tractatus de Propor-*

tionibus. Its Significance for the Development of Mathematical Physics (Madison, Wisc.: University of Wisconsin Press, 1955), pp. 110—117. Also: A. George Molland, 'The Geometrical Background to the Merton School,' *The British Journal for the History of Science, 4* (1968), 108—125; Edith Sylla, 'Compounding Ratios: Bradwardine, Oresme, and the first edition of Newton's *Principia*,' in Everett Mendelsohn (ed.), *Transformation and Tradition in the Sciences. Essays Presented to I. Bernard Cohen* (Cambridge: Cambridge University Press, 1984), pp. 11—43.
40. H. Lamar Crosby (ed.), *op. cit.* (above, note 39), pp. 96—99.
41. Edith Sylla, 'Medieval Concepts of the Latitude of Forms: The Oxford Calculators,' *Archives d'Histoire Doctrinale et Litteraire du Moyen Age, 40* (1973), pp. 264—271. In an earlier section of this article, I discuss some of the material from Part II of the *Summa*.
42. MS Peterhouse 272, f. 23va.
43. *Ibid.*, f. 23vb.
44. *Ibid.*, f. 25va.
45. *Ibid.*
46. *Ibid.*, f. 24ra. This is Dumbleton's second counter-argument.
47. *Ibid.*
48. *Ibid.*, f. 24rb.
49. *Ibid.*
50. This is Dumbleton's first argument against the position; *cf. ibid.*, f. 24ra.
51. This is Dumbleton's third argument against the position (*ibid.*).
52. Since 1984 I have published two other papers relevant to these considerations, namely, 'Mathematical physics and imagination in the work of the Oxford Calculators: Roger Swineshead's *On Natural Motions*,' in Edward Grant and John Murdoch (eds.), *Mathematics and its applications to science and natural philosophy in the Middle Ages. Essays in honor of Marshall Clagett* (Cambridge, London, etc.: Cambridge University Press, 1987), pp. 69—101; and 'Alvarus Thomas and the Role of Logic and Calculations in Sixteenth Century Natural Philosophy,' in Stefano Caroti (ed.), *Studies in Medieval Natural Philosophy* Biblioteca di Nuncius. Studi e Testi, Vol. 1, (Florence: Leo S. Olschki, 1989), pp. 257—98. I also gave a paper at the Rutgers University Colloquium on Late Medieval Science and Technology, 18—19 April 1986, entitled, 'The Oxford Calculators and Aristotelianism: Tradition and Transformation in Late Medieval Science as Exemplified by Paul of Venice and Alvarus Thomas.'

APPENDIX

Uses of 'imaginatio' and related words in Dumbleton's *Summa*, Books II and III, as found in MS Peterhouse 272, with a few variants.

Book II

12ra ... nam quidam varietates rerum sensibilium advertentes tantam instabilitatem et motum in rebus iudicarunt ut dicerent omnia ista subesse motui ut quidlibet ita se habet ut ymaginacio nostra dictat, sicut iteratur 4 *Metaphysice* textu commenti 16^m et 17^m. . .

12ra . . . ita posset ymaginari de positione ipsius Putagoras quod numerus productus est tamquam formam et numeri producentes sunt tanquam materia. . . .

14ra . . . Philosophus non intelligit aliquam unicam materiam sive formam vel privationem necessario requisitam ad generacionem rei naturalis ac si ymaginaretur quod tria certa essent principia ut materia forma et privatio que inessent omni generacione rei materialis.

15ra . . . non est locutio de forma ut est ymaginata abstracta sed de tali que realiter extensa in materia.

16ra . . . quia Philosophus ymaginatur alteracionem esse respectu utriusque extremi in uno subiecto in numero.

16vb . . . Primo arguitur quod nulla est latitudo realis nec ymaginaria cui omnes gradus rerum naturalium sunt applicandi.

16vb Item sic: si esset latitudo perfectionum in speciebus, ergo ista esset infinita respectu utriusque extremi, quod non capit ymaginatione. Et consequentia sic arguitur, quia sit A latitudo talis perfectionis et B terminus, id est non gradus, vel ita se habens (scilicet B ad A) sicut quies ad motus. Et sit C gradus ymaginatus infinitus in perfectione et capiatur perfectio humana . . .

17ra Si ergo nulla pars talis latitudinis ymaginate alicui speciei convenit determinate, nec tota alicui speciei convenit nec potest competere, ergo nulli tali latitudini proprie nec eius partibus specierum perfectiones sunt comparabiles, quod est probandum.

17va Similiter aliud exemplum ponunt quod sicut corpus in vacuo moveretur, si esset vacuum, et nihil pertransiret nisi ymaginarie, quia in vacuo non est spacium, sic est iam de facto quod qualitas intenditur et remittitur per motum alterationis per quem nihil reale motu adquiritur tamquam spacium, quia unica qualitas manet a principio usque ad finem alterationis.

17va . . . iuxta hanc positionem nulla talis distantia adquiritur illi calori nec subiecto in quo est ille calor per illam alterationem nisi secundum ymaginationem, eo quod ista positio dicit illum calorem manere simplicem a principio usque in finem sine alico adquisito vel deperdito. Ergo nullus est in hac alteratione verus motus nisi ymaginarie. Ut licet ymaginati fuerimus simplicem motum extra celum, tamen nullus est ibi verus motus, cum ex hoc non sequitur quod sit extra celum verus motus, quia res non sequuntur ymaginationes. Cum in alteratione data non est nisi ymaginatum spacium, sequitur quod non est verus motus sed motus ymaginatus, ut si intelligamus ymaginando A punctum movere recte et nihil adquirere de facto de distantia que est divisibilis in partes distinctas realiter, tunc A punctus nullo modo vere moveretur nisi secundum ymaginationem, sic in proposito de uniformi alteratione.

17vb . . . Causa deceptionis ponencium istam positionem est hec: ipsi ymaginantur gradum esse solum intensiorem alio secundum quod plus vel minus distat a non gradu latitudinis illius speciei, licet tamen ille gradus in se sit realiter indivisibilis intensive, nec aliquam latitudinem contineat in se. Ac si homo ymaginaretur quod punctus terminans pedalem quantitatem contineret in se magis de quantitate quam punctus terminans semipedalem quantitatem, et sic referendo puncta ad distantias quas terminant, unus magis continet de quantitate quam alius. Puncta tamen de se nullam quantitatem habent.

18ra Item arguitur quod non est ymaginabile quod infinita indivisibilia faciant aliquam magnitudinem vel distantiam, quia si sic, detur quod puncta possent. Contra. Ymaginetur unum punctum in aere et applicentur sibi infinita corpora piramedalia secundum conum quorum unum sit subduplum ad aliud, et tertium in quadruplo minus, et sic in infinitum. Circa punctum datum infiniti coni tangunt se et omnes sunt in eodem situm. Cum ergo infinita indivisibilia possunt esse in eodem situ, inpossibile est quod faciant aliquam distantiam.

18va Qui ergo dicunt distantiam in qualitate ymaginariam, consequenter affirmant alterationem motum ymaginarium esse.

18vb ... ergo motus non esset intensius nisi per partes tales reales vel ymaginarias quarum quelibet inciperet a non gradu.

Pro istis et consimilibus concedendum est quod deducitur, quod quodcumque in natura qualificatum sub certo gradu ratione vel re vel imaginarie ipsum habet realiter omnes gradus remissiores illo. Et quod istud sit in motu patet. Et per consequens in omnibus que ratione solum sunt, et ymaginarie, qualitates. Ut ponamus Socratem moveri A gradu uniformi pro hoc instanti et intendat Socrates motum suum usque ad B gradum intenciorem A gradu in hore. Si ergo Socrates deperderet A gradum immediate post hoc ita quod nec ipsum A nec aliquem gradum equalem A immediate post hoc Socrates haberet, sequitur quod Socrates inciperet intendere motum suum a non gradu, quod est contra positum...

18vb Ex his palam concluditur quod omnis gradus qualitatis, sive fuerit qualitas ymaginaria sive vera sive re sive ratione, quod necessario totam latitudinem sub se continet.

19rb Ut si ymaginemur quod candela agat iam per bipedalem quantitatem lumen remissum sub medio. Si ergo illa candela sic disponatur quod tota sua actio concurrat in unam pedalem quantitatem, istud lumen in illo pedali erit duplum intensive ad illud quod prius fuit actum vel multo intencius. Et causa est quia candela ista agit duo lumina nunc coextensa in illo pedali que prius egit in duabus, et totum aggregatum est intencius lumen. Et per hoc experimentum probatur dicta positio, scilicet qualitates simul coextendi.

19va Item ista positio ponit quod motus difformis est ita intensus sicut aliqua eius pars. Cum ita se habet latitudo in caliditate et in qualitatibus aliis ad subiectum et qualitatem in eadem specie sicut latitudo motus ad suum subiectum, ergo qualitas difformis est ita intensa sicut aliqua eius pars. Consequentia sic arguitur: alteretur aliquid latitudine alterationis ab A gradu ad quietem et extendatur latitudo caliditatis sicut ista latitudo motus quantum ad subiectum. Ille due qualitates, scilicet alteratio et illa caliditas, se habent equaliter quantum ad intenciorem in se. Ergo si una cum omni parte quantitativa intensissime (?) equevalet, ergo et utraque. Consequentia tenet per hoc quod si qualitates ymaginarie difformes sunt ita intense sicut aliqua pars earum, a multo maiori qualitates reales ut caliditas et frigiditas et sic de singulis qualitatibus.

20ra Consequentia tenet cum antecedente nisi dicatur quod difformitas est in qualitate quia qualitas habet partes quasdam intenciores et alias remissiores ymaginarie vel vere, quod non videtur verum quia sic quantitas esset difformis vere iuxta modum loquendi quod non admittitur.

Praetereas omnis qualitas uniformis sive difformis quantum ad intencionem eodem modo est divisibilis in partem remissiorem et intensiorum veram vel ymaginariam, ut omnes quantitates vere equaliter dividuntur, ergo quantum ad intencionem in se omnes sunt eque difformes vel nulla est difformis simpliciter intensive.

20rb Nam dum necessario tante intencioni correspondet tanta latitudo per se et proprie, omnes qualitates habentes equales latitudines ymaginarias sive veras equalis intencionis existunt quod est intentum.

20rb Praeterea si consideremus in quanto alico, ut ponendo quod A habeat latitudinem caliditatis vel alterius qualitatis solum ex hoc quod habet partes difformes quarum una illarum est intencior alia, iuxta istum modum ymaginandi latitudo quam ponimus adquisitum in motu alterationis non est in rerum natura nisi secundum quod extenditur. Et per consequens nulla latitudo adquiritur in alteratione uniformi nisi ymaginaria, eo quod nulla pars est intencior alia, quod requiritur ad hoc quod sit latitudo in qualitate iuxta hanc positionem.

21vb Contra istud sequitur quod nulla caliditas remittitur in infinitum quod sic arguitur: intelligatur tota latitudo intencionis, et sit D. Capiatur pars illius correspondens albedo et sit C. Cum latitudo remissionis sic est infinita sicut latitudo intencionis ymaginata, sequitur quod inpossibile est quod aliqua qualitas remittetur per totam latitudinem, ymmo cuicumque qualitati certa pars D correspondet.

Part III

25ra Et per hoc concluditur quod motus et proportio et omnes tales qualitates sive vere sive ymaginarie componitur ex partibus qualitativis vel in actu vel saltem illo modo quo sunt qualitates . . . .

25ra . . . sed nulla superficies cum corpore alico convenit nec convenire potest eo quod omne corpus omnem superficiem ymaginabilem incomparabiliter excedit.

25ra . . . ita cum Socrates movetur iam A gradu ille motus adquirendus addetur A gradui, A manente continue in Socrate, ex quo motu adquisito partibus manentibus ymaginarie, scilicet re vel ratione, fit motus intensior. Similiter proportio dupla respectu alicuius medii manet inter Socratem et suum medium, licet Socrates adquirat aliam duplam. Similiter est ymaginandum quod latitudo caliditatis non terminatur proprie ad gradum summum, sicut nec pedalis terminatur ad pedalem, nisi aliquis velit ponere quod gradus summus se habet ad latitudinem caliditatis sicut punctus ad lineam quod non est ponendum.

26va Ut ymaginetur infinitos circulos inter circulum DMN et circulum AOQ, quorum circulorum C sit centrum et quod quilibet punctus AM linee sit in alico puncto illorum circulorum ymaginatorum. Si ergo brachium AB elongaretur in levatione A quod A semper esset in puncto alico AM linee, sequitur quod B potest ita distare a C quod quodlibet grave potest levare A per AM lineam rectam. Consequentia tenet et antecedens, quia brachium AB potest ita maiori ut CA sit semidiameter alicuius circuli prius ymaginati inter circulum DMN et AOQ et quod A grave sit in toto ascensu in alico puncto ipsius AM et ita patet quod omne grave quantumcumque fuerit modicum in infinitum grave potest levare motu recto per instrumentum.

28rb Item ponatur quod C ignis intendat B ignem inducendo medietatem latitudinis, et ymaginetur quod terra corrumpat partes qualitativas in B. Et intendetur A uniformiter secundum eandem proportionem secundum quam intenditur B et nihil corrumpitur de A....

29va ... et ita ymaginatur Commentator quod est una forma re vel ratione in mobili qua adquiritur aliquod spacium in alico predicamento. Et ideo predicamenta que non habent vera spacia divisibilia secundum rem, in illis non potest esse verus motus, sicut ponatur a quibusdam qualitatem non habere distantiam veram qualitativam sed ymaginariam.

Diffinitio formalis est hec: motus est actus entis in potentia secundum quod in potentia, cuius diffinitionis hec particula, scilicet quod motus est actus entis iuxta Philosophum sic est intelligenda: motus prout accipitur pro forma motus est forma realis vel ymaginaria in moto tanquam in suo proprie subiecto, per quem motum est vere tale vel quale quousque mobile adquisuerit terminum. Ut albedo in corpore est actus quo corpus est album et non alia forma est corpus albus quam albedine, ita nec mobile est motum nisi per formam ymaginariam vel veram.

29vb ... ideo motus dicitur quodammodo medium inter potentiam et actum quia de utroque habet, sicut medium inter duo contraria habet aliqualiter de utroque contrario. Ac si ymaginaretur Socratem mutari ad sanitatem ipso habente infirmitatem, secundum quod Socrates recipit actionem et transmutationem ut habeat per eas sanitatem, Socrates est subiectum et ens illius facti per motum. Sed secundum quod deficit sibi forma sanitatis est non ens, id est in potentia ens, et non in actu.

30vb Et ideo sive movetur secundum unam proportionem sive aliam, sicut proportio habet partes suas simul eam componentes ymaginarie vel realiter, ita motus habet partes suas et ille simul sic sumpte sunt gradus motus sicut due pedales sunt gradus profunditatis.

30vb Sed cum in C instanti habebit quadruplam secundum quam movebitur in C et omnis proportio quadrupla componitur necessario ex duabus duplis, ergo in C instanti Socrates habebit duas duplas. Ergo habebit duos motus realiter vel ymaginarie quorum alter erit A vel sibi equalis.

31rb Pro isto est dicendum quod, ymaginando motum esse unam formam accidentalem, consequens ponatur, scilicet prius naturaliter esse motum in mobili quam ipsum movetur. Nam isto modo intelligendo oportet tria ymaginari: primo quietem, secundo motum, tertio mobile prout actualiter est in adquirendo spacium, et iuxta hoc motus est forma mediante qua adquiritur privatio quietis que privatio est adquisitio spacii. Ac si ymaginaretur quod motus propter quam fit tenebra in medio luminoso est causa precedens privationem luminis, scilicet tenebram, ita iuxta hanc positionem motus est forma qua fit corpus actualiter adquirens spacium que adquisitio est privatio quietis. Aliter tamen potest motus intelligi, non prout est forma qua est actualis adquisitio spacii, sed prout solum est privatio quietis in mobili, et ut sic motus non est aliud nisi ipsum mobile esse actualiter adquirens, et ita non prius est motus quam pertransitio spacii.

31vb Consequens falsum, et ideo positio in se ymaginationem prioritatis et posterioritatis solum retinet et nihil veritatis in re habet.

32ra quasi [philosophus] diceret quod sicut quantum habet partem extra partem ita motus ymaginatur habere partem extra partem et hoc ex quantitate, quia si quantitas magnitudo non esset, non posset esse motus.

32rb ut cecus ymaginando seipsum mutari alico genere motus intelligit tempus.

32rb Pro isto distinguendum quia tempus potest accipi dupliciter pro materia vel pro forma vel pro composito ex hiis ymaginarie vel ratione.

32va Dico instans esse aliud et aliud secundum materiam ex hoc quod intelligimus mobile esse in alio indivisibili quam prius. Ut Socrates moto et ipso existente in medio puncto linee, instans correspondere illi puncto dicimus, cuius instantis punctus est materia quia intellectus noster non instans intelligit nisi aliud indivisibile extra ymaginatum applicet.

32vb Commentator 12 *Metaphysice* commento 30 dicit quod inpossibile ymaginari motus sine tempore innuendo quod simul natura intelliguntur.

33ra Pro isto nota diligenter quod si Socrates generaret pro quolibet instanti diei novum punctum, solum tamen unicum punctum in quolibet instanti maneret, et ymaginetur Socratem isto modo generare lineam in die, hec generatio linee nec est tarda nec velox. Probatio, quia Socrates non potest plura puncta producere in alico tempore quam facit in qualibet parte illius temporis, quia infinita generat in omni parte illius temporis.

33ra Item fuerit consideratum quod quantitas generata ex generatione indivisibilium, ut ponit casus prior, talis quantitas ymaginata non potest esse ex se maior nec minor nisi quod refertur ad aliud.

33ra Sed commento 98 dicit quod usque istud secutus est dicta expositorum, quare non sunt multa ponderanda usque ad illum locum, quia post dicit quod motus non potest ymaginari sine tempore. Vel posset dici quod accipiendo tempus prout habet esse in actu a nostro intellectu posterius est motu simpliciter.

33rb—va . . . et quia isto modo anima considerat de motu, dicitur tempus habere partes in potentia, id est ymaginatas, ex hoc quod anima refert durationem ad spacium pertransitum et pertranseundum, quod est materia temporis et motus. Si ergo ymaginamur motum et durationem unam respectu AE et aliam respectu EC quarte sicut ymaginamur AC esse maiorem ad partem AE et ad partem EC, sicut dicimus AE minorem AC, ita motum et tempus secundum quod ad partes equales vel inequales referimus motus equales et inequales, et tempora equalia et inequalia dicimus et concedimus. Et ideo propter materiam velocitatis et motus inequalitas et equalitas in eis reperiuntur. Et quia Socrates pro hoc instanti applicatur uni puncto ymaginato, scilicet C cuius puncti non potest esse pars pertransita et alia pertranseunda neque secundum rem neque apud ymaginationem, ideo ymaginatur unum indivisibile, quod est instans, illi C correspondere, quod instans secundum materiam est C punctus.

34ra Hic videtur quod sunt ponenda indivisibilia in qualitate qui sunt termini qualitatis et nullius intencionis sunt sicut puncta respectu linee. Nam si tempus habet esse ex motu alterationis, cum tempus non habet esse sine instanti terminante, ergo motu solo alterationis existente instans est et non nisi respectu alicuius indivisibilis intensive, sicut iam ymaginamur de puncto respectu cuius dicimus instans esse in tempore. Ergo ita vere ponenda sunt indivisibilia accidentia que terminant gradus divisibiles intensive, sicut puncta sunt ponenda terminancia lineas.

Pro isto dico sic: quod non maior ratio artat ponere puncta quam qualitates indivisibiles intensive, et ideo ymaginatio est solum que facit puncta.

34rb Pro isto concedendum est iuxta modum loquendi quod C tempus fuit et tamen neganda est quod B linea fuit. Et causa est quia tempus est tale quod cum suis terminis fuerit in actu, ipsum non est nisi apud ymaginationem. Sed de aliis rebus quarum partes manent simul, ita simul sunt cum suis terminis.

SABETAI UNGURU

EXPERIMENT IN MEDIEVAL OPTICS*

Two caveats: 1. 'Medieval Optics' is construed for the sake of this paper to mean thirteenth century optical thought as reflected in Witelo's *Perspectiva*. 2. My examples are drawn mostly from Books II and III of the *Perspectiva*. The first warning is not as limitative as it may superficially appear. After all, the *Perspectiva* is an encyclopedic treatise, synthesizing the best and most up-to-date information on optical matters available in the West, which it presents in as exhaustive a manner as possible, even when this leads to verbosity and looseness of demonstration, and at a relatively high level of technical sophistication. The second cautionary call is indeed confining but far from lethal. It is my feeling that Books II and III are truly representative of Witelo's attitude toward experiment, as can be seen easily by glancing through the other seven books of this great compilation.

I

Even though *experimentum* and *experientia* were used almost interchangeably throughout the Middle Ages all the way to the Renaissance, a growing sharpening of conceptual distinctions took place gradually, and various scholars distinguished between (1) mere experience, observation, and (2) artificially set-up empirical situations, not occurring normally in nature, for example, searched-for, sought-after experience. The latter is *experiment*. Indeed, Francis Bacon, that much maligned 'naive inductivist,' put it this time too, as in so many other cases, pregnantly and succinctly: 'It remains but mere experience, which, when it occurs, is called a chance happening; if it is sought after, [however], it is called experiment.'[1] The following identity is, then, established:

Experientia quaesita ≡ *experimentum*

Various gropings toward this recognition include Peter Maricourt's 'industria manuum' (it was Peter Peregrinus whom Roger Bacon referred to in the *Opus tertium* as 'dominus experimentorum'), Roger Bacon's own 'scientia experimentalis,' Thomas Aquinas' 'scientia experientie' or its identical twin 'scientia experimentalis,' Nicolas of Cusa's 'experimenti statici,' etc., etc. It is common knowledge that qualitative experiments (and not only qualitative) were already known in ancient times, in medicine and alchemy and mechanics, among other departments of knowledge, while truly quantitative experiments make increasingly their appearance in the eleventh, twelfth, and thirteenth centuries in Islamic science and in its Western counterpart, especially through the use of the balance (Abu-r-Baihan and his determination of specific weights) and other measuring devices (Roger Bacon, Witelo, and Nicolas of Cusa).

Illustrating the variety and fluidity of thirteenth-century attitudes to experiment are the views of Saint Thomas and Roger Bacon. The former, while failing to discriminate between 'scientia experientie sive scientia adquisita' and 'scientia experimentalis,' remained largely within the Aristotelian framework when it came to his understanding of experiment (experience) as arising out of repeated memories:

> Again, Isidore says that devils know many things by experience. But experimental [experiential] knowledge is discursive; for out of many memories comes one experience and from many experiences [experiments] there results one universal, as is said [by Aristotle] at the end of the *Posterior Analytics* and at the beginning of the *Metaphysics*. Therefore the knowledge of angels is discursive.[2]

And: 'Again, Isidore says that devils know many things by experience. For experience consists of many memories, as is said in the *Metaphysics*. Therefore there is in them [i.e., angels] also a potentiality for memory.'[3] This, needless to state, is a far cry from *experiment* and *experimental science* as harbingers of modern science. Such examples can be multiplied: 'experience originates from the senses;'[4] 'in actions [in matters of conduct] experience causes not only knowledge [science] but also a certain habit, as a result of usage [custom] which facilitates action [conduct].'[5] 'It must be said that to tempt [i.e., to put to a test] is properly speaking to make a trial of something. Now we make a trial of something in order to know something about it; hence the immediate goal of every tempter is knowledge.'[6] Finally, for the use of *experimentum* in the sense of 'proof' we have: 'Si ergo ratio humana sufficienter experimentum praebet.'[7] All these examples are drawn from the *Summa Theologiae*. They are only confirmed and strengthened by Saint Thomas's other works.

Roger Bacon, on the other hand, waxes grandiloquent when speaking of experiment and experimental science. He also seems to reach Nirvana when discussing mathematics. The trouble is that it is not easy to reconcile Bacon's sophisticated theory with his extremely simplistic practice in these matters. According to him, experimental science is the most powerful of the sciences in its own right, but also a tool for use in the other sciences. As a tool, it has a double 'dignity': the verification of the conclusions of the other sciences and the attainment of new truths which, though lying within the domains of those sciences, are outside the competence of their methods of investigation. Additionally, in its own right, it discovers marvels in nature and harnesses them to human needs, especially ecclesiastical, through the practical arts.

Following in Aristotle's footsteps,[8] Roger Bacon sees the origin of science in man's inborn curiosity. Furthermore, knowledge (science) is attainable in three ways, by authority, reason, and experience:

> But authority has no savor, unless the reason for it is given, and it does not give understanding but belief. For we believe on the strength of authority, but we do not understand through it, nor can reason distinguish between sophism and demonstration, unless we know how to test the conclusion by works, as I will ... show in the experimental sciences.[9]

Through experience we face literally truth. Our questions are answered directly by experience. Moreover, the fundamental sense in experimental science is vision. Besides, experience is needed in all the sciences as a precondition for achieving understanding with absolute certainty. This is so even in the 'abstract' science of mathematics in which drawing of figures and counting are appealing to the senses and providing the test for the truthfulness achieved by demonstration. Error-free is not the same as doubt-free. Bacon interprets Aristotle's statement that 'proof is reasoning that causes us to know' as necessarily entailing 'provided the proof is accompanied by appropriate experience.'[10] This is the key for Bacon's belief in the value of experience to mathematics. Mathematics contains certifying experience as an *intrinsic* ingredient of its basic operations (drawing of figures and counting). Experience and mathematics are mutually necessary to one another. They are essentially complementary in Bacon's scheme: Mathematics supplies errorless truth while experience furnishes doubtless centainty. Concerning the sequence 'belief — experience (experiment) — reason', Bacon says:

Hence in the first place there should be readiness to believe, until in the second place experience follows, so that in the third reasoning may function. For if a man is without experience that a magnet attracts iron, and has not heard from others that it attracts, he will never discover this fact before an experiment. Therefore in the beginning he must believe those who have made the experiment or who have reliable information from experimenters [experts].[11]

Also, in mathematics itself, 'without mathematical [i.e., astronomical] instruments nothing can be known, and these instruments are not made by the Latins and could not indeed be made for 200 pounds, or even 300.'[12] As a coordinator of knowledge, mathematics specifies the quantifiable elements in nature. It grants the scholar the needed certainty, since it can confirm his observational perceptions. Only through mathematics can he determine what properties and causes in nature are susceptible of scientific investigation. Moreover, mathematics is the basis of all logical categories. If follows, then, that logic, too, is subordinated to mathematics.

Experience, of its own nature, is confined to the concrete, the particular. Mathematics, of its own nature, is confined to the universal. According to Bacon, a knowledge of both the particular and the universal is essential for a complete understanding of science and it was therefore perfectly possible to say both that experience is essential for all science and that mathematics is the 'key to the sciences.'

In practice, Bacon follows his own theoretical precepts very little. Most of the 'experiments' he cites are little more than commonsense observations. Besides, he attached minimal importance to experiment as a tool in induction, thinking of it almost exclusively as a test of preconceived — or prereceived — ideas which he considered essential first steps to scientific understanding.[13]

II

There are numerous instances of appeal to and use of experiments and experience in Witelo's *Perspectiva*. The very first proposition of Book II is a case in point. It reads: 'All luminous rays, as well as the multiplication of forms, stretch forth in straight lines.'[14] The proof is exceedingly long, occupying almost two full pages in the Risner edition of 1572. It starts as follows: 'What is proposed here can be made known not by a demonstration but rather instrumentally; indeed, the diversity [of the attempts] of the ancients to prove this made use of a variety of instruments, while we are using that which we describe below, which we believe to harmonize better and more legitimately with what is proposed here.'[15]

Witelo continues with a very detailed description of the construction of his (Alhazen's) instrument, which in Alhazen is called '*organum refractionis*' and which Witelo uses both for the determination of the rectilinearity of propagation of light and colors and for his measurements of angles of refraction between various media, an outline of which is given by Crombie.[16] What interests us here, however, are simply those passages which show Witelo's 'hands-on,' pragmatic, matter-of-fact approach. Some instances should suffice:

Let there be taken, therefore, a round bronze vessel, sufficiently thick (like the mother of an astrolabe), the width of whose bottom is one cubit . . . and let the height of its edge be equal to the width of two inches . . . and let the vessel be placed . . . in a lathe, and let it be shaped by turning until its periphery be truly round both extrinsically and intrinsically, and let its plane surfaces be levelled, and let the column-like body, which is the middle of the back, be [also] made round. . . . And then let the vessel be brought back to the lathe and let three parallel circles be marked in it. . . . And so let the middle one of these circles be divided in 360 parts and, should it be possible, into minutes. Then . . . let a round opening be drilled. . . . Then let a plane, somewhat thick, bronze plate be taken. . . . And let it be smoothed down . . . and let the plate be bored through with a round opening. . . . Then let the little plate be fastened to the bottom of the vessel. . . . Then let a bronze quandrangular rule be taken . . . and let its surfaces be equalized till they are equal to [the surface] of a rectangle. . . . Then . . . let there be made a round opening, whose size is fit for the body which is in the back of the instrument . . . and let it be such that it revolve in the same instrument. . . . And let two little strips be made . . . which should be consolidated over the extremities of the ruler . . . and let a pin be inserted holding together the ruler with the instrument.[17]

This is more than enough. We can, then, clearly say with Witelo: 'Et hoc est propositum.'[18]

This is also, perhaps, the place to point out that Witelo uses this very '*organum refractionis*' in Book X to prove *experimentally*, as he puts it, his various conclusions about refraction. Thus, in proposition X.4, where he deals with the relationship between the angles of incidence and refraction, he says: 'What is proposed here can be proved instrumentally, so that [there be] a demonstration [that] is expressed sensibly by means of an instrument.'[19] Furthermore, the enunciations of propositions X.5, 6, and 7 read respectively:

To determine experimentally the sizes of the angles of refraction from air to water.[20]

To determine experimentally the sizes of the angles of refraction from air or water to plane or convex glass and vice versa.[21]

To find out experimentally the sizes of the angles of refraction from air or water to concave glass and vice versa.[22]

And concerning propositions X.10 and 11 he states respectively: 'This is clear by experience' and 'What is proposed here is obvious by reason and experience.'[23]

Finally, in X.42 one finds the statement 'What is proposed here is sufficiently clear from the preceding; but it pleases me to show the same experimentally and [to unravel] the universal cause by means of a particular example;'[24] in X.48 the proof ends with the words 'But in experiencing these [things] there is also great latitude, which we, [however], leave to such of curious mind;'[25] and in his study of the rainbow,[26] there is repeated appeal to experience and experiment in proving his conclusions.

I should not omit to point out here that some of the claims made on behalf of Witelo's empiricism, especially by Crombie, are often quite exaggerated, if not outright wrong. A case in point is represented by proposition X.8, in which Witelo reproduces in tabular form results he allegedly obtained experimentally in measuring angles of incidence and refraction between different media — air, water, and glass (all six possible arrangements). It is clear in this particular case that Witelo lied and Crombie was taken in. As shown by Albert Lejeune and David Lindberg,[27] Witelo's table is copied in its entirety from Ptolemy's *Optics* and additionally garnished with absurd data when purportedly measuring angles of refraction from the denser to the rarer medium, which could *not* have been obtained by actual measurements. Witelo's 'empirical' numbers, then, are due to (1) his misunderstanding of the 'reciprocal law' of refraction (which he nevertheless duly states after Alhazen) and (2) his failure to grasp the phenomenon of total internal reflection. Clearly, empiricism has its dangers to which neither perspectivists nor historians of optics are immune.

Pursuing now our survey of Witelo's appeal to experiment, I would like to illustrate it with proposition II.5, which, because of its shortness, lends itself to being reproduced in full:

Lights and colors do not blend in transparent bodies but penetrate [them] separately.

The reason for this thing is to be shown experimentally. Let many locally distinct candles be placed in a certain place and let them all be opposite to one aperture leading to a darkened place and let a certain non-transparent body be placed opposite the aperture in the darkened place. And so the lights of the candles appear over that body separately according to the number of the candles, and any of those [lights] appears opposite to one candle, in accordance with the straight line passing through the aperture and through the middle of the light of the candle. And if one candle is covered completely only one light, [which is] opposite to that candle, will be destroyed and should the candle be uncovered, the light returns. And so it is obvious that in the middle of the aperture, where all or many [lights] penetrate one another in one point, the lights do not blend in the same point but are separate according to their essences; and on account of this, when extended later, they are distinguishable locally according to their diversity, by the places in which they fall. And since light traversing colored things gets colored by those colors, as was supposed, it is obvious that if light penetrates separately, the colors too, which are carried with the light, will penetrate separately. What is proposed is therefore clear.[28]

Since the empirical setup is obvious and the proposition speaks for itself, no further comment is really necessary.

The next use of experiment in Book II is in proposition 24, which reads: 'Every luminous body illuminates a smaller place from which it does not escape more strongly than a greater space.'[29] The reader can be spared the 'proof' this time, but not without first being made aware that, in this case, Witelo calls his procedure *per exemplum*.[30] Our next instance is II.42. In it Witelo proves that a perpendicular ray penetrates a body denser than the one it previously traversed without being refracted. The proof is long and does not really concern us. It is performed 'instrumentally' according to the following statement: 'The proof of this proposition rests more on an instrumental endeavor than on other demonstrations.'[31] Then follows a long and detailed description of the experimental arrangement and procedure involved in establishing the truth of the proposition by means of Alhazen's *obiectum refractionis*, in which the practitioner is referred to as the *experimentator*[32] and which ends with the following words: 'From which it is clear that the passage of light through the body of water is by straight lines, by XI.1 [Euclid], and this is what we intended to show experimentally about the intended proposition.'[33]

In proposition II.43, Witelo discusses the behavior of oblique rays issuing from a rarer into a denser medium, namely their refraction toward the perpendicular. Thus he says: 'This proposed theorem can also be shown experimentally [to be true],'[34] which he indeed does, relying on the procedural setup of the previous proposition.

The following proposition, II.44, occupying almost two full pages in Risner's edition, deals with the lack of refraction of perpendicular rays coming from a denser medium (glass) to a rarer one (air or water). The exceedingly lengthy and loquacious proof, a gem of Witelo's phatic style, involving, this time, too, an experimental arrangement, proceeds according to the initial statement: 'The proposed theorem can be similarly shown [to be true] by instrumental experience.'[35]

The next two propositions of Book II (45 and 46) are both tackled and disposed of experimentally. Thus, II.45 discusses the refraction of oblique rays from a denser to a rarer medium according to the methodological statement 'What is now proposed here is to be shown by instrumental experience in conformity with the preceding [propositions].'[36] II.46 reads: 'It is necessary that every incident and refracted ray be situated in the same plane surface.'[37] The proof, relatively succinct for a change, starts with the declaratory statement 'But even that which is now proposed can be shown experimentally'[38] and proceeds directly by reliance on the experimental set up of II.43, ending effectively with the words 'this experiment [test] is sufficiently evident each time.'[39] Finally, II.47 represents an attempt at summing up and taking stock of the results of propositions II.42—II.45. In it Witelo makes it abundantly clear that he is fully aware of the peculiar experimental approach he adopted in the preceding propositions, which he desires, nevertheless, to supplement here by means of more general, physical, and philosophical considerations, coupled with common-sense inferences applied to the nature of light and motion. Thus he says:

That which was proved so far instrumentally by means of specific trials, we intend to enhance by a natural demonstration. Indeed all natural motions which are made according to perpendicular lines are stronger because they are intensified by the universal celestial virtue flowing into every body lying beneath [the heavens] along the shortest straight line.[40]

And in the course of his discussion, which relies, among other things, on reasoning by *reductio ad absurdum*, one also finds the following remarks: 'And so it is plain, according to the previous reasoning concerning the strength of the perpendiculars and by [various] instrumental trials, [specifically] by the 42nd and 44th [propositions] of this [book], that the ray [which is] incident ... perpendicularly penetrates the whole body,'[41] and 'Natura autem frustra nihil agit.'[42] It is certain, therefore, that Witelo engages in experiments fully aware of their *sui generis* nature in his work, the overwhelming character of which is theoretical and deductive.

Compared with Book II, Book III has little, if any, experiments, but plenty of observations and experience. To illustrate, appeal to the latter meets the eye in proposition 3, where it is proved that the eye is spherical and in which it is said: 'And since this is false and against a notion which is clear to the [visual] sense, because it is possible for a thing greater than the eye itself to be seen, it is plain that it is not possible for the surface of the visual organ to be plane;'[43] in proposition 4, reading 'The eye is the spherical organ of visual power constituted of three humors and four tunics issuing from the substance of the brain [and] arranged spherically,'[44] in which one finds the statements 'In what manner the eye may be the organ of visual power, we leave to the labor of another part of [natural] philosophy. That it is indeed spherical is necessary by the preceding proposition and also [follows] from [the fact] that it is of a watery nature, the property of which it is always to become rounded. ... Moreover the assiduous concern of the anatomists has taught [us] thoroughly that the eye is constituted of three humors and four tunics,'[45] and 'Hence it is thus clear that the humors and tunics of the eye are fixed in spherical channels, and the statement of the proposed definition of the eye is clear, according to the experience of all those who have written thus far about the anatomy of the same';[46] in proposition 6, in which it is established that the visible forms *act* on the eye and in which it is said, 'Indeed the eye suffers from strong light as, [for example], at the sight of the solar body or of another strong light, say the light reflected to the eye from a polished body or from another exceedingly white body';[47] and, finally, in proposition 16, reading, 'Vision does not take place without pain and suffering endured by the substance of the eye. From which it is clear that the eye ought to be of an adequate disposition in its health in order to prosecute completely the [process of] vision,'[48] in which appeal is repeatedly made to observation of the eye's reaction to strong light.

These are not all the instances in Book III, or in the rest of the *Perspectiva*, in which experience and/or experiments make their pivotal appearance. But for obvious reasons, the foregoing examples should suffice.

One other concern that is discernible in this book and that certainly deserves to be pointed out is Witelo's effort to show, primarily in proposition III.60 but with reverberations in the following few propositions, that, in a sense, the naked eye is blind:

In fact the distinction between those two whitenesses does not itself pertain to the sensation of whiteness, because the sensation of whiteness stems from the whitening of the eye's surface, which is accomplished by any whiteness, while the distinction of those [two] whites is achieved on account of the difference of action of those two whites in the same eye. Hence that distinction does not [stem] from the sense alone, but it is [also a result stemming] from another power of the soul, which we call the discerning [power]. *And it is the same with the comparison and discrimination of other sensible forms; indeed nothing of those [forms] is perceived by sight alone, but by the collaboration of reasoning and the discerning virtue. For sight by itself has no discerning power, but [only] the discerning power of the soul distinguishes all those [things] by means of sight.* The proposed [thing] is, therefore, clear.[49]

All this shows, I submit, among other things, Witelo's methodological sophistication. It is clear that he is no blind, naive empiricist.

III

On the whole, Crombie's assessment of Witelo's methods seems to me accurate even if one can discern now and then a tendency to exaggerate Witelo's 'inductivism' and to lend too much credence to his rhetorical 'manual and technical skill' or to his 'quantitative experiments.'[50] It is thus quite all right to speak of 'Witelo's effective combination ... of experiment with geometry, of manual skill with rational analysis and synthesis,'[51] as it is to point out that 'The methods by which the various modes of operation of these [visual] forms were to be investigated were, according to Witelo, by observation and experiment and by mathematics,'[52] even though one might quarrel with this particular enumerative order.

Furthermore, Crombie's thorough study of Witelo's theory of the rainbow led the former to state convincingly:

The precise manner in which the different colours were produced by the incorporation of darkness in the rays Witelo then investigated by means of experiments with refraction through crystals and spherical glass vessels filled with water.[53]

He [Witelo] went on to describe further experiments on refraction [by producing artificial rainbows through light passing through hexagonal crystals and spherical glass flasks filed with water]. The subject was unexplored and experiment was the guide: 'For the colour or visible form is carried to vision only by the nature of light which it contains; and to what has been said the careful inquirer will be able by experiment to add many things.'[54]

However, needless to emphasize, Witelo remained a medieval perspectivist in spite of his thoroughness, erudition, *and* alleged 'experimental-inductivist' bent. His theory of vision is medieval, not modern. In the overwhelming majority of instances, his approach is not that of the inductivist, the experimenter, but, on the contrary, of the deductivist, the theoretician. His use of experiments, which is certainly present, is not, however, undertaken in order to originate previously unknown conclusions, but rather in order to establish, support, or test the accuracy and appropriateness of convictions held beforehand. What Crombie said about Witelo's discussions of lenses applies throughout Books II and III and — I venture the not totally uneducated guess — throughout most of the *Perspectiva*: Witelo's conclusions stem from theoretical considerations alone, while in many cases the requisite experimentation is left to the curious reader (*ingenio perquirentis*).[55] And so it is not at all surprising that even though there were 'careful inquirers' after Witelo who were able, perhaps also by 'experiment to add many things,'[56] what they did not, and could not, add by experiment, as long as they remained thoroughly committed to all the elements of the medieval theory of vision, was its modern modification, the Keplerian theory.

This theory, to state the obvious, is *not* an outgrowth of an experimental, inductivist methodology. As A. Koyré has put it, '*too much* methodology is dangerous and ... more often than not ... results in sterility. ... No science has ever started with a *tractatus de methodo* and progressed in the application of such an abstractly devised method.'[57] Kepler's breakthrough and the beginning of modern optics are no exceptions to Koyré's assessment. On the contrary. David C. Lindberg has shown[58] that Kepler's contribution to optical thought can best be understood by seeing it as the continuation and significant modification of the medieval theory, of which it is the culmination. No role to speak of is played in it either by induction or experimentation as such.

Thus, and I am relying heavily on Lindberg's book, Kepler denied that the impressions allegedly left by light on the lens, which was seen as connected directly to the retina and the optic nerve by means of the aranea, are transmitted to the brain through the intermediary of the retina and the optic nerve. This he did by pointing out, first, that the capsule of the aranea surrounding the lens was not connected to the retina, as was shown by Felix Platter (1536—1614), the lens being in touch only with the uvea, and, second, by negating that sight was a form of touch, light and color traversing the eye instantaneously.[59] Moreover, the shape of the lens differs in its posterior surface from the shape assigned to it by Witelo and the perspectivists, in order to avoid intersection of rays at the center of the eye and, consequently, image inversion. The posterior surface of the crystalline humor is more curved, gibbous than the anterior surface (i.e., its radius of curvature is smaller than that of the anterior surface) and the image thrown on the retina is not only inverted but also reversed with respect to the object. Also, clearly, the eye sees more than the hemisphere of rays perpendicular to it, as can be verified by the images of objects at the very periphery of the visual field that enter the eye obliquely. So very oblique rays do leave an impression in the eye. What, then, about not so oblique rays, but such that while oblique, are very close to the perpendicular? Shouldn't they leave a quite powerful impression, because only slightly weakened? How, then, avoid total confusion on the very basis of the perspectivist theory of the visual image itself?

As shown by Lindberg, in anatomical ocular matters Kepler was, like Alhazen earlier, parasitic on the results of contemporary anatomists. He used primarily the works of Felix Platter (who made the retina and the optic nerve the *principal* seat of sensitivity of the eye) and Johannes Jessen (his Prague friend who, though writing later than Platter, stuck to the old views of crystalline photosensitivity). Concerning the geometry of the eye and its influence on light rays, Kepler bases his conclusions on an analysis of the functioning of spherical transparent lenses, assuming the anterior part of the eye (aqueous humor and the anterior surface of the crystalline humor) to be almost spherical in combination, so that rays entering it almost perpendicularly would undergo practically no refraction after they had already been refracted once at the cornea. Thus, to each point in the visual field, serving as the apex of a cone, there corresponds the base of that cone on the anterior surface of the lens; all the rays forming the base are again refracted when they emerge from the posterior surface of the lens (which is hyperbolic, not spherical) into the vitreous humor, away from the perpendicular, reaching together a new vertex at a point on the retina that serves as the new apex of another cone having the same base as the first. There is thus a one-to-one correspondence between points of the object and points of the *retinal* image by means of two cones having a common base and two vertices, such that the image-vertices are inverted and reversed with respect to the object-vertices. After that the image formed on the retina will be interpreted in the brain by means of the visual spirit carried by the optic nerve, this being a physico-psychological problem and not an optical one, thus not lending itself to geometrical analysis.

Kepler's contribution, then, according to Lindberg, is the result of many factors, among which one must mention seeing the eye as a *camera obscura*, a powerful and skillful talent at ray geometry and, crucially, a complete grasp and penetration of the perspectivist tradition, coupled with the readiness to answer the outstanding optical problems of the seventeenth century within the overall framework set by perspectivist optics. Having rejected, for good reasons, the perspectivist cone of perpendicular rays, Kepler still had to come up with a one-to-one correspondence between object-points and image-points. This he did by the 'two-cone' mathematical analysis of ray geometry and the theory of retinal image, in which he cleaved to the main ingredients of perspectivist optics, which were, in keeping with Lindberg, 'the principle of punctiform analysis, the laws of propagation of light, the requirement that one point in the visual field must stimulate one and only one point within the eye, basic conceptions about the nature of seeing, and the commitment to a methodology that incorporates mathematical, physical, and physiological reasoning.'[60]

Let me conclude this brief excursus on Kepler with the following quotation from Lindberg's book:

Kepler was the culminating figure in the perspectivist tradition. . . . That his theory of vision had revolutionary implications, which would be unfolded in the course of the seventeenth century, must not be allowed to obscure the fact that Kepler himself remained firmly within the medieval framework. The theory of the retinal image consituted an alteration in the superstructure of visual theory; at bottom it remained solidly upon a medieval foundation. Kepler attacked the problem of vision with greater skill than had theretofore been applied to it, but he did so without departing from the basic aims and criteria of visual theory established by Alhazen in the eleventh century . . . his theory of vision was not anticipated by medieval scholars; nor did he formulate his theory out of reaction to, or as a repudiation of, the medieval achievement. Rather, Kepler presented a new solution (but not a new kind of solution) to a medieval problem, defined six hundred years earlier by Alhazen. By taking the medieval tradition seriously, by accepting its most basic assumptions but insisting upon more rigor and consistency than the medieval perspectivists themselves had been able to achieve, he was able to perfect it.[61]

IV

It seems to me, then, that in optics (perspective), as in other branches of theoretical knowledge, experiment plays only a minor, derivative role.[62] Moreover, I tend to agree with Koyré that until the development of scientific technology, theoretical and practical achievements were largely independent in the domain of optical thought. The progress of optics in the thirteenth and fourteenth centuries was not determined by methodological considerations but rather, and crucially, by the availability of Alhazen's *De aspectibus* (*Perspectiva*), with its new departures in optical thinking, and by the sharpening and convergence of issues growing out of its sophisticated analysis.

Witelo's use of experiment, like that of Alhazen, his model, stands neither at the origins of a methodological revolution nor issues from such a revolution. As my examples have abundantly shown — and they represent only a lean chrestomathy — Witelo employs experiment in a plain, straightforward, nondramatic way, an almost pedestrian way. It is clear to the careful reader of his *Perspectiva* that experiment enjoys no special status with him. On the contrary. It seems to crop up when no satisfactory alternative is in sight. I would say that Witelo *prefers* the theoretical, deductive approach and that experiment is his second-best choice. There is no fanfare or particular pride in its use, which is quite natural and normal under the given circumstances. It results in no departure from Witelo's theoretical commitments, which it is rather meant to buttress and authenticate. Witelo seems to me to appeal to experiment *faute de mieux*; he would rather give a mathematical proof if he could and if it were at all suitable. And so I find myself, again, in basic agreement with A. Koyré:

As for myself, I don't believe in the explanation of the birth and development of modern science by the human mind turning away from theory to *praxis*. I have always felt that it did not fit the real development of scientific thought, even in the seventeenth century; it seems for me to fit even less that of the thirteenth and fourteenth. I don't deny, of course, that in spite of their alleged — and often real — 'otherworldiness,' the Middle Ages, or to be more exact, a certain, and even a rather large number of people during the Middle Ages, *were* interested in techniques; nor that they gave to mankind a certain number of highly important inventions.... Yet, as a matter of fact, the invention of the plough, of the horse harness, of the crank, and of the stern rudder had nothing to do with scientific development; even such technical marvels as the Gothic arch, stained glass, the foliot or the fusee of late medieval clocks and watches did not depend on, nor result in, any progress in corresponding scientific theories. Strange as it may seem, even such a revolutionary discovery as that of firearms has had no more scientific effect than it had scientific basis. Bullets and cannon balls brought down feudalism and medieval castles, but medieval dynamics resisted the impact. Indeed, if practical interest were the necessary and sufficient precondition of experimental science — in our sense of the word — this science would have been created a thousand years, at least, before Robert Grosseteste, by the engineers of the Roman Empire, if not by those of the Roman Republic.[63]

The implications of what I have presented here strike me as both obvious and nonseminal. They should not be controversial. My excuse, then, for this evidential presentation is simple: I thought it worth saying. My methodological guide was G. K. Chesterton. In Kingsley Amis' words in his introduction to a selection of Chesterton's stories,[64] it is said that

Chesterton's stance on most matters could be summed up very roughly as follows. What is simple, generally agreed, old and obvious is not only more likely to be true than what is complex, original, new and subtle, but much more interesting as well: a prescription calculated to alienate almost any type of progressive thinker.

As to Witelo's use of an experimental methodology in his *Perspectiva*, my reaction can best be stated in the inimitable words of Ambrose Bierce, the author of *The Devil's Dictionary*: 'Example is better than following it' and 'Where there is a will there is a won't.'

Tel-Aviv University
Tel-Aviv, Israel

NOTES

* An enlarged version of this paper, 'Mathematics and experiment in Witelo's *Perspectiva*', appeared in *Mathematics and its applications to science and natural philosophy in the Middle Ages*, edited by Edward Grant and John E. Murdoch (Cambridge Univ. Press, 1987), pp. 269—297. I thank Cambridge University Press for permission to include here material that originally appeared in the Grant-Murdoch volume.
1. 'Restat experientia mera, quae, si occurrat, casus; si quaesita sit, experimentum nominatur' (*Novum organum* I LXXXII).
2. 'Praeterea, Isidorus dicit quod daemones per experientiam multa cognoscunt. Sed experimentalis cognitio est discursiva; ex multis enim memoriis fit unum experimentum, et ex multis experimentis fit unum universale, ut dicitur in fine *Posteriorem* et in principio *Meta*. Ergo cognitio angelorum est discursiva' *Summa theologiae* I.58, 3, quoted after the Blackfriars Edition, Vol. **9** (1968), p. 150.

3. 'Praeterea, Isidorus dicit quod angeli multa noverunt per experientiam. Experientia autem fit ex multis memoriis, ut dicitur in *Meta*. Ergo in eis est etiam memorativa potentia'; *ibid.*, I.54, 5, p. 86.
4. *Ibid.*, I.64, 1, p. 282: 'experientia a sensu oritur.'
5. 'Experientia in operabilibus non solum causat scientiam, sed etiam causat quemdam habitum, propter consuetudinem, qui facit operationem faciliorem'; *ibid.*, I.II.40, 5, Vol. **21** (1965), p. 16.
6. 'Dicendum quod tentare est proprie experimentum sumere de aliquo. Experimentum autem sumitur de aliquo, ut sciatur aliquid circa ipsum; Et ideo proximus finis cujuslibet tentantis est scientia'; *ibid.*, I.114.2, Vol. **15** (1970), p. 76.
7. *Ibid.*, II.II.2.10. What Thomas says here is that 'if human reason provides sufficient proof,' then 'the merit of faith is altogether taken away.'
8. Πάντες ἄνθρωποι τοῦ εἰδέναι ὀρέγονται φύσει.' (*Met.* A.980a22).
9. 'Tamen auctoritas non sapit nisi detur eius ratio, nec dat intellectum sed credulitatem; credimus enim auctoritati, sed non propter eam intelligimus. Nec ratio potest scire an sophisma vel demonstratio, nisi conclusionem sciamus experiri per opera, ut . . . in scientiis experimentalibus demonstrabo'; *Compendium studii philosophiae*, in J. S. Brewer, (ed.), *Opera quaedam hactenus inedita* (London, 1859), p. 397.
10. *Cf.* N. W. Fisher and Sabetai Unguru, 'Experimental Science and Mathematics in Roger Bacon's Thought,' *Traditio* XXVII (1971): 353—378, passim.
11. 'Unde oportet primo credulitatem fieri, donec secundo sequitur experientia, ut tertio ratio comitetur. Si enim inexpertus magnetem trahere ferum nec audiens ab aliis quod trahat in principio debet credere his qui experti sunt, vel qui ab expertis fideliter habuerunt,' J. H. Bridges, (ed.), *The 'Opus Majus' of Roger Bacon* (London, 1900), II, p. 202.
12. 'Sine instrumentis mathematicis nihil potes sciri, et instrumenta haec non sunt facta apud Latinos, et non fierent pro ducentis libris nec trecentis'; *Opus tertium*, in J. S. Brewer, (ed.), *Opera quaedam hactenus inedita*, p. 35.
13. Fisher and Unguru, 'Experimental Science,' pp. 371—372, 366, 376—77.
14. 'Radij quorumcumque luminum et multiplicationes formarum secundum rectas lineas protenduntur' *Opt. Thes. Wit.*, p. 61, i.e., *Opticae Thesaurus Alhazeni Arabis . . . Item Vitellonis Thuringopoloni Libri X* (1572), reprinted by Johnson Reprint (New York, 1972).
15. 'Hoc quod hic proponitur non demonstratione sed instrumentaliter potest declarari; diversitas tamen antiquorum ad hoc probandum pluribus usa est diversis instrumentis, nos vero utimur isto quod hic subscribimus, quod regularius huic proposito credimus convenire' (*ibid.*).
16. A. C. Crombie, *Robert Grosseteste and the Origins of Experimental Science 1100—1700* (Oxford, 1953), pp. 220—23.
17. *Opt. Thes. Wit.*, pp. 61—63:

   Assumatur itaque vas aeneum rotundum convenienter spissum, ad modum matris astrolabij, cuius fundi latitudo sit unius cubiti . . . et altitudo hore eius sit aequalis latitudini duorum digitorum . . . et ponatur hoc vas . . . in tornatorio, et tornetur quousque periferia eius sit extrinsecus et intrinsecus vere rotunditatis, et adaequentur plane superficies ipsius, et corpus columnare, quod est in medio dorsi, fiat rotundum. . . . Et deinde reducatur vas ad tornatorium, et signentur in ipso tres

EXPERIMENT

circuli aequidistantes. ... Dividatur itaque medius istorum circulorum in 360 partes, et si possibile fuerit, per minuta: deinde. ... perforetur foramen rotundum. ... Deinde accipiatur lamina aenea plana aliquantulum spissa. ... planeturque adeo ... et perforetur lamina foramine rotundo ... deinde consolidetur parva lamina fundo vasis. ... Deinde accipiatur regula aenea quadrangula, ... et adaequentur superficies eius, donec fiant aequales rectangulae. ... Deinde ... fiat foramen rotundum, cuius amplitudo sit capax corporis, quod est in dorso instrumenti ... fiatque taliter, quod revolvatur in ipso instrumentum ... fiantque duae pinnulae ... quae consolidentur super extremitates regulae ... et immittatur cuspis continens regulam cum instrumento.

18. *Ibid.*, p. 63.
19. 'Quod hic proponitur potest instrumentaliter demonstrari, ita ut demonstratio auxilio instrumenti sensibiliter exprimatur'; *ibid.*, p. 407.
20. 'Quantitates angulorum refractionis ex aere ad aquam experimentaliter declarare'; *ibid.*, p. 408.
21. 'Quantitates angulorum refractionis ex aere vel aqua ad vitrum planum vel convexum, et econverso experimentaliter declarare'; *ibid.*, p. 410.
22. 'Quantitates angulorum refractiones ex aere vel aqua ad vitrum concavum, vel econverso experimentaliter invenire'; *ibid.*, p. 411.
23. 'Hoc patet per experientiam' and 'Quod hic proponitur, patet ratione et experientia'; *ibid.*, p. 414.
24. 'Quod hic proponitur, patet satis ex praemissis: sed et idem placuit experimentaliter declarare, et universalem caussam particulariter exemplare'; *ibid.*, p. 440.
25. 'Sed et in horum experimentatione est maxima latitudo quam relinquimus ad talia curiosis'; *ibid.*, p. 444.
26. For which, see Crombie, *Robert Grosseteste*, D. C. Lindberg, *Theories of Vision from al-Kindi to Kepler* (Chicago, 1976); W. A. Wallace, O. P., *The Scientific Methodology of Theodoric of Freiberg* (Fribourg, 1959); and Carl B. Boyer, *The Rainbow: From Myth to Mathematics* (New York, 1959).
27. A. Lejeune, *Recherches sur la catoptrique grecque* (Brussels, 1957), pp. 153—155, and D. C. Lindberg, 'Introduction' to reprint of *Opt. Thes. Wit.*, pp. XX—XXI. XXI.
28. *Opt. Thes. Wit.*, p. 64:

Luces et colores in corporibus diaphanis non admiscentur adinvicem, sed penetrant distincti ... Huius rei experimentaliter declarandae caussa, ponantur in loco aliquo candelae multae localiter distinctae, et sint omnes oppositae uni foramini pertranseunti ad locum obscurum, et opponatur foramini in loco obscuro aliquod corpus non diaphanum. Luces itaque candelarum apparent super illud corpus distincte secundum numerum candelarum, et quaelibet illarum apparet opposita uni candelae secundum lineam rectam transeuntem per foramen et per medium luminis candelae: et si cooperiatur una candela, destruetur unum lumen oppositum illi candelae tantum, et discooperta candela, revertitur lumen. Palam itaque quod luces in medio foraminis, ubi se intersecant omnes vel plures in puncto uno, non admiscentur in eodem puncto, sed sunt distinctae per sui ipsarum essentias: et ob hoc cum ulterius protenduntur, tunc secundum locorum, quibus incidunt, diversitatem localiter distinguuntur. Et quoniam lux res coloratas pertransiens, illarum coloribus coloratur, ut suppositum est: palam, si lumen penetrat distinctum, et colores, qui feruntur cum lumine, penetrabunt distincti. Patet ergo propositum.

29. 'Omne corpus luminosum minus spatium, a quo non egreditur, fortius illuminat quam spatium maius illo'; *ibid.*, p. 70.
30. *Ibid.*
31. 'Huius propositionis probatio plus experientiae instrumentorum innititur, quam alteri demonstrationum'; *ibid.*, p. 76.
32. *Ibid.*
33. 'Ex quo patet, quod transitus lucis per corpus aquae est secundum lineas rectas per 1 p. 11. Et hoc est, quod circa propositam propositionem experimentaliter intendimus declarare'; *ibid.*, p. 77.
34. 'Experimentaliter etiam et hoc propositum theorema potest declarari'; *ibid.*
35. 'Instrumentali similiter experientia propositum theorema potest declarari'; *ibid.*, p. 78.
36. 'Hoc quod nunc hic proponitur, est conformiter prioribus per instrumentalem experientiam declarandum'; *ibid.*, p. 80.
37. 'Omnem radium incidentem et refractum in eadem plana superficie consistere est necesse', *ibid.*, p. 81.
38. 'Sed et id, quod nunc proponitur, potest experimentaliter declarari'; *ibid.*
39. 'Satis evidens est haec experimentatio omni tempore'; *ibid.*,
40. 'Illud, quod particularibus experientijs hactenus instrumentaliter probatum est, naturali demonstratione intendemus adiuvare. Omnes enim motus naturales, qui fiunt secundum lineas perpendiculares, sunt fortiores, quoniam coadiuvantur virtute universali coelesti secundum lineam rectam brevissimam, omni subjecto corpore influente'; *ibid.*
41. 'Palam itaque secundum rationem praemissam fortitudinis perpendicularium et per experientias instrumentales per 42 et 44 huius, quoniam radius incidens ... perpendiculariter, penetrat totum corpus'; *ibid.*, p. 82.
42. *Ibid.*
43. 'Et quoniam hoc est falsum et contra suppositionem, quae patet sensui, quoniam possibile est rem maiorem ipso oculo videri: palam, quia non est possibile, ut superficies organi visivi sit plana'; *ibid.*, p. 85.
44. 'Oculus est organum virtutis visivae sphaericum, ex tribus humoribus et quatuor tunicis a substantia cerebri prodeuntibus sphaerice se intersecantibus compositum'; *ibid.*
45. 'Quomodo sit oculus virtutis visivae organum, negotio alterius partis philosophiae relinquimus: quod autem sit sphaericus, necessarium est per praecedentem propositionem: et etiam ex eo, quod est naturae aqueae, cuius proprietas est semper rotundari. ... Quod autem sit oculus ex tribus humoribus et quatuor tunicis compositus, diligens anatomizantium cura edocuit'; *ibid.*
46. 'Sic ergo patet, quod humores et tunicae oculi sphaerice se intersecant: et patet declaratio definitionis propositae oculi secundum omnium eorum experientiam qui de ipsius anatomia hactenus scripserunt'; *ibid.*, p. 87.
47. 'Laeditur enim visus ex forti luce, ut in aspectu corporis solaris vel alterius lucis

fortis, ut lucis reflexae ad oculum a corpore polito, vel ab alio corpore valde albo'; *ibid.*, pp. 87—88.
48. 'Visio non fit sine dolore et passione a substantia oculi abijciente. Ex quo patet, visum oportere convenientis dispositiones in sanitate esse ad hoc, ut complete exerceat visionem' *ibid.*, p. 91.
49. *Ibid.*, p. 112, my emphasis:

> Distinctio vero inter illas duas albedines non est ipse sensus albedinis: quoniam sensus albedinis est ex dealbatione superficiei visus, quae fit ab utraque albedine: distinctio autem illarum albedinum fit propter diversitatem actionis illarum duarum albedinum in ipsum visum. Non est ergo illa distinctio a solo sensu, sed est ab alia virtute animae, quam dicimus distinctivam. Et similiter est de comparatione et distinctione aliarum sensibilium formarum: nihil enim istorum accipitur solo visu, sed ratione et virtute distinctiva coadiuvantibus: visus enim per se non habet virtutem distinguendi, sed virtus distinctiva animae distinguit omnia illa mediante visu. Patet ergo propositum.

50. Crombie, *Robert Grosseteste* (see note 16, above), p. 218.
51. *Ibid.*, p. 214.
52. *Ibid.*, p. 216.
53. *Ibid.*, p. 230.
54. *Ibid.*, p. 232, quoting Witelo X.83, *Opt. Thes. Wit.*, p. 474.
55. *Opt. Thes. Wit.*, p. 439. This appears in prop. X.40.
56. *Cf.* the quotation appearing in text to note 54, above.
57. A. Koyré, 'The Origins of Modern Science: A New Interpretation,' *Diogenes* **16** (1956): pp. 1—22, at pp. 14—15.
58. *Theories of Vision* (see above, note 26).
59. *Ibid.*, p. 188.
60. *Ibid.*, p. 281 n. 122.
61. *Ibid.*, pp. 207—208.
62. I am struck by this statement as almost tautological.
63. Koyré, 'Origins of Modern Science' (note 57, above), p. 12.
64. *G. K. Chesterton Selected Stories* (London, 1972), p. 12.

PART FOUR

Kepler: Cosmology, Astronomy, and Light

FRITZ KRAFFT

THE NEW CELESTIAL PHYSICS OF JOHANNES KEPLER

I. INTRODUCTORY REMARKS

> Above all, there are three things, the number, size and motion of the planetary orbits, I have always studied in order to find out the reason why they are as they are. I was determined to do so by the wonderful harmony of stationary things, like the Sun, the fixed stars and the space between, with God the Father, the Son and the Holy Ghost.[1]

The endeavour to discover the divine harmony of the cosmos as the reason for number, size and movement of the planetary orbits was henceforth the objective of Johannes Kepler's scientific lifework. This he expressed as the impetus and objective of his studies in the foreword to his first work, the *Mysterium cosmographicum*. However, this harmony, set *a priori*, was not limited to mathematical proportions — in this he differs radically from Plato and the Pythagoreans, whom he took as examples; for him mathematical proportions were only an expression of the divine will of creation, as principles of order in a natural world, which were observed in a *natural* way. In other words, Kepler wanted to make astronomy again into harmonics, as well as into physics. Up to that time the three disciplines were separated from, and conflicting with, each other. He wanted to combine them into one synthesis.

Later, empirical science had shown that this was not possible in the way Kepler *seemed* to have been successful; on the other hand Kepler's lifework makes it clear that his conviction that the synthesis of all three approaches was necessary was also the presupposition for those discoveries which proved to be correct beyond the original context, namely the three so-called Keplerian Laws. Despite their largely empirical basis, they were unaccepted for a long time, not only because they broke with the old demand of the astronomers that all celestial motions be described by uniform circular movements, but particularly because they were so closely connected to Keplerian 'physics,' which could not find recognition, and to Keplerian harmonics, which was rejected as an *a priori* assumption by nearly all astronomers; the rejection of Keplerian physics and harmonics resulted in a rejection of the planetary laws, as long as these could not be consistently integrated into a comprehensive physical system. Kepler's physics cannot yet be seen as such a system, even if it was Kepler who first saw the necessity of a new physics for the heliocentric world system and he who developed the first speculative fundamentals for it.

In what follows we shall concentrate only on Kepler's search for the 'natural' causes of planetary motions (which eventually led him to the discovery of his Laws) and will assess his contribution in the framework of the physics of his time.

II. CELESTIAL PHYSICS BEFORE KEPLER

1. *Aristotelian Physics*

Ptolemy's system of the universe was geocentric and his concept of terrestrial physics was Aristotelian. Aristotle's terrestrial physics allowed the deduction that the Earth was fixed and located at the centre of the Universe; it was closely bound up with his celestial physics, based on Eudoxus' mathematical spheres which had become material ones. These material spheres were made of ether, the only material to have the property of moving in circles, but having no other properties, and thus not subject to change. It had been added as a fifth element to the four terrestrial elements.

The concentric circular motion is as natural for ether as the downward motion is for 'heavy' elements and the upward motion for 'light' ones. It seems that this is why Aristotle let originally the uniform velocity of the concentric ether spheres grow with their size — such as the rate of fall and climb of the terrestrial bodies was supposed to be dependent on their size —, thus seeing the Moon, in the sense of the older vortex theories, as the slowest star, which remains the furthest behind the revolution of the sphere of the fixed stars.[2] Later, after he had become acquainted with Eudoxus' system of homocentric spheres (in *De caelo* he only knows this system and not yet that of Callippus[3]), he decided that there had to be other causes for the varying velocities of the rotary motions. In *De caelo* B 12, he therefore states that stars and planets should not be understood as single and entirely inanimate bodies,[4] but as living organisms that participate in action and life ($\mu\varepsilon\tau\varepsilon\chi\acute{o}\nu\tau\omega\nu$

πράξεως καὶ ζωῆς) and achieve their aim (the single motion of the sphere of the fixed stars) only through multiple actions (motions), or achieve that aim only to a certain degree even through multiple actions.[5] In a lecture, added later to his *Methaphysics* as book Λ, Aristotle gives each simple sphere a noëtic soul to function as its teleological mover, a spirit (νοῦς, Latin: *intelligentia*); and in a later addendum (Λ 8), according to Callippus' improvement of the astronomical system, he determines the number of the Intelligences, and with this the number of the ether spheres. He needed fifty-five such ether spheres to make the mathematical model of Eudoxus and Callippus into a physical system moved teleologically from the outside by an unmoved Prime Mover. For each planet there is, in addition to the spheres which gave it its own motion, an equal number of retrogate spheres, which compensate for each individual component of its motion, so that the motions of these spheres would not be transmitted to the next planet.[6]

The Intelligences (in the Middle Ages identified as angels) which, according to Aristotle's *Metaphysics*, were inherent in the spheres, were usually considered as the movers, although occasionally (based on Aristotle himself[7]) the activity of uniform movement was rejected as inadequate for divine Intelligences, and the 'natural' circular movement came to be explained by means of the 'impetus' which was impressed upon by God.[8] Buridanus[9] introduces this theory with the remark that the concentric spheres are moved without resistance and thus the 'impetus' impressed upon them by God cannot decrease. Due to similar considerations, Galileo Galilei states that only circular-concentric motions remain uniform, because, without additional influence of force, an acceleration takes place only when the distance from the respective centre of gravity is decreased and a deceleration only when the distance is increased.[10] But the idea of Intelligences as movers of the spheres prevailed,[11] and often — as in Nicolaus Copernicus — no further reasons were given for the naturalness of the spheres' uniform rotary motion.

Aristotelian physics remained essentially valid until the end of the 16th century. It was based on the strict concentricity of all celestial ether spheres and on the uniformity of their rotary motions. Consequently, the mathematical theories using eccentric circles and epicycles, which were stated separately by Hipparchus and Apollonius respectively, did not correspond to this physics.

2. *Saving the Appearances*

However, Aristotelian physics had a strong influence on mathematical astronomy, too, in so far as it determined the content of astronomy's physical (metaphysical) principles. The circular form of the celestial orbits, introduced by Anaximander and given a theoretical basis by the Pythagoreans and Plato, had found a physical explanation in the Aristotelian system, which linked it to the uniformity of all the circular motions of the ether. This uniformity was introduced by Eudoxus for the convenience of calculation. As long as the existence of the supralunar ether and the 'naturalness' of the celestial circular motions (for which the same reasons were given as for the circular movement of the ether) were accepted, all celestial motions and their components had to be uniform and circular, even if empirical evidence might perhaps not allow for the preservation of strict concentricity within a mathematical theory describing the appearances (phenomena).

This demand for uniformity and circularity of all celestial motions was usually based on an incorrect interpretation of a source of late antiquity (Simplicius), mistakenly attributed to Plato and, moreover, in contradiction with Plato's own celestial physics.[12] The formulation of this demand as σῴζειν τὰ φαινόμενα (*apparentias salvare*, saving the appearances) is first found with the Stoics at the time of Posidonius.[13] But it must be placed then for other reasons as well: Plato never maintained that the apparent motions of the planets followed uniformly circles; in the Stoics, Aristotelian physics had been renewed and further improved (particularly by Posidonius), leading to the first Renaissance of study of Aristotle's works; certain approaches towards a positivism or instrumentalism, which the demand expresses, are to be found for the first time in the Stoics. Finally in Posidonius' time there appeared, originally in competition with each other, those two mathematical theories of eccentric circles or epicycles, which seemed to explain the same things in completely different ways. For the astronomy of that time both theories could not be correct simultaneously, not even kinematically[14]; and neither agreed with the accepted notions of physics, but both could explain the appearances better than Aristotle's concentric spheres could.

The contradiction between these two theories had two consequences. The first was that later mathematical theories, up to the time of Kepler, generally had the status of pure hypotheses, useful in computing planetary positions, but without physical reality. The second was that these hypotheses at least had to be grounded on the uniform and circular nature of all celestial motions as a fundamental physical fact. This is also expressed in the complete version of the demand which gives the astronomers the task of finding out 'by means of which hypotheses, based on uniform and circular [regular] motions, the appearing nonuniformity [the anomalies] in the planetary motions can be saved.'[15]

Later, mathematical theories and physical systems remained irreconcilable. The fact that this demand was falsely attributed to Plato, or the Pythagoreans, or even to Pythagoras himself, gave it a respectability which caused Copernicus himself to do his best to fulfil it by means of replacing Ptolemy's equants by *uniform* rotary motions.

3. *Modifications of Aristotelian ether physics: Sosigenes and Copernicus*

Since late antiquity, however, there have been some attempts to adapt Aristotelian ether physics to mathematical theories by means of slight modifications motivated through the phenomena.

The annular solar eclipse observed by Sosigenes in the year 164 A.D. had moved him to criticize (in his lost work *About the retrograding spheres [of Aristotle]*[16]) Aristotle's ether physics which until then had obviously been preserved within the *Peripatos* without any restrictions: The ether, and with it the ethereal bodies, were grasped as unchangeable. Therefore an annular solar eclipse, observed by Greek scientists in that year for the first time after observing partial and total eclipses,[17] could only occur, if the Moon or the Sun or both were able to change their *apparent* sizes, that means their distances from the Earth. The strict concentricity of all ether spheres was thus refuted. In this context, Sosigenes gives a critical survey of the older history of astronomy (following Eudemus of Rhodes) and the new mathematical (hypothetical) theories for 'saving the appearances' — without, however, naming Ptolemy, or further dealing with his theory — and criticizes the only celestial physics of his time. With regard to the undeniable phenomenon of the annular eclipse, he attacks particularly the axiom on which Aristotelian ether physics is based and which states that each circularly moved body (each ether sphere) has to move in a circular way around the centre of the universe:[18] As mathematical hypotheses, epicycle and eccentric theories seem to be more applicable than homocentric and retrograding spheres, because

> these hypotheses [eccentrics, epicycles] are at first *simpler* than former ones [the homocentric spheres], because they do not have to invent (ἀναπλάττειν) so many celestial bodies artificially and secondly because they [actually] save the appearances, not only all the others, but particularly those referring to the [motion in] depth and the 'anomaly.' But [according to Aristotle] every circularly moved body should be moved around the centre of the world [...]. However, the axiom which states that each circularly moved body moves around *its own centre* could have a greater degree of truth (ἀληθὲς οὖν ἂν μᾶλλον ἀξίωμα εἴη). For it is right to say that all celestial bodies [spheres], which have the middle of the universe as their centres, move around the middle of the universe, but that which is further away from the middle moves around the centre of its own, namely the stars, epicycles and eccentrics — if there are such things as bodies in the heavens, at all.

In order to maintain the principle of the nonexistence of a vacuum, besides the other Aristotelian principles, Sosigenes will even tolerate the possibility, that such bodies penetrate into each other. He no longer claims, however, the absolute truth which Aristotle had claimed for his system of homocentric spheres (but not for their number); for him his own axiom is merely 'more correct' (μᾶλλον ἀληθές), because it opens the possibility of a physical integration of the new mathematical theories, which truly save the appearances. Of course, it remains a proposal which is not explained in detail; for him the available mathematical theories continue to be hypotheses. It was only on this point that Alexander and Simplicius agreed with him; both of them strictly rejected a new celestial physics. As orthodox Aristotelians, both adhered to the old physics of homocentric spheres. Thomas Aquinas, too, the true originator of Christian Aristotelianism, whose interpretations were largely recognized, agrees with them explicitly in his commentary on *De caelo* and in other works.[19] All of them strictly distinguish between celestial physics (Aristotle) and mathematico-hypothetical astronomy (Ptolemy's *Almagest*), which are almost complementary to each other. Being in the tradition of the orthodox Aristotelians, Osiander was still able to refer to the hypothetic character of mathematical astronomy in his foreword to Copernicus' *De revolutionibus* of 1543 — although he did not reproduce the author's intentions correctly.[20]

Copernicus still proceeds from Peripatetical axioms and his argumentation is still fully Aristotelian;[21] and yet he is not an orthodox

Aristotelian but more of a Peripatetic in the sense of Sosigenes whose thoughts remained known through the commentary of Simplicius, which Thomas Aquinas had already ordered to be translated into Latin.[22] Thus it is not improbable that the *'petitio prima'* in Copernicus' *Commentariolus* originates from Sosigenes. It reads: 'Omnium orbium coelestium sive sphaerarum unum centrum non esse.' His theory, strictly considering the uniformity and circularity of all spherical motions, is also said to be 'more correct' than Ptolemy's theory using the equants; he speaks of *'firmiores* demonstrationes.'[23] Sosigenes' emphasis on the uniformity of all spherical motions was also set against the equant motion of his older contemporary Ptolemy, even if Ptolemy's name is no longer mentioned in Simplicius' excerpts.

In fulfilling the demand of the astronomers, in the sense of the Peripatetic Sosigenes, Copernicus annulled the strict separation of (hypothetical) 'astronomy' and (real) 'physics,' so that the traditional distinction became for him largely irrelevant. For the first time since Aristotle, he connected 'physical astronomy,' which explains reality, with 'mathematical astronomy,' which describes appearances. However, he does not say what he thinks about the nature of the spheres and how the planets and their spheres are to be moved. At least in the *Commentariolus*, he seemed to have in mind *solid* spheres, which was generally imputed to him by Kepler.

The question about the nature of the celestial spheres became a problem, unavoidable for astronomers, only after Tycho Brahe had found out, by means of parallactic measurements, that the *Nova* which appeared in 1572 must have originated in the sphere of the fixed stars and that the comet of 1577 moved unhindered through planetary spheres.[24] Thus, all who put their trust in Brahe's art of measuring (to whom Johannes Kepler belonged[25]) were deprived of the possibility of taking refuge in the auxiliary construction of impenetrable solid spheres. This idea of solid spheres had also been propagated in the Middle Ages, following Alhazen's new 'physical' interpretation of the 'mechanical' model of motion in Ptolemy's *Hypotheses planetarum*.[26]

Copernicus himself had always stressed that his new astronomy was no more accurate than Ptolemy's as a basis for calculations — he had for the most part used the same observational material —, but that his system of the universe was actually more correct. In any case, Copernicus himself thought it was a lesser evil to move the Earth away from the centre of the system than fail to fulfil the demand for uniform circular movement, that is the only 'natural' or 'physical' circular movement. He believed that Ptolemy's and his successors' failure to meet this demand adequately, by using equant motions, accounted for the inaccuracies of Ptolemaic astronomy; and this is what he states, in as many words, as his reason for trying to improve astronomy by finally fulfilling a *pre*-Ptolemaic condition[27]: For him it is obvious that the appearances must be saved, that is, exactly reproduced; but Copernicus lays more stress than Ptolemy or anybody else since Ptolemy, on preserving or 'saving' *uniform* circular motions. He finally manages to replace Ptolemy's eccentric equant motion by a uniform epicyclic motion superimposed on a further epicycle with a homocentric deferent. The first, larger, epicycle has the same sidereal period as the deferent, but its motion is retrograde. It carries the second, smaller, epicycle, which finally reproduces the equant motion, by turning in the same direction as the deferent, but at twice the speed. The combined motions of both epicycles have the effect of doubling the eccentricity. Later, in his main work, Copernicus follows Ptolemy (or rather Adrastus of Aphrodisias and Theon of Smyrna) and uses an eccentric circle as deferent with only one (the second) epicycle.

This deferent, however, does not possess the same eccentricity as in Ptolemy, because Copernicus originally had to distribute the doubled eccentricity in a ratio of 3:1 between the two epicycles, in order to reproduce the equant motion. With this reproduction, Copernicus introduced the use of epicycles into the treatment of sidereal periods, whereas previously they had only been used for synodic ones. However, both periods can hardly be explained simultaneously by means of epicycles, and the eccentric equant motion cannot be reproduced merely by an eccentric circle. And yet, only the use of epicycles could reproduce the equant motion, using only *uniform* circular motions. Copernicus now had to find a new way of reproducing the anomalous synodic periods.

According to his own account, he turned for help to pre-Ptolemaic scholars, and he mentions names such as Philolaus, Ecphantus, Hicetas and Aristarchus, the only real ancient proponent of a (hypothetical) heliocentric planetary system. Finally, there were suggestions to be found in Nicolas of Cusa. All these philosophers thought the Earth moved. Ptolemy had already noticed, and explained as a property of the Sun, that the epicyclic loop-motions of the planets were related to the position of the Sun: the radius vector between the centre of the epicycle and the planet being parallel to the radius vector between the Sun and the centre of the Sun's orbit (in synodic motion). This fact finally convinced Copernicus that the ancient writers mentioned above had been right, and their theories had only to be given a mathematical form: Reverse the positions of the Earth and the Sun and each of the loops in the synodic motions is connected with the position of the Earth, so that they can all be explained at once by the *one* annual motion of the Earth.

Thus, for the first time, the motions of all the planets were seen to be connected, and by taking into account the motion of the Earth, the theories of the motions of individual planets formed for the first time a system. However, the sphere of the fixed stars is now still; because it is inconceivable that God should allow the enormous sphere to turn when he could obtain the same apparent effect by rotating the tiny Earth. The daily rising and setting of the heavenly bodies is thus explained; the planets no longer need to have their own 24-hour rotation, and their real motion is given by what was the sidereal component in the older systems. During the emergence phase of the new theory and the *Commentariolus*, Copernicus did not possess new data in comparison with Ptolemy, and he never maintained that he had achieved a greater numerical accuracy than could be obtained in the Ptolemaic system. For instance, he took over the Ptolemaic solar orbit (where Ptolemy had not made use of an equant motion) and used it unchanged as an orbit for the Earth; therefore the reference point for the planetary motions and the centre of the universe was not the Sun, but the centre of the Earth's eccentric orbit round the Sun: The Copernican system is merely a transformation of the Ptolemaic one; and the motions of the planets (including the Earth) require a large number of eccentric circles and epicycles, not many fewer than were required by Ptolemy.

Copernicus himself was convinced that his system was real. The uniformity of all circular motions had been restored; the loss of geocentricity was more than offset, it seemed to him, by the greater economy of his system and by the way it linked individual motions of the planets. Therefore, mainly those contemporaries and successors who, like Giordano Bruno, Galileo Galilei and Johannes Kepler, were interested in physics, were enthusiastic about the heliocentric system, while pure astronomers regarded it as a useful and in some ways simpler basis for calculations, from which no conclusions should be drawn about the nature of the physical world. Astronomers were at liberty to consider the hypothesis as merely mathematical and kinematical in its implications. The Protestant theologian Andreas Osiander had already done so in his anonymous foreword to *De Revolutionibus*. His intention in doing so was not to make the work easier to accept but rather to write from within the tradition of a belief that any mathematical theory of astronomy could not describe reality.[28]

For Copernicus, physics plays only a subordinate role; he paid little attention to the implications of his system. He was not at all interested in what caused the planets, including the Earth, and the spheres to follow the various circular paths. His physics is completely in the spirit of Aristotle and the Stoics[29]; the Earth's rotation was thought 'natural' to a sphere composed of earth, water and air, as previously the motion of the sphere of fixed stars had been regarded as 'natural' to a sphere made of ether.

III. GRAVITATION

1. *Nicolaus Copernicus: Centres of Specific Gravity*

Copernicus' Aristotelian physics lacked only a new theory of gravitation; because the Earth was no longer considered to be stationary and at the centre of the universe, it was merely a centre of gravitational force. (N. Copernicus, *Commentariolus*, petitio secunda: 'Centrum Terrae non esse centrum mundi, sed tantum gravitatis et orbis lunaris' — 'The Centre of the Earth is not the centre of the world, but merely the centre of gravity and of the Moon sphere.') But even this modification could be made by referring back to Plato's assertion that all material and corporeal things gravitate to their like. This idea of Plato's had already been taken up in ancient times, for example by Plutarchus, who extended it into a general theory of cohesion, in which every heavenly body is a centre of specific gravity.[30] All scholars who abandoned the strictly geocentric system before the pupils of Galileo had worked out the principle of *inertia* used such theories of several specific and individual centres of gravitational attraction. These scholars included Nicolas of Cusa, Nicolaus Copernicus and his followers. Even Galileo and his pupils accepted such a theory, allowing bodies to gravitate towards their specific centres of attraction. Each heavenly body was made of specific elements: lunar, saturnian, and, for the

Earth, terrestrial elements; only lunar things gravitate to the Moon, terrestrial ones to the Earth, solar ones to the Sun, etc. These assumptions explained why all the matter in the universe does not coalesce into one lump, and gave a physical basis for the spherical form of all moving celestial bodies, including the Aristotelian Earth. This view is also to be found later, in the seventeenth century, after the discovery that Venus and Mercury showed phases and thus were dark bodies like the Earth. This view was also held by all opponents of Copernicus, believers in the Ptolemaic geocentric and the Tychonic geoheliocentric systems, so that even this argument cannot diminish the claim of the Copernican system to be a perfect example of a turning-point in the history of natural science.

It was only William Gilbert and, influenced by Gilbert's magnetic theory (only after 1604), Johannes Kepler, who chose to pursue a somewhat different path. They ascribed magnetic properties to all bodies, which therefore do something like attracting each other — if they are of similar nature (as are the Moon and the Earth); but they only do so within a certain limited range, depending on their 'mass' (within their '*sphaera activitatis*' or '*orbis virtutis*'). Nevertheless, this first step towards an idea of universal gravitation did not influence the explanations given for celestial motions, because until Robert Hooke and Isaac Newton no such 'universal gravitation' was proposed as an explanation for the *movements* of the planets.

2. *William Gilbert: Central Forces*[31]

The concept of central forces operating through space without a medium and within a limited '*sphaera activitatis*' or '*orbis virtutis*' has two intellectual roots in the history of ideas, which came together at the end of the 16th century: the first is the Neoplatonic idea of the theo-philosophical concept of the '*sphaera activitatis*' of mental and animistic or divine 'forces' (globe or sphere of virtue), having their origin in a light source (in particular the Sun) and radiating within a limited sphere, which played a significant role in the light metaphysics of late antiquity and the Middle Ages, following Plato's parable of the Sun and Plotinus' emanation doctrine; the second is the doctrine of magnetism, whose action at a distance was originally interpreted as a mediate effect (Aristotle, Simplicius, Averroës, Franciscus de Marchia) or as a '*horror vacui*' effect (atomists) or, according to Plato's and Aristotle's theories of the self-movement of natural bodies, was regarded as a tendency of the lodestone, or of the magnetic needle, to orient themselves to specific points ('magnetic mountains,' Northpole, celestial pole, 'point respective' in Robert Norman' book), Northpole, celestial pole, 'point respective' in Robert Norman's book). From his intensive work with theories and nautical applications of the phenomena of Earth magnetism, Kepler was aware of all these doctrines, long before William Gilbert's work even appeared, though its appearance represents a turning-point in Kepler's conceptions.[32]

The Neoplatonic analogy and interpretation of the 'force' of an inanimate body as a 'mental,' or quasi-living power, enabled scholasticism to transfer this analogy, along with the concept of '*sphaera activitatis*', to magnetism and to speak in this case too of the spherical limit of the area of virtue of a somewhat 'mental' magnetic force. This transfer was accomplished first by Robert Norman,[33] whose terminology was still undeveloped. However, this transfer had not yet been accomplished in the area of the attractive or repulsive effect of a magnet, but only in the area of the conveyability of the faculty to orient itself towards the 'point respective' located *within* the Earth. This concept had arisen from the discovery of magnetic inclination which causes a three-dimensional orientation of the magnetic needle.

In connection with magnetic attraction, and exclusively in this respect, Giambattista Porta was the first who, independently of Norman, spoke of a spherical expansion and limitation of magnetic effects in the 1589 edition of his 'Magia naturalis' which was expanded to 20 books; '*orbis virtutis*' is the term, he coined for the preceeding as an analogy for '*sphaera activitatis*.'[34] This term connects the concept of a spherical limitation of the area of the virtue of a 'force' (*sphaera activitatis*) with the concept of magnetic *effluvia*, diffused from a radiation centre analogous to a light source, so that on account of the dispersion their effect decreases in proportion with the distance. According to Porta, every magnet has two such '*orbes virtutis*,' the centres of which are the two magnetic poles. However, Porta had not yet thought of an 'action at a distance', but obviously of material *effluvia* of the magnetic poles.

By combining older concepts, in particular that of Norman's (induction of the) 'directing force' and Porta's concept of 'attractive force', William Gilbert (in his work 'De magnete'[35]) succeeded in transferring all properties of the older '*sphaera activitatis*' to the new term '*orbis virtutis*' of the magnetic force — in particular the immediate action at a distance, as well. Since the magnetic force penetrates solid bodies unhindered, as he proved by experiment (*De magnete*, II, 4 and 16), a property to which Norman had already referred for his 'directing force', it must (according to Gilbert) be immaterial and its transmission and effect cannot be connected with an immediate contact or with a medium. Contrary to the material 'electric' *effluvia*, Gilbert attributes the magnetic force not to *materia*, but rather to the *forma* in the Aristotelian sense (II, 2), and specifically not to the *causa materialis* and *formalis* of scholastics, but to the *prima forma*, the ἐντελέχεια, *vigor, actus* of each single celestial body, which he interprets as a large global magnet (II, 4); thus magnetic action requires two bodies acting on each other, in the context of which he speaks more precisely of συνεντελέχεια and *conactus*. Later on (V, 12), in an Aristotelian way, he identifies the respectively specific ἐντελέχεια (*vigor*) with the star's soul (*anima*); and thus he remains within the framework of the above outlined cohesion theory, according to which terrestrial matter gravitates within the respective '*orbis virtutis*' to the Earth, solar matter to the Sun etc.,[36] now, however, by reason of mutual 'attraction'. This has the effect that the atmosphere of the Earth, which supposedly consisted of evaporated terrestrial matter, is carried along by the orbiting and rotating 'Earth-Water-Globe' (Copernicus, too, had conceived the 'Earth-Water-Air-Globe' as *one* body, so that the carrying along of the atmosphere was part of the 'natural' motion of the Earth): The parts have the same 'natural' movement as the whole, that is the rotating Earth (VI, 4f.). And so Gilbert obtains a 'force' which operates without a medium through void space, a conclusion obtained from other criteria, above all from the critique of the solid spheres of the *primum mobile* and of the planets.[37]

In contrast to Porta but like Norman, however, Gilbert assumes, that there is only *one* centre of magnetic force which is located at the magnet's centre of gravity, so that the form of the '*orbis virtutis*' is dependent on the outer shape of the magnet: That of a globular magnet is spherical, others are spheroidal corresponding to the surface of the magnet (*cf.*, in particular, *De magnete*, II, 7 and 27f.) He distinguishes between two different '*orbes virtuis* (*magneticae*)', whether in respect to celestial bodies including the Earth or in respect to other, smaller globular magnets. The latter he considered models of the Earth and therefore named '*terrella*'. He defines[38] the '*orbis coitionis* (*magneticae*)' as a limit, within which mutual attraction still takes place (corresponding to the '*orbis virtutis*' of Porta); however, not only the poles of the magnet would attract, but, apart from the equator, the entire surface. Secondly he defines the disproportionately larger '*orbis virtutis*' as that area within which a magnetic needle is still deflected, that is to say directed, but no longer attracted (corresponding to the older '*sphaera activitatis*' and Norman's view)[39]:

orbis virtutis est totum illud spatium, per quod quaevis magnetis virtus extenditur.

The division of magnetic force into 'attracting' and 'directing' faculties became effective particularly in Kepler's concepts. Widening the concept of magnetism to a cosmic level, Gilbert succeeded in giving a first reasonable explanation of diverse phenomena of Earth magnetism, a novel explanation of (mutual) gravitation, including the mutual influencing of celestial bodies (within their own '*orbis virtutis*' and '*orbis coitionis*') as well as of other bodies. In this way, ebb and flow on Earth (because the body of the Earth does not hinder the Moon's force on either side) became subject to an interpretation which Kepler later developed further.[40]

According to Gilbert, the size of the '*orbis virtutis*' and '*orbis coitionis*' and the size of the magnetic 'force' are exactly proportional to the size (or the weight) of the magnet; a very small magnet attracts a large mass only in proportion to its own size (II, 29). There is, however, in the field of magnetism neither an '*attractio*' nor a 'tending towards the whole' — as Gilbert emphasizes repeatedly — but only a '*coitio*,' that means a 'tending towards each other' apportioned both to the whole (the celestial body) and to the part, in proportion to their respective sizes. The centre of the '*orbis coitionis*' is always the centre of gravity, therefore the '*coitio*' is a 'central force,' of limited virtue however, so that in this way only the Moon and the Earth (which are, moreover, similar to each other in the sense of the modified older cohesion theory) are able to have an effect on each other. Finally, according to Gilbert, the Sun moves the planets by means of its '*orbis virtutis*'; but he does not explain how this occurs.[41]

3. *Johannes Kepler: Mutual Attraction*

It seems that Kepler developed his concept of 'gravity' largely following the previously mentioned ideas of Gilbert, only several years after he had studied the work for the first time.[42] The external motive could have been, at the earliest, the writing of the 'Introduction' to his *Astronomia nova*, for one finds no remarks, neither in the printed works nor in the letters, on gravity before the manuscript was preliminarily finished at the end of 1604.[43]

A first report of his views was stimulated by the letter of Herwart von Hohenburg of March 8, 1605,[44] inquiring:

Villeicht würdt der herr ex motu Martis auch ein adminiculum ad probandum motum terrae, gefunden haben. Ich hoffe gentzlich, der herr werde jn disen wichtigsten puncten ein mahl exclamieren εὕρηxα, εὕρηxα.

(Perhaps you could find support for proof of the Earth's movement from the movement of Mars. Indeed, I hope that, in connection with this most important point, you will be able to exclaim: *heureka, heureka.*)

In the context of his view of the Sun as the central mover of the planets, which developed some years before (see section IV, below), Kepler's reply of March 28, 1605, deals critically with the Aristotelian doctrine of 'heaviness' and 'lightness,' which presupposes an Earth resting at the centre of the universe (Letter No. 340: 131ff.[45]):

Terrestrial matter is moved towards the Earth not because it is the centre of the universe but because it is its own centre, and this lies outside of the centre of the universe. In addition, gravity is not an *actio*, but rather a *passio* of the stone being attracted.

Like a magnet (*ut magnes*), the Earth attracts heavy bodies by means of an immaterial efflux (*per effluxum immateriatum*), and therefore this efflux is moved along with the Earth which is in turn moved by its soul. This attractive efflux (*effluxus tractorius*) is disproportionately large compared with the stone's disposition for rest (*dispositio ad quietem*), because the body of the Earth is disproportionately larger than that of the stone. [...] Therefore it surpasses that 'rest' almost infinitely, and, because of this, the Earth pulls the stone as on a chain (*instar cathenae*) just as fast as the efflux itself is moved along with the Earth. Therefore a cannon-ball will fall back into the cannon, if it were possible to shoot it out in an exact vertical direction.

In this way Kepler disproves the objection to the earth's movement, brought up by, for example, Tycho Brahe, namely that in the case of the earth's rotation, the distance a cannon-ball flies should vary depending on whether it is fired to the West or to the East.[46] What Kepler has to add here to the concept of gravity advocated in his time, by those who support the new theory of discrete centres of gravity, is, first of all, the principle of the tending towards absolute 'rest' as a property of matter ('materiae enim proprium est quies')[47] — which he had already developed in his planetary physics — and secondly the interpretation that a heavy body does not actively gravitate towards the Earth, but is passively attracted by the magnetic *vis tractoria* (*effluxus tractorius*) of the Earth which Gilbert had already assumed as a result of his interpretation of the Earth as a globular magnet. Here, however, Kepler does not speak of a *mutual* attraction of 'heavy' bodies, but rather he attributes the 'vis tractoria' to the larger body alone (the Earth) and the 'dispositio ad quietem' to the smaller one alone (the stone) although Gilbert had already pointed out the mutuality of magnetic attraction.

Shortly after finishing this letter, Kepler had also begun to reply to several inquiries by letter of David Fabricius (Letter No. 358[48]). The sureness and brevity of his formulations in this letter show that Kepler had already reflected on these problems on the occasion of Herwart's inquiry; and here, in the first sections, one also finds those further modifications which lead to that theory of gravity which is found in the 'Introduction' to the *Astronomia nova*. (Therefore, this introduction could have been written, at the earliest, in the middle of 1605.) Here in the 'Introduction' Kepler puts forward the following axioms for the 'true doctrine of gravity'[49]:

Every material substance is, in so far as it is material, disposed to rest in every place in which it is located alone, outside of the sphere of force (*extra orbem virtutis*) of a related body.

Gravity consists of the mutual corporeal tendency between related bodies towards uniting and conjunction (*unitio seu conjunctio*) — the magnetic force is also of this order —, so that the earth attracts the stone much more than the stone gravitates to the earth.

Heavy bodies (and that is especially the case if we place the Earth in the centre of the universe) are not set in motion towards the centre of the universe as such but towards the centre of a related globular body, namely the Earth. Wherever the Earth is placed, or wherever it is moved to by means of its animistic faculty (*facultas animalis*[50]), the heavy bodies will always be moved towards it.

If the Earth were not globular, then heavy bodies would not be rectilinearly moved from everywhere towards the centre of the Earth, but from different latitudes towards different points.

If two stones were put in any two places in the universe, near each other and outside of the sphere of force of a third related body, then those stones, similar to two magnetic bodies, would join together at a point located between them, whereby the one draws nearer to the other over a distance that is proportional to the mass (*moles*) of the other.

If the Moon and the Earth were not held in their respective orbits by means of an animistic, or some other equivalent, faculty, then the Earth would ascend towards the Moon one 54th part of the distance between them and the Moon would descend towards the Earth approximately the remaining 53 parts of this distance; at this point they would join together-provided, however, that the substance of both is of one and the same density...

Then follows Kepler's tidal theory.[51] In his letter to Fabricius, Kepler justified the Moon-Earth *Gedankenexperiment* saying, that, 'since the Moon is an earthy body' (*cum sit Luna corpus terrestre*),[52] the Earth and the Moon would mutually attract each other relatively to their sizes: Since the volume of the Moon comprises approximately the 40th part of the Earth's volume and the distance between them contains 60 Earth radii, the Moon would attract the Earth a distance of 1 1/2 Earth radii, so that the Earth and the Moon would join together 1 1/2 Earth radii above the Earth centre.[53]

In the second half of the 1620s, Kepler once more summarized succinctly his concept of gravity in some notes on his *Somnium*. The concept of '*inertia*' (*dispositio ad quietem*), the tendency to rest as an inherent property of matter, played a decisive role here as well, but in Nota 77 Kepler succeeded, by means of a novel *Gedankenexperiment*, to determine the relative sizes of the 'magnetic forces' of the Earth and the Moon, considered to be the only 'related' bodies, without making use of that inherent property[54]:

A body which is located between two celestial globes at that point, where the distance between the two is divided into the same proportion as that which the two bodies have to each other, will remain motionless, since the oppositely directed attractions cancel each other out. This happens when the distance of that body from the Earth comprises 58 1/59 Earth radii and its distance from the Moon 58/59 Earth radii. If the body should move only a little bit closer to the Moon, it would then be attracted by the Moon, since the Moon's force would then predominate on account of its being nearer.

Kepler does not indicate here, either, which measurement data were used to produce this new result. The equal density is not specifically mentioned as a prerequisite. However, he said that this only takes place assuming that a certain part of the Earth is equal to the lunar globe and that both possess the same power of attraction.

Under the assumption of equal density, the chronological listing of the results leads, as far as mutual attraction is concerned, to an even decreasing 'Lunar force' in proportion to 'Earth force':

Letter No. 358 to Fabricius (mid-1605):   1 1/2:58 1/2;
Introduction to *Astronomia nova* (after mid-1605):   1:53;
Somnium (1620s):   58/59:58 1/59.

Thus for Kepler the size of the *vis attractoria* depends on the volume and the weight (the 'density') of the body possessing it, as well as depending linearly on the distance, so that he is able to determine by means of the statical moment the point towards which two bodies could mutually attract each other, if their *animae* did not hold them back (thereby interpreting the line between both centres as a balance), although the *orbis virtutis attractivae* (*coitionis*) would have to expand spherically and its force would have to decrease proportionally to the square of the distance, as Kepler himself had shown for light. In section IV below it is shown that, for some time, Kepler had assumed the like in regard to the *orbis virtutis* of the Sun (the Earth) which carries the planets (the Moon) along. However, he also imagined that the Sun's 'virtue' expands only over the plane of the ecliptic, so that in connection with that the static derivation is justified, whereas here at least the Earth together with its *orbis virtuosus* is rotating under the Moon and heavy bodies are falling to it from all sides. And so the 'attractive force' must expand spherically for Kepler, as well. Perhaps it will be impossible to ascertain, whether Gilbert's concept of the attracting force corresponding to the *weight* of the magnet, or Kepler's own conception resulting from the expansion of the motorial *vis magnetica* within the plane of ecliptic, had brought Kepler to this erroneous conclusion. Probably both considerations came together in Kepler's view.

This is also another reason to believe that Kepler dealt with the interpretation of 'gravity' not earlier than 1605, and only on a certain occasion, and that he never thought again about the then developed concept. At that time his planetary physics had been concluded to a large degree and, even regarding the terminology, it would have been impossible for Kepler to combine both 'physics' — and such a combination was not at all necessary for his 'physical' explanation of planetary movement. Moreover, only the Earth and the Moon, by virtue of their 'relationship', would fulfil the requirements for a mutual gravitational effect.[55] In this regard Kepler still remained within the framework of the older cohesion theory; his imperfect concept of *inertia* was not yet sufficient to go beyond this theory. However, he supplemented this theory with the concept of the mutuality of attraction which takes place only within a limited '*orbis virtutis*' (taken from magnetism), instead of considering the gravitating of the parts, separated from the whole, towards the whole. But perhaps Kepler took over 'mutual gravitation' as a magnetic central force from Gilbert. New, however, were his attempts to determine its relative sizes for the celestial bodies Earth and Moon, at least. In contrast to Gilbert, who assumed that magnetism belongs to the *forma* (the soul) of a body, Kepler's concept, that 'gravity' results from an *'affectio corporea,'* whereas the movements of bodies are caused by a *'facultas animalis'*, is also new.

For Kepler, the *vis (facultas) corporea* is the 'force' of a body, a 'body force' in contrast to a 'soul force,' but not a 'corporeal' force, which is a 'force' linked to a body, to corporeal (material) *effluvia*, an opinion of other authors of the 17th century, who distinguished between the *vis corporea* and the *vis incorporea (immateriata)* of a soul or of a body depending on whether the *'orbis virtutis'* (the *'sphaera activitatis'*) is filled up with *effluvia materiata* or *effluvia immateriata*.[56] In contrast to this concept, Kepler maintains that a *'vis* (or *facultas) corporea'* is always immaterial (*immateriata*), as well as acting immediately at a distance[57] as a body's force.

IV. JOHANNES KEPLER'S CELESTIAL PHYSICS

1. *Mysterium Cosmographicum*

Kepler was the first astronomer to recognize that far-reaching physical conclusions must be drawn from the heliocentric system. It was no longer possible to believe in Aristotle's celestial physics with its concentric spheres, which carry the planets and are moved teleologically, since Tycho Brahe, by his measurements of the parallaxes of the comet of 1577 and the nova of 1572, had shown that the ether spheres were neither impenetrable nor immutable; the old medieval Ptolemaic structure of fixed spheres and the strict division between celestial and terrestrial events had to be abandoned. Such, at least, was the opinion of Kepler and of those who accepted the correctness of Tycho's measurements.

Already in his first work, *Prodromus dissertationum cosmographicarum, continens Mysterium cosmographicum* (Tübingen, 1596), it is clear that Kepler was a convinced and enthusiastic advocate of Copernicus' heliocentric system. He wrote that Michael Maestlin, his teacher in Tübingen, had introduced him to this doctrine, and he also wrote that, during his student years in Tübingen, he had already defended the Copernican doctrine at several disputations, and that he had written such a *disputatio* himself (which has been lost) in which he 'had already attributed the Sun's movement to the Earth by means of physical or rather metaphysical reasons, as Copernicus had done by mathematical reasons' (Telluri motum Solarem, ut Copernicus Mathematicis, sic ego Physicis, seu mavis, Metaphysicis rationibus ascriberem).[58] Above all, the reasons presented in Georg Joachim Rheticus' *Narratio prima* convinced him of the correctness of the Copernican theory; and it was especially his own speculations about the set of six in regard to the planets, which had incited him to search for a better causation of just that number.[59] On July 9/19, 1595, he finally discovered the *Mysterium cosmographicum*[60]: By explaining the synodic loops in the motion of the planets as parallax effects, caused by the annual motion of the Earth, Copernicus had been able to estimate the distances of all the planets from the Earth and thus also their distances from the Sun. (No parallaxes could be seen for the fixed stars, and this left a great void between the orbit of Saturn and the sphere of the fixed stars, which Tycho Brahe, for instance, considered to be evidence against the heliocentric system.)

From an astronomical standpoint this was the essential advantage of the heliocentric system compared with the kinematically equivalent ones of Ptolemy and Brahe; for, once again reversing the positions of the Earth and the Sun, Brahe had taken over the greater economy and simplicity, which Copernicus had attained by systematically connecting several components of motions (the second anomalies), and used it in his own geocentric system. In this system the Sun, now being the centre of all planetary orbits, once again revolves annually. In the first chapters of the *Astronomia nova*, written between 1601 and 1605, Kepler proves the kinematic and geometrical equivalence of all three systems.

Copernicus himself had said that his system gave the same planetary positions as the Ptolemaic, being based, for the most part, on the same observations. Therefore no empirical methods could be devised to decide between the two systems. And even when Kepler learned of Tycho's minute observations, he realized that these would only result in modifications of the periods attributed to the eccentrics and epicycles in each of the two systems. The same was true of the empirical and astronomical results Maestlin had taught him.

But greater simplicity is still no proof that the system is correct. In the tradition of ancient and scholastic natural science such proof can be obtained only by identifying *causes* and deducing the phenomena from these *causes*. This is also true for modern science, in so far as it does not pursue the positivist ideal of mere description of the phenomena, or of their reduction to an axiom system. But naturally, modern science differs in what it regards as an acceptable cause. Since Kepler, reasoning has been supposed to be merely *a posteriori*, working from empirical data, but up to his time it was mainly *a priori*, or subject to rules taken from other fields. Kepler begins his studies with *a priori* reasoning and is led by that to *a posteriori* reasoning; he perfects the old type of science and founds the new type, even if he himself is not yet able, or willing, to make a sharp distinction between the two methods. For him *a priori* causes always underlie the *a posteriori* ones.

In the sense of 'saving the appearances' (σῴζειν τὰ φαινόμενα), the Ptolemaic and Copernican theories were equivalent *hypotheses* without any claim to physical reality. Osiander had also expressed himself in that sense. Kepler, however, completely disagreed with this view. Like Copernicus, he was convinced that only the heliocentric system corresponds to reality. Therefore for him its reality can and must be deduced from other realities, and these must necessarily be superordinate to the phenomena and by that to the reality of Divine Creation expressing itself in these phenomena.

Such a reality, transcending the material Creation and thereby even its Creator, is seen by Kepler — following the Neoplatonists — to be the kingdom of quantities. He explicitly refers to the saying ascribed to Plato, 'God always works geometrically' (θεὸς ἀεὶ γεωμετρεῖ). He thus sees geometry as the archetypal *a priori* basis of God's Creation. It served as a model for God's Creation and therefore it alone leads man, made in God's image, to understand the plan of Divine Creation written in the 'Book of Nature'. Understanding God's will, as expressed in this Creation, becomes simultaneously the praising of God. Consequently, natural science on a quantitative basis is for Kepler at once the study and the service of God; and it is in this sense that he writes in an letter of October 3, 1595, to his teacher Maestlin in connection with his *Mysterium cosmographicum*[61]:

I want to publish this work as soon as possible, to the Glory of God, who makes himself known through the Book of Nature [. . .]. I wanted to be a theologian; for a long time my mind was troubled; but now, see how by my striving, God is glorified in astronomy as well.

Of course, Kepler was not the first to give a theological justification for the study of Nature. He was influenced in such a view by the natural theology of his time, which was also practiced by the pietistic theologians at Tübingen while he was a student there. Where he differs from this contemporary theology, however, is in linking this idea with that of a geometrical archetype, whereby he went back to Plato and the Christian Neoplatonists (especially Proclus), as he himself clearly acknowledged. He believes, as Plato believed, that the universe is built on a geometrical plan. And it is also to the work of Plato and his commentators that Kepler turns in his search for the *archetypus* of the heliocentric system, for the law which governs the newly discovered order in the structure of the cosmos; for according to *prisca theologia* the nearer one is to the time of the universe's creation, the nearer and more immediate one is to Divine wisdom, the report of which has to be interpreted.

Investigating the distances between the six planets in the Copernican system, Kepler, in his search for the real archetypes in the works of ancient authors, first tried to fit these distances within Plato's fundamental scale of planetary harmonic intervals, expressed as ratios of the numbers 1, 2, 3, 4, 8, 9, 27.[62] However, the disagreements showed themselves to be too great. He had no better success even when he used the technique of completing the correspondence with the archetypal series of numbers by introducing principally invisible objects into the visible world (such as the 'Counter-Earth' of Philolaus), a technique still known from the older Pythagoreans; however additional planets, assumed to be so small as to be invisible, were not sufficient to fill in the gaps in the series.[63] From ratios of *integers* Kepler then moved on to ratios of *lengths* geometrically attained, and of *areas* enclosed by the orbits, until, as he believed at least, he discovered the reason why there are *six* planets, which simultaneously determined their distances: Since the cosmos is three-dimensional, his proportions should be determined by means of extraordinary three-dimensional solids. Whereas until then there had been thought to be seven Planets with six gaps in-between them, now in the Copernican system the Moon is a satellite of the Earth and so there are only six planets with five gaps in-between them — as many gaps as there were extraordinary solids, namely the five regular polyhedra, the so-called Platonic solids.

Because he was convinced that the structure of the cosmos was based on *a priori* geometry and harmony, Kepler immediately suspected an inner relation between these two extraordinary facts — in a way similar to the ancient Pythagoreans and Plato who had assumed the idea of a musical harmony in the structure of the universe. On July 9/19, 1595 in discovering his *Mysterium cosmographicum*, he finally had success. Using Copernicus' values of the largest and smallest distances of each of the planets from the Sun, Kepler found, at a first attempt, that the eccentric orbits could be fitted amazingly, and almost accurately, between the five Platonic solids, placed one inside the other, using an arrangement of their circumscribing and inscribing spheres.

The relatively small differences which were still found when he — with the help of Michael Maestlin — calculated newly the distances of the planets in relation to the Sun, as the real physical centre instead of in relation to the centre of the eccentric orbit of the Earth (as Copernicus had done), did not cause Kepler any doubts about the archetype; instead he questioned the capacity of astronomy to make sufficiently accurate observations. He had not yet come across Tycho's exact data; the idea of the archetype still had a clear priority over an exact correspondence to experience.

Kepler was forced to disagree with Copernicus on yet another point. There was no room between the polyhedra for the epicycles Copernicus used to replace Ptolemy's equant-point motions. As Maestlin pointed out to Kepler, these epicycles jut into the orbital spheres whereby they and the archetype could not both exist simultaneously. So Kepler followed Ptolemy once more in using equant-point motions, and thereby not uniform movement around the eccentric circle that Copernicus had already eliminated. But what made him justify this step?

According to the physics of that time, a spherical shell of ether surrounding the Sun as its centre could only rotate uniformly. That was the reason why Copernicus had replaced the non-uniform equant movement with the movement of uniformly rotating spheres, to reattain the physical reality of planetary movements. As was usually assumed, such spheres contain the motive source for their self-rotation within themselves or within the planetary body. Kepler endeavours to demonstrate that there are harmonic proportions within the orbital velocities or periods as well. To this aim he compared, in chapter 20, the relative distances of the planets from the Sun with their relative (angular) velocities, and found out that this is on the one hand a confirmation of the order of the planets in the Copernican system — whereupon Rhaeticus had already taken into consideration that the Sun is the mover of the planets[64] —, and that, on the other hand, the orbital periods stand in fact in a similar proportion to the respective distances from the centre (the Sun). Thereby Copernicus' triumph over the ancient view is already guaranteed.[65]

But if one wanted 'to come closer to the truth (*praecisius ad veritatem accedere*) and to hope for any *equality* of the proportions (*proportionum aequalitas*)' then this would call upon deciding between two alternatives: 'Either the motive souls (*motrices animae*) [of the planets or the spheres respectively] become weaker with increasing distance from the Sun' — the existence of such souls being the common view[66] — 'or there is only one motive soul in the centre of all spheres, namely the Sun, which impels the nearer body more strongly, but somehow becomes weaker in regard to more distant bodies because of the length of the distance and the decreasing virtue (*virtus*).'

By calculating, Kepler recognized that the planets do not move with equal linear velocity, because the orbital periods do not increase linearly with the radius (with the length of the orbit), but that the linear velocities also become slower with increasing distance from the centre. A greater distance of the planet from the Sun therefore leads to a twofold increase of the period, which is 'double' in relation to the difference of the distances ('*incrementum periodi duplum esse ad ἀποστημάτων differentiam*'[67].

It must *not* be understood that this 'increase which is twofold' in proportion to the distance is a reverse quadratic ratio to distance ($1/r^2$) — Kepler had objected strongly to this idea his entire life; rather the velocity decreases proportionally both to the length of path over which the 'motive force' is dispersed, and over which the planet is moved so to speak by the same 'quantity of force,' and to the distance from the centre, as well.[68] Proceeding from this additional decrease in velocity, Kepler then decided in favour of the alternative of a central *anima motrix* of the Sun as the motive source: Similarly to the lighting capacity of a light source, its motive capacity decreases with distance.

Kepler says here that light becomes weaker 'in proportion to the circles,' that means to the radius of the light sphere; he speaks of the '*circulus*' but not of the '*sphaera*' (*orbis*) that is to say the surface of the globe. He corrected this mistake in the first part of his *Astronomiae pars optica* where he showed that light becomes weaker by the square of the distance. However, it is curious that this realization exerted no influence at all on his concept of the *vis animalis* and its *orbis virtutis*. On the contrary, the relatively good results of chapter 20 of the

Mysterium cosmographicum later induced him to have the *vis motrix* of the Sun spreading only over the plane of the ecliptic, so that the linear decrease is thus mathematically justified. (As has already been mentioned above, the correct calculation of the spherical decrease of the light had no influence on the spherically extended *vis attractoria* of 'gravity,' either.) The analogy of the *sphaera activitatis* of a soul (*anima: vis animalis*) with a light sphere is very old and has its roots in Plato's parable of the Sun, as was shown in section III, 2. New however, is the idea of the Sun's *vis **motrix** animalis*.

Now when Maestlin had drawn Kepler's attention to the fact that Copernicus' epicycles run into the 'orbital spheres' of the planets (the eccentric deferents) embedded between the polyhedra, Kepler did not surmount this difficulty, as we saw, by abandoning the archetype already otherwise approved; on the contrary, its beauty and suitability confirmed his opinion that the planets obviously move merely around Copernicus' eccentric deferents as their orbits, whereas the epicycles give only the degree of latitudinal nonuniformity. Thus Kepler followed once more Ptolemy in using equant-point motions (nonuniform movement around the eccentric circle) to explain the First Anomaly, and by that he once again reversed Copernicus' transformation of the equant into an epicycle. The necessity to give up the Copernican epicycles because of their incompatibility with the archetype corroborates additionally his concept of the centricity of the motive soul, derived from the proportion of orbital lengths and velocities (periods); in this way the soul becomes the 'physical' cause of the larger orbital velocity at perihelion and of the smaller at aphelion[69] — corresponding to Ptolemy's equant. Thereby Kepler obtains eccentric planetary orbits, determined by the regular polyhedra thought to be the model for Creation, whereas the varying orbital velocities are determined by the greater or lesser remoteness of the Sun's *anima motrix*.

While Copernicus had wanted to explain that anomaly of Ptolemy's equant-point motion by means of epicycles and not equant movements, especially for 'physical' reasons, and thereby had been forced to abandon geocentricity, Kepler's new physics, deduced from the Copernican system by using the idea of the *a priori* archetype, forced him to bring back the Ptolemaic equant-point motions, in direct contradiction to Copernicus' original intentions to establish a heliocentric system.

The history of science sometimes progresses in strange ways, and its progress is not always unambiguous, as this historical example shows in particular: the original reason for adopting the heliocentric system is no longer valid for the already adopted system. For Kepler, an older and stronger archetype appears as the criterion for accepting a heliocentric system, which it is supposed to corroborate while simultaneously condemning the geocentric one. Today we know that equant-point movements correspond to the anomaly later described by Kepler's so-called second law of planetary motion, the Area Law. Kepler was the first who gave 'physical' reasons for this anomaly, taken from his central motive source, deduced from the proportions of velocities.

And thus Kepler was also the first who returned to the point which astronomy had ignored since the time of Eudoxus. In the *Nomoi (Laws)* Plato had already rejected the idea of this astronomy which divided motions up into individual autonomous mathematical components, stressing that the planets only *seem* to describe several circles[70]: 'They do not really describe many circles but rather only one.' Only with Kepler do the planets again describe this one and only circular orbit (which afterwards became an elliptical one).

In this context it should be borne in mind that up to this time, in the changes that the history of astronomy underwent, no empirical data were used which had not already been available to Ptolemy: Copernicus had merely carried out transformations of the Ptolemaic elements — he frequently confirms that — and Kepler was led to abandon Copernicus' additional epicycles and to replace them by a single eccentric circle, around which the planet moves nonuniformly (without using self-moved solid spheres), merely by deduction based on his *a priori* archetype and on Copernican data. The result was, however, an entirely novel celestial *physics*.

2. *Astronomia nova seu Physica coelestis*

At the end of chapter 22 of his *Mysterium cosmographicum* Kepler summarizes his results once again[71]:

The entire world is filled with a soul which carries along every star and comet it seizes, and this is done with that velocity which corresponds to the distance of the position from the Sun and to the relative intensity of the force (*fortitudo virtutis*). Furthermore we give every single planet a special soul, by means of which the star ascends on its orbit.

It was to this last thought that he devoted himself a long time thereafter. However, in regard to the central *anima motrix* of the Sun he remarks in the second edition of the *Mysterium* (1621) that here *anima* is to be replaced by *species immateriata* of the Sun, providing thereby in brief the essential content of his celestial physics (*physica coelestis*), which will be repeated in the *Astronomia nova* and in the *Epitome astronomiae Copernicanae* (Liber IV). But it was still a long way from the speculative idea to its detailed development. The stages of this development can partly be learned from his informative letters, an impossible task however, in the present context.[72]

As is well known the eccentric circular orbits were soon changed by Kepler into ellipses; because, intending to find further confirmation for his discovery of the archetypal *Mysterium cosmographicum* in the more accurate measurements made by Tycho Brahe, Kepler accepted Tycho's invitation to become his assistant in Prague, and, after Brahe's death in 1601, became *Mathematicus* at the Court of Prague. His hopes were already partially realized in 1600, when, with the help of Tycho's observations, he was able to show that the Earth also had an equant-point motion, in contrast to the theories of Ptolemy (for the Sun, respectively) and Copernicus — a fact on which he had already insisted in his *Mysterium cosmographicum* of 1596, based on the conviction that cosmic regularities were uniform and universal.[73]

In Prague Kepler was given the task of working out a new (geocentric) orbit of Mars, based on Tycho's observations. This involved him in heavy computation and trying out of various hypotheses, starting with a simple eccentric circular orbit, then an oval and a 'chubby-faced' orbit, and finally an ellipse. Because of his conviction that each planet describes only *one* orbit (resulting from his novel 'physics'), he was led in 1605 to propose the first two of what were later to be known as Kepler's laws of planetary motion.

Two factors, apart from his idea of the archetype, were crucial in this discovery: (1) Kepler's unlimited confidence in Tycho Brahe's observations, the best ever to be made without a telescope. Kepler soon learned to value them, and they led him again to abandon, first his circular and then his oval orbit, because in the final analysis of the depth movement there was a discrepancy of 8 minutes of arc between the values they predicted and those found by observations, an accuracy which for any other astronomer — and for Kepler himself in former times — would have been considered quite sufficient. Kepler was the first who had taken into consideration the astronomical implications of these depth movements at all (the attempt to confirm the archetype had induced him to that). In calculating the consequences of the various hypotheses with regard to the orbital shape derived by him from physical causes, he repeatedly encountered discrepancies between the orbital shapes and Brahe's measurement data, without possessing criteria for deciding, which to prefer. Therefore he interrupted his astronomical studies and proved, as a matter of priority, the optical prerequisites for achieving exact observational data in astronomy. The result was the work *Astronomiae pars optica* (*The Optical Part of Astronomy*), which appeared in 1604. In this book it is proved that light propagates rectilinearly even within a pinhole camera, so that Brahe's methods of observations carried out with specific dioptres could deliver accurate data. Therefore the discrepancy must have had its cause in the traditional orbital shapes. (2) Kepler recognized that the new astronomy and its consequences also demanded a new physics, to bridge the old gap between astronomical hypothesis and physical reality by means of a synthesis of mathematical astronomy and physics.

Here some former speculations and deductions of Kepler found confirmation in William Gilbert's ideas put forth in his work *De magnete* of 1600. Gilbert had assumed that the Earth and all other celestial bodies were large magnets and had shown through experiments that a magnet has an effect on iron or another magnet even when penetrating through other matter and thus that the magnetic 'force' is a 'force acting at a distance' which is not linked to a medium as a carrier.

During an extended illness while he was writing the *Apologia Tychonis contra Ursum*[74] Kepler had become acquainted with this work sometime at the end of 1600/beginning of 1601 — and had also read it, superficially, at that time.[75] Thereupon he replaced the effect of his central '*anima motrix*' of the Sun, originally thought to be 'animistic,' with 'magnetic central force' which disperses itself within a limited '*orbis virtutis*.' Later on, besides Nicolaus Copernicus and Tycho Brahe he cited William Gilbert as the third to whom he was indebted for most of the stimuli,[76] and he always mentioned Gilbert with the greatest of praise.

But already in his letter from the beginning of July 1600 to Archduke Ferdinand, the cousin of Emperor Rudolph II, which was a letter of application for the post of court-astronomer,[77] Kepler had, independently of Gilbert, spoken of a *virtus motrix* instead of an *anima motrix* of the Sun, which disperses itself from the centre into the space of the universe; thus the 'force' becomes weaker the further it is, because it is spread out over a more expansive globe (*in spaciosiorem orbem*). Therefore the 'motive impulse' (*motus impressio*), which is transmitted to the planets, is weaker, when the planet is moved by another motive principle further away from the seat of the force (*a sede virtutis*).

It was only the 'how' of the effect and transmission of the force which had remained open and which now had been classified as being 'magnetic' by Gilbert. Thereby the question asked in chapter 16 of the *Mysterium cosmographicum*[78]: 'by means of which lever, which chain, which brazen bonds is our Earth held fast in its orbit in the heavens', is answered. In 1596, Kepler had still thought the air (atmosphere), in the form of an Earth heaven, to be this medium through which the influences of the stars are exerted — in the sense of a contact theory. The 'other motive principle' in the planets constituted thereafter the essential problem of Kepler's physical reasoning.

On December 10/20, 1601, he reports on progress in the theory of Mars; in regard to 'physics,' he writes, the most important result is the realization that the line of the apsides of the Earth runs through the *punctum aequans* and the Sun, and that the eccentricity has to be halved. Thereupon, from the research on Mars and Earth, it follows that the cause of the equant-point motion is entirely of physical nature ('causam aequantis esse mere physicam'), but appears in geometric dimensions; 'for the proportion in which any two distances are to each other is like the dwell (*mora*) of the planet in the point of one distance to the dwell in the point of the other distance.'[79]

This early place is important because it introduces the concept of (virtual) momentary movements; here also lie the roots for the later distance law, afterwards developed into the Area Law.[80]

In November, 1602, Herwart von Hohenburg had brought Gilbert's work to the attention of Kepler again[81]; from Kepler's answer follows that in contrast to Gilbert, he already saw the 'magnetic force' as a 'body force' (*virtus corporalis*), because it assumes corporeal dimensions ('quia dimensiones accipit corporeas').[82] And so the 'facultas naturalis' of the Sun should be understood in this sense; with its help the Sun carries the planets along, something already mentioned in the letter of October 1, 1602,[83] to David Fabricius. On the other hand, the ability of the Earth (or of any planet) to move away from, or nearer to, the Sun corresponding to its longitudinal position, so that the correct velocity can be transmitted to it by the Sun's force, is still thought to be a '*facultas animalis*' or rather '*intellectualis*' of the Earth (or of the planet): Its approaching and moving away in relation to the Sun occurs fully without 'forces' (*vires*) but merely by means of intellect and tendency, 'because the celestial globes are *not heavy versus the Sun*' (quia globi caelestes versus solem graves non sunt). This intellect is able to calculate the distances from the quantity of light received from the Sun. The degree of the depth movement to and from the Sun is determined by Copernicus' epicycle, once again introduced subsidiarily for purposes of calculation.

At the beginning of 1605, Kepler wrote to Longomontanus regarding the 'vicarious hypothesis' that he had dealt mainly with physics in the last years[84]:

I had already constructed a *hypothesis* four years ago, which to the exact second determines any eccentric positions of the planets. But I am not happy with this because it is not physical but rather what is called a hypothesis.

In his letter of March 5, 1605, Kepler eventually reports to M. Maestlin, who had, finally, just answered his many letters, about the new physics (contained in the first 50 chapters of the *Astronomia nova*), which in his opinion had enabled the solution of the astronomical problems.[85] After a lot of effort he had found that the planetary orbit was an oval and that the planet was not carried along an epicycle, 'but rather librates on the diameter to and fro'. This is how he puts it:

The Sun is a circularly magnetic body and rotates in its place [by virtue of a *facultas animalis* as already found in Plato], whereby it carries around its *orbis virtutis* with it[86]; this *virtus* is not attractive but rather motorial (*non attractoria sed promotoria*). The planetary bodies, however, tend towards rest by themselves wherever they are placed in the universe. Therefore power is to be employed so that they are moved by the Sun, whereupon they are moved more slowly, when they are more removed from the Sun, and more quickly, when they are nearer to the Sun — that means so that the eccentric is moved uniformly around the equant-point. Now every planetary globe must be considered magnetic or quasi-magnetic (because I desire similarity and not exactly the thing itself), of course with a straight 'line of force' (*linea virtutis*), which has two poles, the one fleeing the Sun, the other following it. This axis is held by an animistic force (*vis animalis*) in almost the same direction relative to the universe. Therefore, if a planet is carried around by the Sun, it alternately turns its fleeing and its following pole to the Sun. This causes the distance to fluctuate to and fro.

The calculations of the distances then soon resulted in the elliptical shape of the planetary orbits.[87] (At the time of this letter the first 52 chapters of the *Astronomia nova* had been completed.) Since the corporeal force globe of the '*species immateriata*' propagates instantly — therefore passing through every body without being linked to a medium as carrier[88] — and rotates together with the solar body, it has the same angular velocity everywhere. Besides the motorial faculty's decrease in proportion to greater distance, there must be — in contrast to the conceptions in the *Mysterium cosmographicum* — another circumstance which causes as well a reduced velocity of the moved body.[89] Following the Aristotelian view of violent motion as caused from the outside, Kepler sees this circumstance in the tendency towards absolute rest of any material body, that is in the 'inertia' — which Kepler then combines with his theory of gravity as well[90]: The velocity proportions of the planets show, in his words[91]: 'that in all planets, even in the lowest one, Mercury, there is a 'faculty' inherent in matter (*vis materialis*) to flee the influence of the *orbis virtutis* of the Sun.'

Therefore with increasing distance the motions of the planets remain behind the rotation of the Sun which takes place on an axis, the poles of which are directed to the poles of the ecliptic. Kepler then numerically deduces the rotary velocity from the approximate equality of the proportion of the lunar orbit to the Earth's diameter and the proportion of the Mercurial orbit to the Sun's diameter: The Sun's rotary period must be in the same proportion to Mercury's orbital period as the Earth's rotary period is to the Moon's orbital period. From his calculations he gets a solar rotary period of three days — taking into consideration a daily rotation as well.[92] Later, after the discovery of sunspots, Kepler altered, in the *Epitome*, the rotary period following the observations of Galileo Galilei and Christopher Scheiner; he then supposed it to be shorter than Mercury's orbital period (25 or 26 days).[93]

In regard to the carrying along of the planets by the '*orbis virtutis*' of the Sun, Kepler uses a bold analogy to the magnet[94]:

The magnet does not attract everywhere, but it possesses, so to speak, 'filaments' (*filamenta*) or rectilinear 'fibres' (as the seat of the motorial force [*fibrae rectae*]), which extend longitudinally, so that it does not attract an iron needle placed at the side between the magnet's poles, but only directs it parallel to its fibres (*parallelon suis fibris dirigat*). It can be assumed in the same way that, in the Sun, in contrast to the magnet, there is no power attracting the planets (*vis Planetarum attractoria*) at all (because all the planets would continue to approach the Sun, until they would combine with it), but only a directing force ([*vis*] *directoria*), and that the Sun therefore possesses circular fibres (*fibrae circulares*) which extend themselves round about within the plane of the ecliptic.

Here at last it turns out that Gilbert's distinction between two different 'magnetic forces,' with respectively different '*orbes virtutis*,' was fateful for Kepler — and prevented the combining of the celestial physics, thereupon modified, with the gravity theory, established afterwards. The 'motorial force' of the Sun is seen in analogy to the magnetic 'directory force'; the planets are not attracted by the Sun, they are 'not heavy versus Sun'. Here Kepler speaks of 'so-called filaments/fibres,' but as with the 'quasi-magnetic' of the cosmic forces, he drops the restriction at once. The analogy to magnetism soon becomes reality and thereby exposes the possibility of criticism, because magnets do not behave in such a way at all[95]:

The circular fibres in the plane of the ecliptic (the plane of planetary motions) are thought to be almost strings on which the planets are drawn around the Sun — along concentric orbits and with uniform velocities. It was not until the 57th chapter of the *Astronomia nova*, that Kepler passed on his reasoning, as to how the originally assumed 'planetary spirit,' which should cause the deviation from these circles (which in turn causes the variability of the velocity), can be replaced by a 'natural' or even a 'corporeal force' (*naturalis* or *corporalis facultas*).[96] Since this force must have contrary effects according to the planet's orbital position, it could not have its seat within the exact resting centre of the Sun. (Earlier, in chapter 39, he had already condemned an attractive force of the Sun with good reasons.) Since the rotary axis of the Earth, because of its obliquity in respect to the plane of the ecliptic, in its annual orbit alternately turns the one pole and the other pole to the Sun, and since, following Gilbert, the Earth is also thought to be a large magnet, it is then obvious, that the different effect has to be attributed to the different magnet poles of the Earth themselves. Thereupon these poles either draw themselves to the Sun or push themselves away from it; the Sun remains neutral 'like iron not yet magnetized,' also because of its circular magnetic fibres, whereas the Earth (the planets) needs in this case rectilinear (radial) fibres.[97]

But Kepler immediately expresses some doubts for this solution because, in the case of the Earth, the aphelion and the perihelion almost converge with the solstitial points but not with the equinoxes (where the magnetic effect reverses itself) as the theory requires — in addition to that, these points wander. In order to explain that, Kepler again referred to the Earth spirit — with all the consequences of the distances, points, angles or apparent solar diameter to which it has to conform on the oval orbit of the Earth without being able to do so. Later on, Kepler no longer concerned himself with this problem, however. He makes no mention of it in the theses of his *Tertius interveniens* nor in succeeding letters.[98]

However, later in the *Epitome* (1620), Kepler raised, among other, the following objections to the view that it is the planets alone that draw themselves to the Sun and push themselves away from it by means of their own magnetic corporeal force[99]:

(1) Since the Sun lies eccentrically to the orbital centre of the planets, their movement to and fro would take place along a line which runs outside of the Sun.
(2) It would be improbable that the small '*orbes virtutis*' of the planets would extend themselves up to the Sun if the 'orbis' of the much larger Sun did not extend out to the planets as well.
(3) The fibres of the planetary bodies would be slightly inclined by the Sun.
(4) When several planets are in conjunction, they should jointly be able to pull the Sun out of its position.

All of this could be avoided, if the mutual attraction and repulsion were avoided. Therefore the depth movement of all planets must have also been caused by the disproportionately larger Sun *alone*. However, in order to render possible the attraction and the repulsion of the corresponding magnetic poles of the planets, the magnet of the Sun must be constructed differently[100]: Unlike a magnet (and the planetary dipoles of the older view), the Sun's faculty to attract or to repulse the planets, depending on the positions of their axes, must not be found on one side alone but rather on all sides of the Sun. Therefore it had to have a homogeneous effect outwards. Kepler thought that one pole was located in the centre of the Sun and the other formed the entire solar surface (*superficies tota*). The explanation is that the monopole magnetic attraction of the Sun interacts with the planetary dipoles, which have their rotational and magnetic axes inclined to the ecliptic and are therefore drawn towards the Sun or pushed away from it, depending on whether the like or unlike pole is nearer the Sun. Depending, then, on how far they are from the Sun, the planets are carried round more quickly or more slowly by the Sun's '*orbis virtutis*' which rotates with it.

In this theory the planetary movements of longitude as well as of depth[101] are caused only by corporeal forces of the Sun. The aim to reduce the causes of the entire planetary movements to the *vis corporealis* of the Sun as a 'single most simple force' (*vis una simplicissima*), which drives the 'celestial machine' as a weight drives a clock — expressed for example in his letter of February 10, 1605, to Herwart[102] — is thus fully reached. It led Kepler to the laws of planetary motion; but at the same time it shows clearly that the idea of a general mutual gravitation, or of a corresponding force, as a motive principle never came into Kepler's mind. In addition, 'gravity' is something that must not be linked to the motorially acting magnetic forces of the Sun.

Kepler's physics itself proved a failure, and on the whole it found no advocates.[103] Therefore, despite better approximation of experience, Kepler's laws of planetary motion, deduced from this physics, could not meet with approval before they had been integrated in another further-reaching physics, namely that of Isaac Newton. What was valid in Kepler's confirmation of Copernican astronomy is also valid here: The results finally succeeded; the initial and motivating ideas were rejected.

Philipps-Universität
Marburg, Germany

NOTES

1. J. Kepler, *Gesammelte Werke* (hereafter quoted as: KGW), Vol. I (Munich 1938), p. 9: 33—36: *Mysterium cosmographicum*, 'Praefatio ad lectorem.'
2. *De caelo* B 8, 289^b 34ff.; *cf.* F. Krafft, *Dynamische und statische Betrachtungsweise in der antiken Mechanik* (*Boethius*, Vol. **107**), (Wiesbaden, 1970), pp. 67—68; the argumentation is taken from the *Quaestiones mechanicae*.
3. *Cf.* H. J. Easterling, 'Homocentric Spheres in 'De caelo,'' *Phronesis*, 6 (1961), 138ff.
4. See *De caelo* B 1, 284^a24ff., where Aristotele still rejected the existence of stellar souls.
5. According to Eudoxus the planets require four spheres each, the Sun and the Moon require three each but, in the case of the latter, he did not take into consideration the nonuniformity within the sidereal period. Probably, Aristotle alludes to this here.
6. See *Metaphysics* Λ8. — Since Simplicius, Aristotle's numbering of the spheres has often been criticized; see *e.g.*, G. E. R. Lloyd, *Aristotle: The Growth and Structure of His Thought* (Cambridge etc., 1968), p. 151: 'The number of spheres seems too high, since on analysis it turns out that the first primary

sphere of each planet exactly reduplicates the movement of the last reacting sphere of the planet above.' This objection, however, is not justified; for every rotating sphere is moved only by itself (respectively by its *intelligentia*). Since the points on the outer sphere holding in place the poles of the inner sphere are removed from the outer sphere's own poles, the resulting motion for the poles of the inner sphere will be circular. This is the only possible influence of an outer sphere upon a neighboring inner sphere. The retrograte spheres of a planet's system of spheres only bring the axis of the first sphere of the next following planet once again into alignment with the axis of its own first sphere (respectively of the sphere of the fixed stars), and no redublication of velocities occurs.

7. See n. 4, *supra*.
8. For the impetus theory see M. Wolff, *Geschichte der Impetustheorie: Untersuchungen zum Ursprung der klassischen Mechanik* (Frankfurt, 1978). For a critique of Wolff's attempt at deriving the impetus theory from the socioeconomic situation in 6th century Alexandria (the roots of this theory simply lie almost a thousand years in the past), see F. Krafft, 'Zielgerichtetheit und Zielsetzung in Wissenschaft und Natur: Entstehen und Verdrängen teleologischer Denkweisen in den exakten Naturwissenschaften,' *Berichte zur Wissenschaftsgeschichte, 5* (1982), pp. 53—74, especially pp. 58—61 and 65—67.
9. Buridanus, 'Questiones de caelo et mundo,' II, questio 12.
10. G. Galilei, *Dialogo*, first day: *Le opere di Galileo Galilei*. Edizione Nazionale, Vol. 7 (Florence, 1965), pp. 43ff.
11. *Cf.* G. Riccioli, *Almagestum novum* . . . (Bologne, 1651), Pars II, pp. 247—270, where Riccioli compiles a list of the advocates and opponents of this conception; see also H. A. Wolfson, 'The Problem of the Souls of the Spheres from the Byzantine Commentaries on Aristotle through the Arabs and St. Thomas to Kepler,' *Dumbarton Oaks Papers, 10* (1961), pp. 67—93, and note 66 below.
12. See F. Krafft, *Geschichte der Naturwissenschaft, I: Die Begründung einer Wissenschaft von der Natur durch die Griechen* (Freiburg, 1970), pp. 347—354.
13. *Cf.* F. Krafft, 'Der Mathematikos und der Physikos: Bemerkungen zu der angeblichen Platonischen Aufgabe, die Phänomene zu retten,' in *Alte Probleme — Neue Ansätze. Drei Vorträge* . . . , *Würzburg 1964, (Beiträge zur Geschichte der Wissenschaft und der Technik, 5)*, (Wiesbaden, 1965), pp. 5—24. This paper is an outgrowth of a review (*Sudhoffs Archiv, 49* (1965), pp. 221—223) of J. Mittelstraß, *Die Rettung der Phänomene: Ursprung und Geschichte eines antiken Forschungsprinzips* (Berlin, 1962). Discussions of the more recent literature are found in F. Krafft, 'Physikalische Realität oder mathematische Hypothese? Andreas Osiander und die physikalische Erneuerung der antiken Astronomie durch Nicolaus Copernicus,' *Philosophia naturalis, 14* (1973), pp. 243—275; E. J. Aiton, 'Celestial Spheres and Circles,' *History of Science, 19* (1981), pp. 75—114, especially pp. 78—81; U. Charpa, 'Zum Verhältnis von Historie und Philosophie in der Wissenschaftsgeschichte — Ptolemaica als Didactica,' *Philosophia naturalis, 20* (1983), pp. 179—191.
14. Adrastus' of Aphrodisias and Theon's of Smyrna proof that both theories are kinematically equivalent has been sometimes erroneously attributed to the older

Apollonius of Perga, citing Ptolemy (*Almagest*, XII, 1). The transformation of the theorems of the epicycle theory into the equivalent eccentric theory was probably carried out by Ptolemy himself.
15. *Simplicii in Aristotelis de caelo libros commentaria*, edidit J. L. Heiberg (*Commentaria in Aristotelem Graeca*, Vol. 7), (Berlin, 1894), p. 488 — which was erroneously declared to be a fragment (No. 121) of Eudoxus by François Lasserre, *Die Fragmente des Eudoxos von Knidos* (Berlin, 1966); see F. Krafft: 'Physikalische Realität...' (n. 13), pp. 250f. and *passim* (248ff. and 274f.).
16. See M. Schramm, *Ibn al-Haythams Weg zur Physik* (Wiesbaden, 1963), especially pp. 15—63. Excerpts from this work have survived in the *De caelo* commentary of Simplicius, who had taken them from the commentary of Alexander of Aphrodisias; both Simplicius and Alexander supported the strict Aristotelian point of view.
17. The zone of annularity did not include Alexandria (see M. Schramm, *op. cit.*, (n. 16), pp. 23—25) where Ptolemy worked.
18. *Simplicii in Aristotelis de caelo libros* (n. 15), pp. 509: 16—510: 3.
19. See J. Mittelstraß, *Die Rettung der Phänomene*... (n. 13), pp. 173—178.
20. See F. Krafft, 'Physikalische Realität' (n. 13).
21. *Cf.* H. Guerlac, 'Copernicus and Aristotle's Cosmos,' *Journal for the History of Ideas*, 29 (1968), pp. 109—113.
22. See also Henricus Bate, *Speculum divinorum et quorundam naturalium* (written between 1280 and 1300), caput 16: 'Expositio Sosigenis in praemissis iuxta Simplicium.' For a general discussion of the problem, see E. J. Aiton (n. 13).
23. N. Copernicus, *De revolutionibus orbium coelestium libri VI* (Nürnberg, 1543), Praefatio; N. Kopernikus, *Gesamtausgabe*, Vol. **2**: *De revolutionibus*... Hanc editionem curaverunt F. Zeller, C. Zeller (Munich, 1949), p. 6.
24. *Cf.* C. D. Hellman: 'Was Tycho Brahe as influential as he thought?', *The British Journal for the History of Science*, 1 (1962/63), pp. 295—324.
25. However, the Aristotelian doctrine of the 'meteora' still had advocates for a long time, who mistrusted Brahe's measurements and interpreted inconsistent data in favour of Aristotle. See my commentary on book 5, Appendix in *Otto von Guerickes Neue (sog.) Magdeburger Versuche über den Leeren Raum, nebst Briefen, Urkunden und anderen Zeugnissen*, ... übersetzt und hrsg. von H. Schimank unter Mitarbeit von H. Gossen, G. Maurach und F. Krafft, (Düsseldorf, 1968), pp. (296)ff; also related literature cited by C. D. Hellman (see n. 24).
26. For details see F. Krafft, 'Johannes Keplers Beitrag zur Himmelsphysik', in F. Krafft, K. Meyer, B. Sticker, (eds.), *Internationales Kepler-Symposium Weil der Stadt 1971* (*arbor scientiarum*, Series A, Vol. **1**), (Hildesheim, 1973), pp. 55—139, especially pp. 68—73.
27. See F. Krafft, 'Copernicus retroversus, I: Copernicus Fulfills Greek Astronomy', in *Colloquia Copernicana, Toruń 1973*, Vol. **III** (*Studia Copernicana*, 13), (Wrocław etc., 1975), pp. 113—123, and F. Krafft, 'Copernicus retrogradis: Die 'Copernicanische Wende' als Ergebnis absoluter Paradigmatreue', in A. Diemer, (ed.), *Die Struktur wissenschaftlicher Revolutionen und die Geschichte der Wissenschaften* (Meisenheim, 1977), pp. 20—48.

28. See F. Krafft, 'Physikalische Realität' (n. 13).
29. See H. Guerlac (n. 21). Extensive materials on Copernicus' usage of the Aristotelian manner of argumentation are included in the comment of A. Birkenmajer in Nicolaus Copernicus, *Über die Kreisbewegungen der Weltkörper* (*De revolutinoibus orbium caelestium*), *Erstes Buch, zweisprachige Ausgabe*, hrsg. und eingeleitet von G. Klaus (Berlin, 1959), pp. 89ff. In the translation of the title the term '*Weltkörper*' is misleading, because '*orbes*' are not the planets, but rather the individual spheres (circles).
30. See F. Krafft, 'Copernicus retroversus, II: Gravitation und Kohäsionstheorie,' in *Colloquia Copernicana, Toruń 1973*, Vol. 4 (*Studia Copernicana*, 14), (Wrocław etc., 1975), pp. 65—78.
31. For more details see F. Krafft, 'Sphaera activitatis — orbis virtutis: Das Entstehen der Vorstellung von Zentralkräften,' *Sudhoffs Archiv*, 54 (1970), 113—140, and F. Krafft, 'Johannes Keplers Beitrag' (n. 26), pp. 79—95; for the prehistory of the conception see also F. Krafft, 'Kreis und Kugel', in *Historisches Wörterbuch der Philosophie*, Vol. 4 (Basel/Stuttgart, 1976), col. 1211—1226, especially No. 7a.
32. See S. Günther, 'Johannes Kepler und der tellurisch-kosmische Magnetismus', *Geographische Abhandlungen*, 3/2 (1888), 19—43.
33. R. Norman, *The newe Attractive, Containyng a short discourse of the Magnes or Lodestone [sic!], and amongst other his vertues, of a newe discovered secret and subtill propertie, concernyng the Declinyng of the Needle, touched there with under the plaine of the Horizon. Now first founde out by Robert Norman Hydrographer* (London, 1581), reprinted in G. Hellmann, *Rara Magnetica 1269—1599* (*Neudrucke von Schriften und Karten über Meteorologie und Erdmagnetismus*, No. 10), (Berlin, 1898), pp. 29—40.
34. G. Porta, *Magiae naturalis libri XX, in quibus Scientiarum Naturalium divitiae, & deliciae demonstrantur. Iam de novo, ab omnibus mendis repurgati, in lucem prodierunt* ... (Frankfurt, 1597) (the edition which was available to me; first edition 1589). In book VII, 15 there are certain methods treated regarding the finding of the poles of a magnet, *i.e.*, those points which diffuse the magnetic force in its 'circuit,' concluding on p. 305 with: 'Punctum in orbem suam vim diffundere cognovimus, ut a centro ad circumferentiam. Et sicut lumen candelae spargitur undecunque, & cubiculum illuminant, & quanto ab ea longius recesserit, eo languidius splendet, & paulo longius deperditur, & quanto proprius accesserit, eo vividius lucet. Eodem modo vis illa ex eo puncto emanat, & si proprius haeserit, vegetius trahet, & quo longius, remissius, immo & si multum recedet, evanescit, & nulla fit, ob id pro iis, quae dicenda sunt, punctus ille suae virtutis diffusivus cognoscendus erit, et signandus, et orbem suae virtutis vocabimus virium longitudinem.' This is also the meaning of the term '*orbis virtutis*' in the following chapters of book VII (especially chapter 22 and 26). The word 'vocabimus' makes it probable that Porta himself coined that term in this context.
35. W. Gilbert, *De magnete, magneticisque corporibus, et de magno magnete tellure; Physiologia nova, plurimis et argumentis, et experimentis demostrata*' (London, 1600).
36. See especially II, 4 and VI, 5, and the posthumously published book by W.

Gilbert, *De mundo nostro Sublunari Philosophia nova* (Amsterdam, 1651), I, 20 and II, 21.

37. *Cf. De magnete*, VI, 3; in *De mundo* (n. 36) the density proportion of the optical media water to air to the alleged optical medium ether, led Gilbert to the conclusion that this ether did not exist but that there was an interplanetary vacuum instead. A similar optico-astronomical reasoning led Kepler later to the same conclusion; but for 'physical' reasons he rejected it, although he did not need to use ether as a medium for his own celestial physics. *Cf.* J. Kepler, *Epitome astronomiae Copernicanae*, Liber I, pars III (KGW VII, pp. 53 f., especially p. 54: 5 f.): 'Itaque si per Physicam liceret, astronomus totum aetheris spacium plane Vacuum posset supponere'; he reasons similarly in KGW II, pp. 119 f. (*Astronomiae pars optica*, Caput IV, propositio X, problema III) and KGW IV, p. 293: 37 ff. (*Dissertatio cum nuncio sidereo*).

38. *De magnete*, 'Verborum quorundam interpretatio,' at the end of the Praefatio, p. (VIa), and *passim*.

39. *Ibid*. In his widely read handbook on magnetics (definition 7), Athanasius Kircher uses the same differentiation, which in the meantime had become canonical. (A. Kircher, *Magnes sive de Arte Magnetica . . . Editio secunda post Romanam multo correctior* (Cologne, 1643), p. 34; this is the edition which was available to me; first edition Rome 1641.

40. See, *e.g.*, Nota 202 of J. Kepler's *Somnium* (*Kepler's Somnium: The Dream, or Posthumous Work on Lunar Astronomy*, translated with a Commentary by E. Rosen (Madison etc., 1967), pp. 123 f.); see also below note 54 and the Introduction to his *Astronomia nova* (KGW *III*, 26: 1—23) as well as his Letter No. 409 of late 1607 (KGW *XV*, 386—388).

41. In his *De mundo* (n. 36) published in 1651, posthumously, there is a woodcut which shows the Copernican System surrounded by fixed stars, between which there appears the inscription (p. 202): 'Stellae extra orbem virtutis Solis sive formam effusam non moventur a Sole, sed fixae nobis apparent'. ('The Stars outside of the *orbis virtutis*, or rather the effluent *forma* of the Sun, are not moved by the Sun, but appear to us as being fixed.')

42. See below. The letters quoted by No. and line in the following are found in KGW *XIV—XVI* (Munich, 1949—1954).

43. The stages in which Kepler wrote and published his works are now definitively treated by F. Seck, 'Johannes Kepler und der Buchdruck: Zur äußeren Entstehungsgeschichte seiner Werke', *Archiv für Geschichte des Buchwesens, 9* (1970), coll. 610—726; for *Astronomia nova* see coll. 643—647.

44. Letter No. 336: 18—21.

45. See also Letter No. 358: 223 ff.

46. See the dispute in the correspondence between Tycho and Chr. Rothmann in *Tychonis Brahe Opera omnia*, edidit J. L. E. Dreyer, Vol. **6** (Hauniae, 1919), p. 219. In a letter to Kepler, at the end of 1604, David Fabricius had mentioned this argument (Letter No. 315: 25—32); Kepler entered into the particulars of this argument in his letter No. 358 (see also Letter No. 508). Here (No. 358: 31—36) he also responded to Fabricius' question concerning magnetic aberration using W. Gilbert's explanation (with explicit reference to his name).

47. Thus Kepler transfers the Aristotelian conception of 'violent' (artificial-mechanical) motions to 'natural' motions, which he no longer explains as being the result of an inner drive of the moved body but rather as being caused by another 'natural' body, from the outside, in a 'natural' manner (in the same way as Aristotle explained 'artificial' motions). In letter No. 358: 50—58 he specifies this: 'Like darkness, resting is a *privatio*, which does not require creation but is attached to that which is created like a kind of nothingness (*ut nullitas aliqua*). Motion is, on the contrary, something positive like light. Thus, when a stone is moved from its place, it is not moved in so far as something material, but in so far it is a passive receptor of an outside impulse or attraction (*extrinsecus impulsus vel attractus*) . . .'.
48. For the entry dates of the several passages of letter No. 358 (not finished until October 11, 1605), see M. Caspar's notes in KGW *XV*, 527. The first passages (1—12) were written after Easter, and the 13th passage was written in May after a few weeks' break; the first part was sent off after June 4, 1605.
49. KGW *III*, 25: 18—26: 3.
50. M. Caspar (Johannes Kepler, *Neue Astronomie, übersetzt und eingeleitet von Max Caspar* (Munich/Berlin, 1929), p. 26 and *passim*) translated *animalis* always as 'animalisch' which means 'bestial'/'animal' (adjective from the Latin noun *animal*), whereas *animalis* is an adjective for *anima* ('soul' *not* 'spirit'); therefore it was translated here as 'animistic'.
51. See note 40, *supra*.
52. Letter, No. 358: 132.
53. In the letter to Fabricius, Kepler also deals with the problem of different densities regarding his concept of the Earth's '*orbis virtutis*' which carries the Moon along from his almost developed celestial physics (see section IV): Since the Earth together with its '*orbis virtutis*' rotates once daily and since this *orbis* extends out to the Moon and there, corresponding to the Moon's distance, is sixty times weaker than on the Earth's surface, the Moon must cover the same distance in 60 days that it would have covered in one day on the Earth's surface, namely 5265 German miles. 'If the Earth only rotated and there were no proportion of density and size between the bodies of the Moon and the Earth, the Moon would simply follow the Earth which, together with its *orbis* containing the Moon, rotates daily without remaining the Earth, the Moon would simply follow the Earth which, together with its *orbis* containing the Moon, rotates daily without remaining even the slightest fraction behind the Earth. However, since it does remain 57 out of 59 parts behind, it follows thé *virtus* only by two parts, so that after 29 1/2 revolutions of the Earth's *orbis virtuosus* the Moon is revolved only once.' Therefore the ratio of its density to the 'force' (*fortitudo*) of the *orbis virtuosus* (the ratio of which is 1:60 to the Earth) would come to 2:59, hence to the density of the Earth it would come to 1:1770. Since the volumes stand in proportion of 1:40 to each other, the ratio of the densities of the Earth and the Moon would be 1:44. Later, before sending the first part of the letter, Kepler corrects himself in a note in the margin. There would remain a density ratio of 1:57/80; and therefore the densities of the Moon and the Earth come close to 4:3 (like Gold: Silver). The point at which the Earth and the Moon would meet,

must, corresponding to their 'densities' (here: 'mass'), be located at a point of division which divides the distance of the centres of both into a ratio of approximately 4:56, but not into a ratio of 1 1/2:58 1/2. However, this calculation was not carried out any further by Kepler, either. His uncertainty shows, that he actually had carried out these calculations for the first time on the occasion of his letter to Fabricius, and it stimulated him, not even to include corresponding reflections on density in the introduction to his *Astronomia nova* at all — which he therefore could only have written after finishing this letter together with the notes in its margin. In the *Epitome* (see KGW *VII*, pp. 324—326) one finds a completely different numerical connection between the Earth's rotation and the Moon's orbit. In reference to the 'relationship' between the Earth and the Moon see *ibid.*, p. 319: 23ff.: 'Quemadmodum igitur, ut Magnes Magnetem aut ferrum trahat, cognatio corporum efficit: sic etiam de Luna non est incredibile, ut illa moveatur a Terrae cognato corpore: licet nec hic nec illic intercedat aliquis contactus corporum.' In answering the question why the Moon does not rotate, he goes on to say: 'nisi quia circa Lunam nullus amplius planeta circumire cernitur; nullum igitur habet Luna planetam, cui motum inferat, gyratione sui corporis: gyratio igitur in Luna, ut supervacua, fuit omissa.' — The *Epitome* (*Cf.* KGW *VII*, 75ff.) contains no new ideas concerning the interpretation of gravity.

54. *Somnium* (n. 40), Nota 77, see also Notae 62, 66, 67, 74—77. Recent astronomical data would result in a ratio of 1:49 (see E. Rosen, *Somnium* (n. 40), p. 74). As the density of the Moon is approximately 3/5 that of the Earth, Kepler's data would result in a ratio of 0.88:59.12.
55. In the *Somnium*, Nota 202, Kepler also takes into consideration that, at the time of spring tides, both the Moon *and* the Sun have an effect on the tidal current (so that the Sun too had to be 'related' to the Earth); but this idea remains isolated and even within the 'Notae' on the *Somnium* there is no repercussion on his theory of gravity.
56. *Cf.*, for example, Otto von Guericke (see n. 25), *IV*, 1 *passim*.
57. See, for example, J. Kepler, *Epitome*, KGW *VII*, 89ff.
58. KGW *I*, 9: 16—19.
59. KGW *I*, 10: 32ff.
60. For details see M. Caspar's 'Nachbericht', KGW I, pp. 403ff.; also the introduction to his German translation (J. Kepler, *Mysterium Cosmographicum — Das Weltgeheimnis, Übersetzt und eingeleitet von M. Caspar* (Augsburg, 1923)); E. J. Aiton, 'Johannes Kepler and the *Mysterium cosmographicum, Sudhoffs Archiv, 61* (1977), pp. 173—194, and his introduction and commentary to J. Kepler, *Mysterium Cosmographicum — The Secret of Universe*, transl. by A. M. Duncan (New York, 1981); and F. Krafft, 'Johannes Keplers Beitrag' (n. 26), pp. 96—101.
61. Letter No. 23.
62. See F. Krafft, *Geschichte der Naturwissenschaft* (n. 12), pp. 347ff.
63. It is interesting that the minor planets discovered since 1800 are in fact debris of a former planet, which once had an orbit in the gap between Mars and Jupiter to which Kepler drew attention. These planets were discovered because the search was carried out in a region where, according to Titius' and Bode's law of

planetary distances, a planet should be found, but where no planet had been found.

64. To which Kepler KGW *I*, p. 70:33 refers to, as well.
65. KGW *I*, 70: 18ff.
66. In the sense of the *intelligentiae* of the spheres (see above note 11) or of the concept of the (moving) souls of stars which is already found in Plato's *Timaeus* and which is represented in all of its nuances in Julius Caesar Scaliger's *Exotericarum exercitationum liber XV de subtilitate ad Hieronymum Cardanum* (Paris, 1557). In his early years, Kepler was strongly influenced by this work which was widely circulated at that time. See M. Caspar in KGW *I*, p. XXIIf., and Kepler's note to chapter 20 in the second edition of the *Mysterium cosmographicum*.
67. KGW I, 71: 15f.
68. Through this Kepler came to the formulation, that the (mean) orbital radii of two planets ($r_1 < r_2$) relate to their orbital periods (u_1, u_2) as follows: $r_1:r_2 = u_1:(u_1 + u_2)/2$. In the second edition of the *Mysterium*, Kepler corrects this by means of his Third Law.
69. KGW *I*, p. 76 (*Mysterium Cosmographicum*, chapter 22).
70. Plato, *Nomoi* VII, 22, 822 A: καὶ οὐ πολλὰς ἀλλὰ μίαν ἀεὶ κύκλῳ διεξέρχεται. This has clearly been directed against the several 'circles' of Eudoxus (Callippus) which were still multiplied in the time thereafter. The demand for an astronomy without hypotheses, that means without eccentrics and epicycles, in the Platonic sense, was made emphatically by Petrus Ramus in 1569 (*Scholarum Mathematicarum Libri XXXI* (Basel, 1569), Liber II, pp. 49f.): According to him, astronomy must only be based on logic and mathematics which was the case with the Chaldeans (Babylonians) and the Egyptians; the trail blazed by Eudoxus had to be broken off, and with it the strict orthodox Aristotelian celestial physics, maintained in its continuing tradition by Alexander of Aphrodisias/Simplicius/ al-Bitruji/Thomas Aquinas/ G. Fracastoro. Later on, that is in the same year in which Kepler's *Mysterium cosmographicum* also appeared, Ramus' demand became well known within a larger circle of astronomers, because Tycho Brahe included in his collection of letters, (Uraniburg, 1596), p. 60, a letter to Chr. Rothmann in which he reported about a discussion concerning this theme that he had with Ramus in Augsburg in 1570. However, for Tycho an astronomy without these hypotheses is unthinkable; as astronomy had to be based on numerical data and measurements, the appearing planetary motions could only be rendered by such circles. Afterwards, Kepler justifiably saw himself as the man who had fulfilled Ramus' demand; and he placed a quotation from the above mentioned work (p. 50) together with a reply at the beginning of his *Astronomia nova* (KGW *III*, p. 6). See also E. J. Aiton, 'Johannes Kepler and the Astronomy without Hypotheses', *Japanese Studies in the History of Science*, **14** (1975), pp. 49—71.
71. KGW *I*, p. 77; see also Letter No. 23 to M. Maestlin, October 3, 1595.
72. See for further details F. Krafft, 'Johannes Keplers Beitrag' (n. 26), especially pp. 101—131.
73. See Letter No. 168; *Mysterium cosmographicum*, KGW *I*, p. 77: 31ff.
74. Edited by Chr. Frisch in *Johannis Kepleri astronomi Opera omnia*, Vol. I

(Frankfurt/Erlangen, 1858); here Gilbert's work is mentioned with praise on p. 243. [Proof addition 1991: See new edition by Volker Bialas in KGW XX/1 (1988), pp. 17–62, especially p. 24.]
75. See Letter No. 242 [January 12, 1603]: 219–221. E. Rosen, *Kepler's Somnium* (n. 40), p. XVIII is of another opinion. See also letters quoted in F. Krafft, 'Johannes Keplers Beitrag' (n. 26), pp. 104–106, 110, and note 74.
76. J. Kepler, *Epitome*, KGW VII, p. 254.
77. Letter No. 166: 63–67; see also Letter No. 168 to Herwart von Hohenburg, July 12/22, 1600 and No. 183 (72–77) to Maestlin, February 8, 1601.
78. KGW I, 56: 30ff.
79. Letter No. 203: 64–68.
80. Cf. E. J. Aiton, 'Kepler's Second Law of Planetary Motion', *Isis, 60* (1969), pp. 75–90.
81. Letter No. 235.
82. Letter No. 242: 361ff. From this letter one discovers, as well, that Kepler had studied the work of G. Porta (see n. 34) in the edition of 1597 and had repeated the magnetic experiments mentioned therein (No. 242: 136ff. and 207ff.).
83. Letter No. 226, especially line 618ff.
84. Letter No. 323: 175ff.
85. Letter No. 335: 10ff.
86. More details concerning this are found in KGW III, pp. 242–246 (*Astronomia nova*, chapter 34).
87. See E. J. Aiton, 'Kepler's Path to the Construction and Rejection of His First Oval Orbit of Mars', *Annals of Science, 35* (1978), pp. 173–190; C. A. Wilson, 'Kepler's Derivation of the Elliptical Path', *Isis, 59* (1968), pp. 4–25.
88. Cf. KGW III, pp 236–242 (chapter 33), pp. 247 f. (chapter 35).
89. Cf. KGW III, pp. 248–252 (chapter 36), pp. 254–262 (chapters 38 f.); see also chapter 33 (see n. 88) and 34 (see n. 86).
90. See section III, 3, *supra* especially n. 47 on 'violent' motion.
91. KGW III, p. 244: 25–27.
92. KGW III, pp. 244f.
93. J. Kepler, *Epitome*, KGW VII, pp. 298: 20ff., 290: 16f.; the idea had already been mentioned in Letter No. 658 to Odo Malcutius of July 18, 1613.
94. KGW III, p. 246: 4ff.
95. It was A. Kircher in particular, who criticized Kepler's magnetic physics; see F. Krafft, 'Johannes Keplers Beitrag' (n. 26), pp. 131ff.
96. KGW III, p. 348: 27f.
97. KGW III, p. 355.
98. See, *e.g.*, Letters No. 424 (150ff.), 448, but also Letter No. 340: 108ff.
99. J. Kepler, *Epitome*, KGW VII, 300 and 336.
100. KGW VII, 300: 30ff.
101. The latitudinal motion is a special problem (see in particular *Astronomia nova*, chapter 63), which cannot be discussed here.
102. Letter No. 325: 55ff.
103. See F. Krafft, 'Die Keplerschen Gesetze im Urteil des 17. Jahrhunderts', in R.

Haase, (ed.) *Kepler Symposion zu Johannes Keplers 350. Todestag, 25.—28. September 1980 im Rahmen des Internationalen Brucknerfestes '80 Linz Bericht* (Linz: Linzer Veranstaltungsgesellschaft, 1982), pp. 75—98, to some extent opposing J. L. Russell, 'Kepler's Law of Planetary Motion, 1609—1666', *The British Journal for the History of Science, 2* (1964), pp. 1—24; see also C. A. Wilson, 'From Kepler's Laws, So-called, to Universal Gravitation: Empirical Factors', *Archive for History of Exact Sciences, 6* (1970), pp. 89—170.

DAVID C. LINDBERG

KEPLER AND THE INCORPOREALITY OF LIGHT*

INTRODUCTION

It is my purpose in this paper to explore Johannes Kepler's ideas about the nature of light. It is my premise that we can succeed in this venture only if we go to the trouble of viewing Kepler's theory against the background of the ancient and medieval optical tradition, which formed its immediate historical context. One reason for past failures is that we have inspected Kepler from the perspective of Newton or the present, rather than that of Aristotle or Plotinus or Roger Bacon. We have been so impressed with Kepler's optical successes, in devising a new theory of vision, solving the ancient problem of the *camera obscura*, and formulating a geometrical theory of the telescope, that we have expended little effort in discovering whence they came or what they meant in the context of the optical tradition. Yet surely it was by grappling with the past — attempting to come to terms with the ideas of his predecessors, to separate the gold from the dross in ancient and medieval optical theory, to introduce clarity and precision, and to adjust existing theory to the teachings of sense, intellect, and philosophical presupposition — that Kepler fashioned his own theory. In short, Kepler's achievement can be fully appreciated only if viewed within a disciplinary tradition. This will not strike historians as a novel thesis; the novelty will come if we put it into practice.

The first necessity is to identify and define the basic categories of ancient speculation about the nature of light. Others have classified ancient theories of light according to their degree of mathematization, the direction of transmission (that is, by the intromission or extramission of light in the act of vision), or their level of sophistication (some theories being primitive, others advanced).[1] Such efforts are not without value, but I believe they miss the heart of the matter. In this paper, I wish to propose what I hope will prove a more useful scheme: a first division of ancient theories of light into corpuscular and noncorpuscular, and a second division of noncorpuscular theories into those that view light as corporeal (that is, as the modification of a corporeal medium) and those that take light to be wholly incorporeal.[2] We thus obtain three working categories: theories that view light, respectively, as corpuscular, the modification of a corporeal medium, and incorporeal. Such a classification scheme, I believe, confers the significant benefit of enabling us to see the alternatives as they presented themselves to Kepler.

THE CORPUSCULAR THEORY OF LIGHT

The atomistic theory, as developed by Leucippus and Democritus (late fifth century B.C.), was more concerned with visual perception than with light as a substance in its own right. However, Aristotle certainly inferred a theory of light from the writings of the atomists, and we may do the same.[3] According to the atomists, an effluence passes from visible objects to the observer. This effluence consists of atoms, thrown off from the surface of the object, which maintain more or less their original configuration as they pass to the eye of the observer, thus communicating to soul atoms more or less faithfully the visible qualities of the object. Epicurus (ca. 341–270 B.C.) elaborates the theory as follows:

Again, there are outlines or films, which are of the same shape as solid bodies, but of a thinness far exceeding that of any object that we see. For it is not impossible that there should be found in the surrounding air combinations of this kind, materials adapted for expressing the hollowness and thinness of surfaces, and effluxes preserving the same relative position and motion which they had in the solid objects from which they come. To these films we give the name of 'images' or 'idols.' Furthermore, so long as nothing comes in the way to offer resistance, motion through the void accomplishes any imaginable distance in an inconceivably short time....

For particles are continually streaming off from the surface of bodies, though no diminution of the bodies is observed, because other particles take their place. And those given off for a long time retain the position and arrangement which their atoms had when they formed part of the solid bodies, although occasionally they are thrown into confusion....

These films are responsible for vision:

We must also consider that it is by the entrance of something coming from external objects that we see their shapes and think of them. For external things would not stamp on us their own nature of colour and form through the medium of the air which is between them and us, or by means of rays of light or currents of any sort going from us to them, so well as by the entrance into our eyes or minds, to whichever their size is suitable, of certain films coming from the things themselves, these films or outlines being of the same colour and shape as the external things themselves.[4]

Lucretius would later compare these thin corpuscular films to the skin of a snake or cicada.[5] What is important in this theory is the claim that impressions of light and color result from the transmission through space of minute bodies or corpuscles.[6]

ARISTOTLE AND THE MODIFICATION THEORY OF LIGHT

It is difficult to estimate how widely the corpuscular theory, in one version or another, was held in antiquity. Presumably it flourished within Epicurean circles, but its further spread and development were hampered by Aristotle's firm rejection of it. Aristotle (384—322 B.C.) maintained repeatedly that light and color are not emanations of body. One of his principal arguments rested on the immeasurably swift passage of light and color from horizon to horizon, an impossibility for bodies undergoing local motion. Light, he wrote,

> is neither fire, nor in general any body, nor an emanation from any body (for in that case too it would be a body of some kind) . . . Empedocles, and anyone else who has argued on similar lines, is wrong in saying that light travels, and arrives at a certain time between the earth and its envelope, without our noticing it; this is contrary both to the light of reason, and to observed facts; it would be possible for it to escape our observation in a small intervening space, but that it does so all the way between east and west is too large a claim.[7]

If the corpuscular theory is wrong, what is Aristotle's noncorpuscular alternative? Before we can answer this question, there are some important conceptual and verbal distinctions to be made. It is customary in contemporary English to employ the adjectives 'material' and 'corporeal' and the nouns 'matter' and 'substance' synonymously.[8] But the Latin cognates from which these three terms are derived (and the Greek terms of which they are translations) denote three quite different conceptions — in the technical senses, that is, given them by Aristotle. *Matter*, in Aristotle's metaphysics, is an indeterminate substrate. Matter is not itself the 'stuff' (clay, lumber, bronze) out of which anything is constructed, for such 'stuff' exists and bears forms or properties before anything is constructed out of it. Aristotle is very clear about this in his *Metaphysics*:

> By matter I mean that which in itself is neither a particular thing nor of a certain quantity nor assigned to any other of the categories by which being is determined. For

there is something of which each of these is predicated, whose being is different from that of each of the predicates.... Therefore the ultimate substratum [matter] is of itself neither a particular thing nor of a particular quantity nor otherwise positively characterized....[9]

Matter, devoid of definition, does not exist independently of form.

That which has independent existence is *substance* (this piece of lumber, that chair or person). Substances, normally composites of matter and form, are the beings that make up the sensible universe. At least in the sublunary realm they occupy space, as tangible, three-dimensional objects. They may therefore be designated *corporeal*, for to be corporeal is to be a dimensional thing, a body or *corpus*. It follows from these definitions that matter, of itself, is incorporeal. What is corporeal in Aristotle's world is not matter or form, but the composite substance. Matter, the substrate, has no quantity or dimensionality (as the quotation above makes clear); in the absence of form it lacks even existence.[10]

If there are corporeal substances, are there also incorporeal substances? Here we must proceed with special care. Aristotle grants the existence in the superlunary world of nonspatial beings, such as the prime mover and the intelligences that move the planetary spheres.[11] If these are independently existing, nonspatial beings, they are by definition incorporeal substances. Note, however, that they are not only incorporeal, but also immaterial. Although substances, they are devoid of matter and consist, therefore, of bare form. Can there be substances that are not composites of matter and form? Apparently. Aristotle's successors looking for a better solution and building on Aristotle's own obscure remarks about the intelligible matter of mathematical objects, would develop the notion of a spiritual matter that can receive spiritual forms to produce spiritual (and therefore incorporeal) substances.[12]

We are now in a position to appreciate Aristotle's theory of light. We have seen how Aristotle attacked the idea that light is body. But if not corpuscular, what is light? The answer is found in an examination of the notion of transparency. There are transparent substances such as air and water, Aristotle argues, that owe their transparency to their participation in the same nature as the uppermost heavenly body. Light is a state of such transparent substances — to be specific, their actualization, whereby they become actually transparent, so that colored bodies separated from the observer by the medium become visible.

Such actualization takes place in the presence of fire or something else that shares the nature of the outermost heavenly sphere:

Of this [transparent] substance light is the activity — the activity of what is transparent insofar as it has in it the power of becoming transparent; where this power is present, there is also the potentiality of its contrary, darkness. Light is as it were the proper colour of what is transparent, and exists whenever the potentially transparent is excited to actuality by the influence of fire or something resembling the uppermost body.[13]

As the state of a medium, rather than the local motion of a body, light requires no time for passage, since the entire medium may have its state transformed in an instant: 'Generally speaking, change of state and travel in space are different; for spatial movements naturally first reach the intervening space, but with things that change their state it is not the same; for it is possible that such a change of state should occur in a thing all at once, and not in half first; for instance, a body of water may all freeze at one time.'[14]

Color, according to Aristotle, is that characteristic of the surface of bodies which has the capacity to move the actually transparent. Therefore, the transparent must first be brought to actuality by the presence of fire or its like; the actualized transparent is then moved to further change by the presence of a colored body, and this change is communicated to an observer, who thus perceives the colored object. 'It is the essence of colour to produce movement in the actually transparent; and the actuality of the transparent is light. . . . Colour moves the transparent medium, e.g., the air, and this, being continuous, acts upon the sense organ.'[15]

It should be obvious from the foregoing analysis that light belongs not to the realm of matter, but of its correlative: form. There is, of course, a material substrate; but if, in the sublunary world, one can always distinguish the substrate from its properties, light clearly belongs among the properties. It must be equally apparent that light is no substance. As a state of something else, it has no independent being; if there is no transparent medium to be actualized, there can be no actualization of it and, therefore, no light. Now if light is not a substance, then clearly it cannot be a corporeal substance. Light is not a body and of itself possesses no dimensionality. Nonetheless, light participates in corporeality by virtue of being the state of a corporeal substance, the transparent. Similarly, it has a derived dimensionality through the dimensions of the actualized medium. In short, light belongs to the realm of the corporeal even though not itself a *corpus*.[16]

Aristotle's theory of light became the prototype of many to follow, which viewed light as a state or property of the transparent medium. Aristotelian commentators typically adopted the Aristotelian theory, perhaps with variations.[17] But others of different or mixed philosophical allegiance also agreed with Aristotle on the necessity of a medium and viewed light as its state or qualification.[18] Some historians (including myself) have employed the term 'mediumistic' to characterize such theories.[19] I concede that this term is an aesthetic disgrace (not to speak of its unfortunate associations with modern spiritualism), and I resolve, therefore, to desist from its further use, even while recognizing that it nicely captures the Aristotelian stress on the medium. In this paper, I employ the more civilized (but also more ambiguous) expression 'modification theory' or, on certain occasions when an adjective is required, the term 'corporeal' (secondary or derived corporeality being understood).

THE INCORPOREALISM OF PLOTINUS[20]

The third class of ancient theories of the nature of light is most clearly represented in the work of Plotinus (d. 270 A.D.), the founder of Neoplatonism. Drawing from Platonic and Stoic sources, Plotinus developed a metaphysical system based on radiation or emanation. He identified the source of all being as the transcendent, self-sufficient One, from which proceeds all lesser being by an overflowing or emanation of its esence, just as rays of light emanate from the sun. Emanation from the One gives rise to *nous*; further emanation carries us down the scale of being to soul and ultimately to the world of sensible being. The image of the One becomes progressively weaker through successive emanations; when we reach matter, we have 'an image which has escaped from being and truth,' total privation or negation, absolute non-being.[21]

If the One is truly self-sufficient, how can it 'stir' itself to the production of any secondary being? 'What are we to think of,' Plotinus inquires, 'as surrounding the One in its repose?' His answer is that 'it must be a radiation from it while it remains unchanged, like the bright light of the sun which, so to speak, runs round it, springing from it continually while it remains unchanged.'²² Plotinus has thus far been discoursing on a metaphysical plane, about the source of being. But in order to elucidate the process by which being emanates from the One, he now introduces an analogy from the physical realm:

> All things which exist, as long as they remain in being, necessarily produce from their own substances, in dependence on their present power, a surrounding reality directed to what is outside them, a kind of image of the archetypes from which it was produced: fire produces the heat which comes from it; snow does not only keep its cold inside itself. Perfumed things show this particularly clearly. As long as they exist, something is diffused from themselves around them, and what is near them enjoys their existence. And all things when they come to perfection produce: the One is always perfect and therefore produces everlastingly; and its product is less than itself.²³

Plotinus has thus stated a principle of universal activity: everything that exists produces an image or likeness of itself, which it directs into its surroundings. This idea would become one of the hallmarks of Neoplatonic emanationism.²⁴ Visible light, of course, is the most accessible instance of this activity and therefore the paradigm case — the door to an understanding of the universal principle of emanation.

If we are to understand Plotinus's theory of the nature of visible light, however, we must begin by glancing at his conception of matter. Plotinus adopts Aristotle's notion of matter as the correlative of form; matter is a passive receptacle, on which forms are impressed and in which they reside. Form, Plotinus argues, is in matter as images are in a mirror — or, for a better analogy, as illumination is in the transparent medium, for in the latter case the illumination is visible, while its substrate, the medium, remains totally unperceived.²⁵ Plotinus inquires:

> Is matter, then, the same thing as otherness? No, rather it is the same thing as the part of otherness which is opposed to the things which in the full and proper sense exist, that is to say rational formative principles. Therefore, though it is non-existent, it has a certain sort of existence in this way, and is the same thing as privation, if privation is opposition to the things that exist in rational form.²⁶

Matter is total privation and thus identical with evil: '... that which has nothing because it is in want, or rather is want, must necessarily be evil.'²⁷ Such matter, of course, is of itself without dimensions and consequently incorporeal.²⁸ What is *corporeal* is the composite of matter and form in the lower, sensible realm.²⁹

Plotinus dealt with the nature of light on several occasions. Discussing beauty in the first *Ennead*, he remarks that light functions as form in relation to matter: 'And the simple beauty of colour comes about by shape and the mastery of the darkness in matter by the presence of light which is incorporeal and formative power and form.'[30] This statement was to prove enormously influential, giving rise to the idea of light as the form of corporeal substance, which became a major theme in the work of later writers.[31] Armstrong has argued, however, that when he wrote this Plotinus did not intend light to be viewed solely as the formal correlative of matter; for Plotinus continues: 'This is why fire itself is more beautiful than all other bodies, because it has the rank of form in relation to the other elements; it is above them in place and is the finest and subtlest of all bodies, being close to the incorporeal.'[32] The close connection drawn here between light and fire (both given the status of form) and the admission (see just above) that fire, itself body, though the least substantial of bodies, is the form that brings corporeal substance into existence has prompted Armstrong to conclude that Plotinus intends, at least at this stage in his philosophical career, to give 'light a very special status on the frontier of spirit and matter.'[33]

Plotinus returns to the nature of light in the fourth *Ennead*, now employing the Aristotelian notion of activity or actualization (*energeia*). Light, he argues, is the activity of a luminous body — an activity that has the capacity to radiate itself outwardly into its surroundings:

But the activity within the luminous body, which is like its life, is greater and is a kind of source and origin of its [outward] activity; that which is outside the limits of the body, an image of that within, is a second activity which is not separated from the first. For each thing that exists has an activity, which is a likeness of itself, so that while it exists that likeness exists, and while it stays in its place the likeness goes far out, sometimes a longer, sometimes a lesser distance. . . . The light from bodies, therefore, is the external activity of a luminous body.[34]

Plotinus then adds, returning to the idea of form, that light in such bodies corresponds 'to the form of the primarily luminous body. When a body of this kind together with its matter enters into a mixture, it gives colour.'[35] Thus in every luminous body there is a form (*eidos*) or actualization (*energeia*) which is the source of the body's activity, sending forth its power or image into the surroundings.

Two points made in the immediately preceding paragraphs require further comment: first, the claim that light is incorporeal; second, the idea of two activities or lights, the one internal to the luminous body, the other radiating outside it. Both claims are clarified in a passage from the second *Ennead* on the fiery nature of the sun. After quoting Plato on the origin of the sun, Plotinus continues:

> ... by fire he [Plato] does not mean either of the other kinds of fire but the light which he says is other than flame, and only gently warm. This light is a body, but another light shines from it which has the same name, which we teach is incorporeal. This is given from the first light, shining out from it as its flower and splendour.[36]

The first light, representing the luminosity of the sun or other fiery body, its form or activity, is held to be corporeal. The other light, offspring and image of the first, is light radiating spherically outward from the luminous body; this light, Plotinus maintains, is incorporeal.

In what sense is the first light corporeal? In the same sense as Aristotle's light in the transparent medium is corporeal. Just as Aristotelian light has corporeal being by virtue of its presence in the transparent medium, so Plotinian first light is corporeal by virtue of bodily existence in the sun or other luminous body; it participates in corporeality through its host.

What, then, of the incorporeality of Plotinus's second or radiating light? Simply stated, radiating light neither issues forth as a corporeal object (a corpuscle) nor utilizes or qualifies the surrounding corporeal medium. Plotinus's denial of the corpuscularity of light, the first of these alternatives, will come as no surprise. Nonetheless, Plotinus goes to the trouble of arguing the point in a discussion of mirror images. After defining the image in a mirror as the activity or *energeia* of the visible object acting on a recipient suited to receive it, he proceeds to maintain that in the process nothing actually flows from the visible object to the mirror.[37] While 'the object stands there, the image also is visible, in the form of colour shaped to a certain pattern, and when the object is not there, the reflecting surface no longer holds what is held when the conditions were favorable.'[38]

The second alternative, Aristotle's claim that radiating light has corporeal existence as a state or property of a corporeal medium, was a more serious rival. This, too, Plotinus emphatically rejects. In the radiation of light, there is no necessity of a corporeal medium:

> But if [light] is only a quality, and a quality of something, since every quality is in a substrate, one must look for a body in which light will be. But if it is an activity from something else, why should it not exist and travel to what lies beyond without the existence of an adjoining body, but with a kind of void in between (if that is possible)?[39]

Radiating light employs no medium because 'no other object can possess it.'[40] 'Light therefore is not a modification of the air, but a self-existent in whose path air happens to be present.'[41] The actualization which is light goes forth from the luminous body, leaps over the intervening space, and actualizes a suitable recipient, bringing forth color or a mirror image according to the nature of the recipient; the actualization 'must be thought of as being lodged, both in the active and powerful source and in the point at which it settles.'[42] It is never lodged in the medium.

THEORIES OF LIGHT IN THE MIDDLE AGES

These ancient theories had diverse fates during the Middle Ages. The corpuscular theory of the atomists was vigorously attacked by Aristotle, as we have seen, and became a standard whipping boy among his commentators; after the decline of the Epicurean school, it had no defender until the revival of atomism at the end of the sixteenth century.[43] The incorporealism of Plotinus also found little support before the Renaissance. During the Middle Ages it was the modification, or corporealist, theory of Aristotle that overwhelmingly prevailed. This was true even in nominally Neoplatonic circles and among scholars who borrowed heavily from Neoplatonic philosophy.

A good illustration of these claims is al-Kindi (d. ca. 873), the 'first philosopher of the Arabs,' who developed a natural philosophy pervaded with Neoplatonic emanationism. Al-Kindi argued that every being in the world radiates in all directions, 'so that every place in the world contains rays from everything that has actual existence.'[44] He has thus picked up the part of Plotinus's theory of light that emphasizes the active nature of every being, the self-radiating property of everything that is. But when it comes to the nature of light, al-Kindi is firmly in the Aristotelian camp. He refers repeatedly to light as an 'impression' on the medium; a ray of light, he argues, is 'the impression of luminous bodies in dark bodies, denoted by the name 'light' because of the alteration of accidents produced in the bodies receiving the impression.'[45] Whereas Plotinus's light 'leaps over' the medium to actualize a recipient beyond, al-Kindi's light is a quality embodied in the medium and requiring the medium for its very existence.

The most forceful statement of this position produced during the entire Middle Ages came from the pen of Roger Bacon (d. ca. 1292).[46] Bacon's *De multiplicatione specierum* painstakingly explores the physics of radiation — or, as it is referred to there, the multiplication of species. The term 'species' goes back to classical Latin, with the primary meaning of 'aspect' or 'appearance.' Augustine employed it to denote the likeness in the senses and intellect of a perceived object, and this sense of the term remained current throughout the Middle Ages.[47] With Robert Grosseteste (d. 1253) and Bacon, however, the meaning of the term was broadened to denote the similitude emanating from an object, whether or not a percipient being was present to receive it. Species were the forces or powers by which any object, including a luminous object, acted on its surroundings. Bacon defined a species as:

the first effect of an agent; for all judge that through species [all] other effects are produced. Thus the wise and foolish disagree about many things in their knowledge of species, but they agree in this, that the agent sends forth a species into the matter of the recipient, so that, through the species first produced, it can bring forth out of the potentiality of the matter [of the recipient] the complete effect that it intends.[48]

Light is the one visible instance of a species.

Bacon's *De multiplicatione specierum* is a ringing statement of the corporealist position, or modification theory. After considering various alternatives, Bacon concludes that species are generated by a 'bringing forth out of the active potentiality of the recipient matter.'[49] Moreover, this occurs only by direct contact between agent and recipient. Thus an agent produces its species in the part of the medium immediately adjacent to itself: that first part of the medium then acts as agent, replicating itself in the immediately adjacent second part of the medium, and so forth, until resistance of the medium or loss due to secondary radiation brings the process to an end. Bacon summarizes in his *Opus maius*:

But a species is not body, nor is it moved as a whole from one place to another; but that which is produced [by an agent] in the first part of the air [or other medium] is not separated from that part, since form cannot be separated from the matter in which it is unless it should be mind; rather, it produces a likeness to itself in the second part of the air, and so on. Therefore, there is no change of place, but a generation multiplied through the different parts of the medium; nor is it body which is generated there, but a corporeal form that does not have dimensions of itself but is produced according to the dimensions of the air; and it is not produced by a flow from the luminous body, but by a drawing forth out of the potentiality of the matter of the air.[50]

Bacon has been influenced by the Neoplatonic doctrine of emanation, holding that all things are sources of activity, radiating their likeness in all directions. But he explicitly and forcefully repudiates the Plotinian position on the relationship of light to the medium, allying himself with Aristotle and al-Kindi. Whereas Plotinus had rejected any connection between light and material media, Bacon reinstated light as the actualization of potentialities present in the medium. While acknowledging that species are not themselves bodies, Bacon insisted that species are embodied in the medium and through the medium acquire dimensionality. They play corporeal form to the matter of the medium. 'Therefore,' Bacon writes, 'since the medium is the material cause, in which and from the potentiality of which a species is generated by the agent and generator, this species cannot have a corporeal nature distinct from the medium.' This is evident because an effect, whether complete or incomplete, 'does not have a new corporeal dimension, but that which belongs to the medium or body in which [it] is generated.'[51] Since the opinion that species are incorporeal 'cannot be saved by any rational judgement, nor does it have any probability, as is evident to any man who wishes to dismiss the foolishness of the vulgar and to follow reason, I therefore state unconditionally that the species of a corporeal thing is truly corporeal and has truly corporeal being.'[52]

Bacon's defense of the corporeality of light, along with Aristotle's own remarks on the subject, firmly established this as the prevailing opinion through the end of the Middle Ages and into the Renaissance. There was an occasional objector — William of Ockham in the fourteenth century and Marsilio Ficino in the fifteenth.[53] But when Johannes Kepler turned to the nature of light about the beginning of the seventeenth century, corporealism (that is, the modification theory) was firmly entrenched.

KEPLER ON THE NATURE OF LIGHT

Light makes an appearance in almost every book Kepler wrote, but received fullest analysis in his *Ad Vitellionem paralipomena* (1604). From a reading of this work it is clear that Kepler had drunk deeply from the springs of Neoplatonic emanationism — specifically, that he regarded light as one of the basic archetypes and agents employed by the Creator in the design and operation of the sensible world. Echoing

Plotinian themes, Kepler argues that light is the offspring and image of the divine Trinity. When God 'conceived the corporeal world, He gave it a form as like Himself as possible.' This was the sphere, whose central point, circumference, and intervening space represent God the Father, Christ the Son, and the Holy Spirit: 'For in forming the sphere the all-wise Creator produced for His pleasure the image of His Holy Trinity.'[54] Whatever aspires to perfection must imitate this most perfect shape. Therefore all bodies are endowed with virtues that 'issue forth and strive for sphericity.'[55] Magnetic virtue is such an emanation; motive virtue emanating from the sun to drive the planets is another. But above these in beauty and importance is light, 'source and most excellent thing in the whole corporeal world, origin of animal faculties, and link between the corporeal and spiritual worlds.'[56] This is a clear statement of the Plotinian principle of universal activity.

We have thus glimpsed how light fits into Kepler's metaphysics and cosmology. But what is its nature? Chapter 1 of the *Paralipomena* is entitled 'On the Nature of Light,' and here we must begin our investigation. In the opening propositions we learn that light by nature emanates from each point of its source in all directions.[57] We also learn that light is suited for propagation to infinity, since it is weightless and consequently offers no resistance to the power of the luminous body that propels it.[58] A similar argument is used to demonstrate that light is propagated with infinite swiftness and that light is attenuated laterally (because the rays from a given point spread out) but not longitudinally.[59] Light is propagated in straight lines because it 'seeks to produce a spherical figure,' and 'if light were to utilize a curved path, there would be no equality in its diffusion and consequently no resemblance to a sphere.'[60] The straight line along which light is propagated is called a 'ray.' Rays are not light, and certainly not corporeal, but simply geometrical lines representing the motion of light. Light itself is a geometrical surface — an expanding, spherical two-dimensional surface. On several occasions Kepler maintains that it is devoid of matter,[61] and, as a two-dimensional entity, it is of course incorporeal.

All of this is very tersely expressed in the first eight propositions of the *Paralipomena*. For elaboration we must turn to an appendix attached to the first chapter, in which Kepler assembles and systematically refutes a series of Aristotelian theses about light and vision. The attack is very wide ranging, but its focus is on the Aristotelian notion that light is a state (*habitus*) of the potentially transparent medium, induced by the presence of a luminous body, whereby the transparent achieves actuality; and the further notion that color moves or alters the actually transparent, thus sending a qualitative change through the transparent medium to the eye of the observer. In short, Kepler makes an all-out attack on the modification theory of light.

One of Aristotle's mistakes, Kepler believes, was to define transparency in terms of light and light in terms of transparency. Kepler argues that transparency is associated with the 'internal disposition of the body,' which is totally independent of the presence or absence of light.[62] Light, on the other hand, is an emanation from a luminous body or illuminated opaque body and has no need of a transparent medium for its existence. It is a 'species,' which not only requires no medium for its propagation but which, in the absence of all media, would be more perfectly propagated.[63] As for color, Kepler maintains that its visibility does not depend on illumination of the medium (so as to bring it to a state of actual transparency), but on illumination of the colored object itself. Experience teaches that illumination of the medium is irrelevant; colored objects become visible if and only if illumination is directed onto them.[64]

One of the most important clues to the nature of light is the term 'species,' employed sometimes by Kepler as a synonym for *lux* or *lumen* or *virtus*.[65] I am not in the least persuaded by those who have argued that Kepler continued to employ the term 'species' while rejecting its traditional meaning.[66] Kepler was extremely sensitive to the meanings of words; he carefully defined terms that seemed to him problematic and coined new ones when existing terminology was insufficient. In the absence of arguments to the contrary, I think we must suppose that Kepler intended the word 'species' in its traditional sense, to denote the likeness emanating from everything in the natural world — an expression of the Neoplatonic notion that all existing things are centers of activity, radiating similitudes and influences in all directions about themselves.

Within the Neoplatonic tradition, however, there were unresolved issues. Both the corporeality and the incorporeality of species had been defended, and Kepler would be compelled to make a choice. On one side was the Baconian tradition, allied on this issue with the pronouncements of Aristotle and ratified by centuries of scholastic commentary, proclaiming the corporeality of light or species. On the other side were Plotinus and a few followers (including by this time, Marsilio

Ficino and Francesco Patrizi), urging the incorporeality — or, in some cases, the quasi-incorporeality — of light. Kepler, confronted by these alternatives, broke with Bacon and Aristotle and cast his lot with the incorporealists.

To understand Kepler's position, we must first note that the concept of matter had undergone gradual modification since antiquity, so that 'matter' (for many) had become synonymous with 'corporeal substance.' It is abundantly clear that Kepler employed the term in this new sense.[67] Kepler asserts the incorporeality of light when he defines it as a two-dimensional geometrical surface and maintains that it has 'neither matter nor weight nor resistance.'[68] He defends this claim by pointing to the infinite swiftness of light. We know from Aristotle that time in motion is dependent on the ratio between the applied motive force and the weight of the mobile. 'But here the ratio of the motive power to the light that it moves is infinite, since light has no matter and therefore no weight.'[69] Moreover, we can infer from the infinite swiftness of light that the medium offers no resistance, and therefore that 'light lacks matter by which resistance could occur.'[70] Finally, the phenomena of refraction demonstrate the incorporeality of light. If light were corporeal substance, it would encounter resistance not only upon crossing a transparent interface, but also while traversing the new transparent substance; consequently, it would be continuously retarded and continuously bent. But such does not occur; once within the new medium, light assumes a new, but still rectilinear, course. Refraction is apparently a surface phenomenon, and from this Kepler feels entitled to infer that light is but a geometrical surface.[71]

One might well ask how light, an incorporeal surface, is able to interact with the surface of a corporeal medium to produce refraction at all. Kepler replies that both light and the surface of the medium participate, each in its own way, in density; and that whatever moves (including light) must participate in the accidents of motion, including impact.[72] This may seem unsatisfactory, but it was no doubt the best way Kepler could find of reconciling the obvious fact that light is deflected as it passes from one transparent medium to another with the equally obvious fact that, once within the second medium, light is subject to no further deflection.

We observe Kepler's incorporealism, finally, in a comparison of light and planetary *virtus motrix*. The motive power or species emanating from the sun to move the planets in their orbits, Kepler argues, will not be some geometrical body, but a certain surface, exactly as light. Thus in general the species emanating immaterially from things do not proceed according to corporeal dimensions, even though they issue from a body (in this case the body of the sun). This follows, indeed, from the law of emanation, for emanation is terminated not in and of itself, but by the surfaces of illuminated bodies; and therefore, as light is considered a certain surface because its emanation is received and terminated by surfaces, so the *virtus motrix* is considered a certain geometrical body because it is terminated or received by the entire bodies of the things it moves. Thus the *virtus motrix* cannot be or subsist anywhere in the entire world except in the moved bodies; consequently, it can never *be* in the medium between the source and the mobile, although, so to speak, it *was* there, exactly as light.[73]

Neither light nor *virtus motrix* (nor any other species) has being in the medium. The medium represents intervening space or substance, which must be passed through or leapt over, without being fundamentally involved in the process of transmission. The medium is not a necessity, but an obstacle. Kepler repeats this point, arguing that although light and the *virtus motrix* 'crossed over the space between the source and the illuminated body,' they do not 'have being there.'[74] This is precisely the view of Plotinus and the incorporealists.

CONCLUSION

In this paper I have attempted to identify a triumvirate of ancient theories of the nature of light, all known to Kepler and providing him with his basic alternatives. No simple threefold classification can hope to capture all of the nuances, of course, but I have argued that the central issue facing all of those who considered the nature of light down to, and including, Kepler was its relationship to body — the need to determine (1) whether radiating light is itself body, as the corpuscularians maintained, (2) whether transparent body mediates between luminous source and illuminated objects, as upholders of the modification theory affirmed, or (3) whether intervening body is but an obstacle, as the incorporealists believed. In Kepler's day, the modificationists (whether of Aristotelian or of Baconian persuasion) were clearly in command, at least numerically, but the other possibilities were beginning to attract more notice. The corpuscular theory was in the process of being revived as a genuine possibility after almost two millenia of abuse,[75] and renewed attention to the source documents of Neoplatonism had made a few converts to the incorporealism of Plotinus. Faced with these alternatives, Kepler joined the incorporealists.

Although we will never know all that went into this choice, it should come as no surprise to scholars sensitive to the Neoplatonic tendencies that motivated other aspects of Kepler's scientific work — particularly his search for cosmic harmony. It is apparent, that is, that Kepler's commitment to Neoplatonic philosophy was general — more general than has been acknowledged by those who believe that Kepler repudiated Neoplatonic light metaphysics and took the first halting steps toward mechanization.

I am not meaning to suggest that Kepler merely appropriated a theory of light from the available ancient and medieval alternatives — that he walked through a philosophical cafeteria line, assembling a meal of ready-prepared Neoplatonic dishes. Kepler was certainly responding to ancient theories and profoundly influenced by them, but his response was both critical and creative. His own theory of the nature of light surely belongs to the Neoplatonic incorporealist tradition, but more than the right pedigree was required. It also had to satisfy a variety of rational and empirical criteria: internal consistency, compatibility with observational data, and general congruity with the rest of Kepler's physics, metaphysics, and cosmology. The acknowledged swiftness of light and the observation that refraction was exclusively a surface phenomenon seem to have been decisive, testifying powerfully against a corpuscular or modification theory of light. This was not blind traditionalism, but reasoned choice.

While choosing, Kepler also creatively modified. By virtue of its incorporealism, Kepler's theory must be judged fundamentally Plotinian, but it was not merely Plotinian. Kepler combined the Plotinian understanding of the relationship between light and the medium with the geometrical optics of the Baconian (or perspectivist) tradition.[76] But he did more than this: he not only maintained that the behavior of light can be described mathematically (as all good perspectivists believed); he also made the much more extreme claim that light is mathematical in its very essence. While admitting that the 'innermost nature' of light may lie beyond the reach of human knowledge, Kepler insists on the legitimacy of the quest.[77] To the best of his understanding, light is, in essence, a mathematical surface. Devoid of all matter or corporeal substance (though, of course, capable of interacting with bodies), it belongs inevitably to the intelligible realm of mathematics. This is a most revealing position, when viewed within the history of attempts to understand the relationship between mathematics and nature: it carries the mathematical reductionism of the Platonic tradition to the very heart of the science of optics. Plenty of Kepler's predecessors were eager to submit light to mathematical analysis; nobody since Plato had been so bold as to claim for light a purely mathematical nature.

University of Wisconsin
Madison, Wisconsin, U.S.A.

NOTES

* Research for this paper was supported by grants from the National Science Foundation and the Graduate School of the University of Wisconsin. Certain sections of this paper bear a resemblance, more or less close, to parts of my 'The Genesis of Kepler's Theory of Light: Light Metaphysics from Plotinus to Kepler,' *Osiris*, 2 (1986), pp. 5—42. It will clarify the relationship between the two papers to note that the *Osiris* paper, though published first, was written second. However, a few passages and some of the citations in the present paper have been revised since publication of the *Osiris* paper.

1. On ancient theories of light, see Arthur Erich Haas, 'Antike Lichttheorien,' *Archiv für Geschichte der Philosophie*, 20 (1907), pp. 345—86; Vasco Ronchi, *The Nature of Light*, trans. V. Barocas (London: William Heinemann, 1970), chap. 1; David C. Lindberg, *Theories of Vision from al-Kindi to Kepler* (Chicago: University of Chicago Press, 1976), chap. 1 (or see the German translation, published as *Auge und Licht im Mittelalter*, trans. Matthias Althoff, Frankfurt: Suhrkamp, 1987).

2. I would prefer not to speak of corpuscular, corporeal, and incorporeal theories of light, for such phraseology, strictly construed, suggests that what is corpuscular or corporeal is the theory rather than the light (though 'corpuscular theory of light' does have the sanction of usage). Nonetheless, the alternatives are extremely cumbersome. In this paper I will employ a variety of locutions, some of which sidestep the problem, some of which do not.

3. Aristotle, *De anima*, II.7. On the atomistic theory, see Cyril Bailey, *The Greek Atomists and Epicurus* (Oxford: Clarendon Press, 1928), pp. 103—4, 165—70, 406—13; Kurt von Fritz, 'Democritus' Theory of Vision,' in *Science, Medicine and History: Essays on the Evolution of Scientific Thought and Medical Practice, Written in Honour of Charles Singer*, ed. E. A. Underwood, 2 vols. (London: Oxford University Press, 1953), *I*, pp. 83—99; Lindberg, *Theories of Vision*, pp. 2—3.

4. Diogenes Laertius, *Lives of Eminent Philosophers*, X, trans. R. D. Hicks, 2 vols. (London: William Heinemann, 1925), *II*, pp. 575—79.

5. *De rerum natura*, IV. 54—61.
6. Empedocles' and Plato's association of light with fire should be viewed as variations of the corpuscular theory; see Lindberg, *Theories of Vision*, pp. 4—6.
7. *De anima*, II.7.418^{b}14—27, trans. W. S. Hett (London: William Heinemann, 1936), p. 105; *cf. De sensu*, II.440^{a}16. On Aristotle, see also Lindberg, *Theories of Vision*, pp. 6—9.
8. The same, I believe, is true of the counterparts of these terms in most other European languages.
9. VIII.3.1029^{a}20—25, trans. W. D. Ross, 2nd ed., in *The Works of Aristotle Translated into English*, ed. J. A. Smith and W. D. Ross, 12 vols. (Oxford: Clarendon Press, 1908—1952), *VIII*.
10. For a good, recent discussion of Aristotelian substance, see Abraham Edel, *Aristotle and His Philosophy* (Chapel Hill: University of North Carolina Press, 1982), chap. 8.
11. W. D. Ross, *Aristotle: A Complete Exposition of His Works and Thought* (New York: Meridian Books, 1959), pp. 175—80; Edel, *Aristotle*, pp. 129—33.
12. Aristotle, *Metaphysics*, VII.10.103^{a}2—12, VII.11.1036^{b}32—1037^{a}5. For a discussion of Aristotle's meaning, see Ross's note in Aristotle, *Metaphysics*, ed. David Ross (Oxford: Clarendon Press, 1924), *II*, 199—200; Thomas Heath, *Mathematics in Aristotle* (Oxford: Clarendon Press, 1949), pp. 213—14.
13. *De anima*, II.7.418^{b}9—13, trans. J. A. Smith, in *The Works of Aristotle Translated into English*, II, with several changes.
14. *De sensu*, VI.446^{b}28—447^{a}3, trans. W. S. Hett (London: William Heinemann, 1936), p. 269, with several revisions.
15. *De anima*, II.7.419^{a}9—15, trans. Hett, p. 107.
16. Similar claims can be made for color.
17. For an early Aristotelian commentator, see *The De anima of Alexander of Aphrodisias*, trans. with commentary by Athanasios P. Fotinis (Washington, D.C.: University Press of America, 1979), pp. 236—49. Such commentaries continued to appear through the Middle Ages.
18. See, for example, al-Kindi, below.
19. David C. Lindberg, 'The Science of Optics,' in *Science in the Middle Ages*, ed. Lindberg (Chicago: University of Chicago Press, 1978), p. 340.
20. This section bears a close resemblance to a section (which it sometimes condenses and sometimes expands) in my 'The Genesis of Kepler's Theory of Light: Light Metaphysics from Plotinus to Kepler,' *Osiris*, 2 (1986), pp. 5—42, esp. pp. 9—12.
21. *Enneads*, II.4.15, trans. A. H. Armstrong, 7 vols. (London: William Heinemann, 1966—1988), *II*, p. 147.
22. *Ibid.*, V.1.6, trans. Armstrong, *V*, 31.
23. *Ibid.*, *V*. 31—33.
24. For a sketch of the history of this doctrine, see David C. Lindberg, *Roger Bacon's Philosophy of Nature: A Critical Edition, with English Translation, Introduction, and Notes, of De multiplicatione specierum and De speculis comburentibus* (Oxford: Clarendon Press, 1983), pp. xxxv—lvi.
25. *Enneads*, III.6.13.
26. *Enneads*, II.4.16, trans. Armstrong, *II*, 147.

27. *Ibid., II*, 149.
28. See *Enneads*, III.6.16—18.
29. On the superlunary world, see II.1. On intelligible matter (the receptacle of eternal forms in the intelligible realm), see II.4.4—5.
30. I.6.3, trans. Armstrong, *I*, 241.
31. For example, Robert Grosseteste. See James McEvoy, *The Philosophy of Robert Grosseteste* (Oxford: Clarendon Press, 1982), pp. 151—52, 160—61; Lindberg, *Bacon's Philosophy of Nature*, pp. li—lii.
32. *Enneads*, I.6.3, trans. Armstrong, *I*, 142.
33. A. H. Armstrong, *The Architecture of the Intelligible Universe in the Philosophy of Plotinus* (Cambridge: Cambridge University Press, 1940), p. 55. Armstrong has noted the strong Stoic flavor of these claims. This chapter of the first *Ennead* was a very early Plotinian work; see Armstrong's note, *Enneads, I*, 231.
34. *Enneads*, IV.5.7, trans. Armstrong, *IV*, 305—7.
35. *Ibid.*, trans. Armstrong, *IV*, 307—9.
36. *Ibid.*, II.1.7, trans. Armstrong, *II*, 29—31.
37. *Ibid.*, IV.5.7.
38. *Ibid.*, IV.5.7. Here I prefer the translation of Stephen Mackenna, 2nd rev. ed. revised by B. S. Page (London: Faber and Faber, 1956), p. 336. *cf. Enneads*, VI.4.7, VI.4.10.
39. *Ibid.*, IV.5.6, trans. Armstrong, *IV*, 303.
40. *Ibid.*, trans. MacKenna, p. 334.
41. *Ibid.*, p. 335.
42. *Ibid.*
43. For a possible exception, see Julius R. Weinberg, *Nicolaus of Autrecourt: A Study in 14th Century Thought* (Princeton: Princeton University Press, 1948), p. 145.
44. *De radiis*, ed. M.-Th. d'Alverny and F. Hudry, in *Archives d'histoire doctrinale et littéraire du moyen âge*, 41 (1974), p. 224. On al-Kindi's emanationism, see also Lindberg, *Bacon's Philosophy of Nature*, pp. xliv—xlvi.
45. *De aspectibus*, prop. 11, in Axel Anthon Björnho and Sebastian Vogl, (eds.) 'Alkindi, Tideus und Pseudo-Euklid. Drei optische Werke,' *Abhandlungen zur Geschichte der mathematischen Wissenschaften*, 26, pt. 3 (1912), 13. See also Lindberg, *Theories of Vision*, pp. 24—26, 30—31.
46. These paragraphs on Bacon offer an abbreviated version of matters treated more fully in my *Bacon's Philosophy of Nature*, pp. liii—lxxi; and my 'Genesis of Kepler's Theory of Light,' (see full reference in n. 20 *supra*). Occasionally I have borrowed wording as well as substance.
47. Pierre Michaud-Quantin, 'Les champs sémantiques de *species*. Tradition latine et traductions du grec,' in Michaud-Quantin's *Etudes sur le vocabulaire philosophique du moyen âge* (Rome: Edizioni dell'Ateneo, 1970), p. 113; Lindberg, *Bacon's Philosophy of Nature*, pp. liv—lv.
48. *De multiplicatione specierum*, I.1.75—80, trans. Lindberg, *Bacon's Philosophy of Nature*, p. 7.
49. *Ibid.*, I.3.52, trans. Lindberg, p. 47.
50. *The Opus Majus of Roger Bacon*, ed. John Henry Bridges, 2 vols. (Oxford:

Clarendon Press, 1897), *II*, 71—72; translated in Lindberg, *Bacon's Philosophy of Nature*, p. lxiii.
51. *De multiplicatione specierum*, III.1.37—42, trans. Lindberg, p. 181.
52. III.2.76—80, trans. Lindberg, p. 191.
53. On Ockham and the response to his doctrine, see Katherine H. Tachau, 'The Problem of the *Species in medio* at Oxford in the Generation after Ockham,' *Mediaeval Studies, 44* (1982), pp. 394—443. On medieval discussions, see also Anneliese Maier, 'Das Problem der 'Species sensibiles in medio' und die neue Naturphilosophie des 14. Jahrhunderts,' *Freiburger Zeitschrift für Philosophie und Theologie, 10* (1963), pp. 3—32; reprinted in Maier's *Ausgehendes Mittelalter: Gesammelte Aufsätze zur Geistesgeschichte des 14. Jahrhunderts*, 2 vols. (Rome: Edizioni di storia e letteratura, 1964—1967), *II*, pp. 419—51. On Ficino, see my 'Genesis of Kepler's Theory of Light,' pp. 22—29.
54. *Ad Vitellionem paralipomena, quibus astronomiae pars optica traditur* (Frankfurt, 1604), ed. Franz Hammer, in *Gesammelte Werke*, ed. Walther von Dyck and Max Caspar, 18 vols. (Munich: C. H. Beck, 1937—1963), *II*, p. 19 (hereafter cited as *Paralipomena*). I have also benefitted greatly from the French translation of Catherine Chevalley: Johann Kepler, *Les fondements de l'optique moderne: Paralipomènes à Vitellion (1604)* (Paris: J. Vrin, 1980). My views on Kepler's theory of the nature of light are also expounded, with much greater attention to his metaphysics and cosmology of light, in 'Genesis of Kepler's Theory of Light,' pp. 29—42. Several paragraphs of the present paper are extracted from that account.
55. *Paralipomena*, chap. 1, p. 19.
56. *Ibid.*, pp. 19—20.
57. *Ibid.*, props. 1—2, p. 20.
58. *Ibid.*, prop. 3, p. 20.
59. *Ibid.*, props. 5—7, p. 21.
60. *Ibid.*, prop. 4, p. 20; *cf.* Chevalley's translation, p. 109.
61. *Ibid.*, props. 3—5, pp. 20—21. On the immateriality of light, see Catherine Chevalley, 'Sur le statut d'une question apparement dénuée de sens: la nature immatérielle de la lumière,' *XVIIe Siècle, 34* (1982), pp. 257—66.
62. *Paralipomena*, p. 46; trans. Chevalley, p. 150. On transparency, see prop. 11, p. 22.
63. *Ibid.*, chap. 1, reply to Aristotelian thesis 16, p. 42; trans. Chevalley, p. 145.
64. *Ibid.*, reply to thesis 10, p. 40.
65. For a useful discussion of Kepler's terminology, see Catherine Chevalley's introduction to *Les fondements de l'optique moderne*, pp. 69—71. For examples of Kepler's use of the term 'species,' see *Paralipomena*, pp. 37, 45; *Astronomia nova*, III.33, in *Gesammelte Werke*, *III*, 240.
66. Ronchi, *Nature of Light*, p. 92. Stephen M. Straker's 'Kepler's Optics: A Study in the Development of Seventeenth-Century Natural Philosophy,' Ph.D. dissertation, Indiana University (1970), p. 504. On the significance of Kepler's term 'species,' see Catherine de Buzon [Chevalley], 'La propagation de la lumière dans l'optique de Kepler,' in *Roemer et la vitesse de la lumière* (Paris: J. Vrin, 1978), pp. 75—78.
67. See Ivor Leclerc, *The Nature of Physical Existence* (London: George Allen &

Unwin, 1972), chaps. 9—10; James A. Weisheipl, O. P., 'The Concept of Matter in Fourteenth Century Science,' in *The Concept of Matter in Greek and Medieval Philosophy*, ed. Ernan McMullin (Notre Dame, Ind.: University of Notre Dame Press, 1963), pp. 150—56. On Kepler's usage of the term, see my *Osiris* paper (reference in n. 20 *supra*).

68. *Paralipomena*, chap. 1, props. 3, 8, pp. 20—21.
69. *Ibid.*, prop. 5, p. 21.
70. *Ibid.*
71. On Kepler's theory of refraction, see Straker, 'Kepler's Optics,' chap. 7.
72. *Paralipomena*, prop. 20, p. 27.
73. *Astronomia nova*, III.33, *Gesammelte Werke, III*, p. 240. This passage is quoted by Alexandre Koyré, *The Astronomical Revolution*, trans. R. E. W. Maddison (Paris: Hermann, 1973), pp. 200—1.
74. *Astronomia nova*, III.33, *Gesammelte Werke, III*, p. 241. In this passage Kepler once again admits that there is a sense in which the light '*was* present in the medium.' He also points out that the same claim was made earlier in the *Paralipomena*; for that passage, see Kepler's reply to the fourth Aristotelian thesis, *Paralipomena*, chap. 1, p. 40.
75. Straker, 'Kepler's Optics,' pp. 507—8, has called attention to Thomas Harriot's attempt to persuade Kepler of the truth of the corpuscular theory.
76. Elsewhere, I have argued that Kepler was the culminating figure of this Baconian tradition, and none of that do I wish to retract. But I do wish to restrict its applicability to Bacon's conception of the optical enterprise and to his more technical, mathematical achievement (laws of radiation and the like); on the nature of light, Kepler repudiated the Baconian position. See Lindberg, *Theories of Vision*, pp. 205—8; and 'Laying the Foundations of Geometrical Optics: Maurolico, Kepler, and the Medieval Tradition,' in David Lindberg and Geoffrey Cantor, *The Discourse of Light from the Middle Ages to the Enlightenment* (Los Angeles: William Andrews Clark Memorial Library, 1985), pp. 41—53.
77. *Paralipomena*, p. 39.

PART FIVE

Science, Religion, and Political Power

JOHN D. NORTH

ONE TRUTH OR MORE?
DEMARCATION IN THE UNIVERSE OF DISCOURSE

There emerged in the course of cosmological discussion in the Middle Ages a strategy of argument that may be described as 'intellectual pacifism.' This took several forms. The best known was the doctrine ascribed, rightly or wrongly, to Averroes and his followers, namely that there are two sorts of truth, one philosophical, the other theological. This was supposedly worked out to reconcile propositions that were at first sight inconsistent. There were indeed, then as now, a number of different sorts of intellectual demarcation with this end in view, and they deserve closer attention than they are usually given. Why did some sorts of conceptual inconsistency or incompatibility seem relatively unimportant, while others were thought so serious that communication was deemed more or less impossible across a subject boundary? Two themes in particular will be used here as illustration, namely creation and eternity.

'Demarcation' came into the English language from a good Spanish word that gained currency after Pope Alexander VI had divided up the Americas as between Spain and Portugal (Bull of 4 May 1493). The territorial metaphor is too tempting for me to be able to resist it, although its dangers are manifest. It does at least have the merit of reminding us of a very strong motive for intellectual demarcation — the desire to avoid conflict. It shows, too, that one may divide territory without knowing quite what falls on either side of the dividing line. (Thus Alexander was in a different situation from Jehovah's at the time He gave Israel to the Jews.) The great disadvantage of the metaphor is that it fails to do justice to the more interesting sorts of intellectual divisions, those between not parts of a world, but *kinds* of world. To illustrate the point with a well known example: there is a topological boundary between celestial and terrestrial regions, that is, a boundary between the sphere of the Moon and the sphere of fire, in the Aristotelian world. Topologically considered, this is not particularly interesting. What makes it so is the fact that different *principles* for the behaviour of matter operate on the two sides of the boundary. A reasonable parallel might seem to be that between the different *mores* operating in the Spanish and Portuguese dominions, but this analogy is potentially misleading as soon as we turn to examples where there is no topological boundary involved. Despite my cosmological concern, I shall not often have spatial divisions, as such, in mind.

KANT, WITTGENSTEIN AND DEMARCATION

Use of the word 'demarcation' in the context of philosophy must surely bring two philosophers in particular to mind, namely Kant and Wittgenstein, and it will be useful to begin with a brief reference to both, in order to show potential ambiguities in the notion. The whole drift of Kant's criticism of metaphysics is that it falsely pretends to be an extension of (as we should say) scientific thinking. The contradictions in the antinomy of pure reason, for instance, were said to stem from a false premiss, namely that the world of the senses (the scientific world) is a world of things-in-themselves (the world of metaphysics). Kant insisted that these must be treated as fundamentally *different* worlds. Do so, and all will be well. He wanted to determine the limits of thought. By the nature of his exposition of the categories of thought, he is today often seen as 'really' having had in mind the limits of language. One way or another, he can be seen as one who made explicit a division implicit in much medieval writing — and it is no accident that the arguments on the two sides of the antinomy reached him after a long journey passing through medieval cosmology *en route*.

The early Wittgenstein, the Wittgenstein of the *Tractatus*, took an optimistic line over the possibility of a certain sort of metaphysics — a metaphysics of language, based on the essential nature of propositions, a metaphysics that could be shown, but not put into words, and thus not characterised as true. This is a form of demarcation without, as far as I know, any medieval counterpart, although it bears a certain resemblance to the view that religious knowledge can be intuitively grasped, without the use of anything analogous to scientific reasoning. The distinction Wittgenstein drew in his later philosophy, as elaborated in the *Philosophical Investigations*, between scientific and philosophical thinking, was based on his new-found belief that there is something essentially mistaken about the act of generalising in philosophy. His view of philosophy, as a pursuit of illumination through particular facts about language, rather reminds one of certain histordans' view of their craft — on no account must one generalize! Whether either they or he could live up to the ideal is a moot point. Needless to say, most recent linguistic philosophers have been prepared to embrace the systematic methods of logic and the sciences, and to the extent that they see logic (in whatever form) as setting limits to the range of linguistic possibilities, they are in a Kantian line of succession (where the limits were limits set to knowledge), and going back yet further, are of the same philosophical family as those who insisted that God's very actions were limited by the demands of logical consistency.

Taking stock of our demarcations, we have Kant's, marking off knowledge from the rest — the rest, as it happens, included morality and religion — and we have Wittgenstein's — marking off the realm of language from all else. (Near the end of the *Tractatus* he explained somewhat obscurely how morality and religion fall within the realms of factual language, without forming a part of it.)

As is well known, he argued that the meaning of even a simple sign, such as an ordinary name, is its role in a language-game, in a rule-governed activity, a form of life, autonomous, with no external aims and goals. It was this that led him to say that the commonest form of nonsense arises when words are used outside the language-games appropriate to them, a thing easily done, since the dress of language hides the diversity of its 'games' — describing, judging, giving commands, guessing, praying, and so forth. What of the rules of the game? We say that the rules of the game are arbitrary when they have *no effect outside the game*. (The goals of grammar being those of language as a whole, he took its rules to be arbitrary.) He thus insulated — or encouraged others to insulate — various modes of thought, modes of discourse, each being granted its own objective basis. He provided some, as I said earlier, with their 'live and let live' attitude to philosophy and religion and science. That his work could be taken as a charter of non-agression had something to do with his failure to supply guidelines as to what counts as an 'effect outside the game'.

This is not the place to discuss how successfully Wittgenstein carried out his own programme, but at the risk of discrediting my demarcation metaphor, I should like to mention his views on the distinction between religious and factual discourse. Physics has a place in human life; religious propositions do not express factual possibilities or hypotheses, but they do make a difference to the lives of those who hold to them, and their meaning is a function of this effect. (For this reason, Wittgenstein did not like the idea that religion was a form of rudimentary empirical science, but this seems to me to be no reason at all.) So two language-games may overlap in their effects. My topological metaphor seems inappropriate, which is of no great moment; but it also seems inappropriate to have supposed that religion and science — on his view of them — could not be in conflict. There are close parallels to this sort of tension in the Middle Ages, as when the science of astrology and the dictates of religion made different recommendations. (As for metaphors, more suitable might have been comparison with a locomotive with two engines, one electrical, the other driven by steam. Their effects may be compared, and one might be preferred to the other, so there is potential conflict, not in the head-on sense, but in the sense that they compete for our favour, just as alternative scientific theories compete.)

TEMPIER, AVERROISM AND TWO TRUTHS

Whatever justification a modern philosopher might find for setting apart different sorts of discourse, one would not have thought that this was likely to have presented much of a problem in the Middle Ages. In principle, since teaching was then the prerogative of the Church (or churches, as I should say, since I shall also introduce Jewish and Moslem ideas), there should have been no obstacle to a monolithic truth. In practice, there were confrontations of pagan learning with Holy Scripture and the Church Fathers. In the West, despite a rearguard action by such representatives of the older monastic teaching as Peter Damian and Bernard of Clairvaux, this became especially serious after the twelfth century, and the possibilities for confrontation multiplied after 1255, when the arts faculty in the University of Paris stipulated the study of all of Aristotle's works, to the consternation of many within the theology faculty. Étienne Tempier's notorious condemnations of suspect theses in 1277 were aimed at members of an arts faculty guilty of ever greater vanity — even intellectual complacency — as regards the sort of rationality they purveyed. Reconciling Scripture and the sciences was not new, but it became more urgent, now the spotlight of debate was upon it. Some Aristotelians felt too strongly about their Philosopher to expose his faults. Siger of Brabant told them they should not scruple to do so. But what if the scriptural text should seem to be getting the worst of the argument? Even to contemplate issuing a directive on this point would have been to court disaster. There were danger areas enough — such as in problems of nature and grace, the immortality of the soul, and the resurrection of the body; and then, with the coming of the Averroistic debates of the thirteenth century, the problems associated with the working of the spheres, the nature of creation, and the eternity of the world took on a new lease of life.

The two greatest critics of the Averroist interpretation in the thirteenth century were of course Albertus Magnus (note especially his *De xv problematibus*) and Thomas Aquinas. Giles of Rome (*Errores philosophorum*) and the many others were usually on a distinctly lower plane. Our concern is not with the discussion in general, but with the manner in which theologico-philosophical reconciliation was achieved, especially on a select number of cosmological points. The problem seems to have been making itself felt in Paris from the 1240s, when Bonaventure was a student, for as he later wrote:[1]

When I was a student, I heard it said that Aristotle posited the world as eternal, and when I heard the reasons and arguments quoted to that effect, my heart began to beat, and I asked myself: how is this possible? But now, all this is so public that no hesitation is permitted.

By 1268 he was attacking a number of specifically Averroist positions, including this one concerning eternity. He had earlier disagreed with Aristotle on the question (that is, when commenting on the *Sentences*), but now he was opposing his contemporaries. It is reasonable to suppose that the short list of condemned propositions put out by Tempier in 1270 was prompted, at least in part, by Bonaventure's attack. That year, 1270, saw Giles of Rome's *On the errors of the philosophers*; and the turmoil in Paris was such that by January 1277 Pope John XXI (Peter of Spain) had written to Tempier on the question, and had thus presumably thereby prompted the mammoth list of 219 condemned propositions. Of all the thorns in the bishop's side, the most troublesome were Siger of Brabant and Boethius of Dacia. They must certainly have been in Tempier's thoughts when, in the preamble to the list of propositions, he wrote of persons who

say that [the propositions] are true according to philosophy but not according to the Catholic faith, as if there were two contrary truths, and as if, against the truth of Holy Scripture, there could be truth in the sayings of the damned gentiles, of whom it was written: 'I will destroy the wisdom of the wise,' because true wisdom destroys false wisdom.[2]

This is an attack on a doctrine, real or supposed, since known as the doctrine of double truth.

Did anyone profess a doctrine of double truth? Gilson said that he knew of no text wherein Siger himself stated that necessary philosophical conclusions were true when they contradicted Christian faith, but that on the contrary, he held that in such cases truth was on the side of faith.[3] The problem of deciding whether Siger was being prudent or sincere is interesting, but it is not my problem. The most one can say is that whenever philosophy contradicts theology, Siger refrains from asserting the truth of philosophy; and from this fact, Gilson concludes that 'unless he were deliberately lying, which is possible but cannot be proved, Siger must have thought that certain conclusions could be philosophically irrefutable without being true'.[4] F. van Steenberghen, Siger's editor and biographer, disagrees with this interpretation, just as he dislikes Gilson's appeal to God's miraculous intervention as a potential justification for a situation in which faith tells us that philosophy is wrong. (The idea, for which Gilson could quote contemporary opinion, was that God may override the causes as they are known to philosophy; and that by reason of God's miraculous intervention, even the predictions of the Prophets were sometimes wrong.) For van Steenberghen, Siger's avoidance of an attribution of falsehood to philosophy was because he would have found it disagreeable, as is evidenced by his oblique way of speaking whenever Aristotle's conclusions conflict with the faith.[5] To suppose that Siger could have imagined that God can reconcile contradictory things is, according to van Steenberghen, out of the question. At all events, one thing is clear, namely that if anyone believed in a doctrine of double truth, it is hard to pin this on Siger, and I see little purpose in considering further the very numerous discussions of his views — discussions which take their starting point, as often as not, from the fact that Dante puts Siger in Paradise, with praise from the lips of his erstwhile adversary Thomas Aquinas.[6]

Boethius of Dacia, who, after Tempier's condemnations, fled Paris with Siger and appealed to the pope, is as innocent of a doctrine of double truth as Siger. Surviving texts are fewer in the case of Boethius, but much the same has to be said of him as was said of Siger: philosophical conclusions contrary to the faith were not called true by either, in an unequivocal and explicit way. Even Averroes himself (Ibn Rushd, A.D. 1126—1198), to whom the doctrine of double truth has been ascribed, seems not to have accepted anything of the sort, at least in a form that would make for an epistemologically interesting thesis. It appears that he said only that when the Koran touches on the same subject as philosophy, it is philosophy that should take the lead, and that the sacred text must be so interpreted that it will agree with the dictates of demonstrative reason. When philosophy is silent, instruction must come from the Koran.[7] There seems to be the idea in Averroes' writing of a theology that uses allegorical methods for the common man's sake, whereas philosophy can penetrate to a more exact understanding of the truth. This conceding of precedence to philosophy is characteristic of some of the fourteenth century Latin Averroists. On occasion they even managed to make moral capital out of their stance: faith carries merit, and there is no merit in believing what can be proved by reason![8] In John of Jandun's case there is a certain insouciance, if not truculence, in his tone: 'Sed demonstrare nescio; gaudeant qui hoc sciunt' — 'I do not know how to prove this [proposition contrary to Aristotle]; bully for those who do.'[9] But this does not look like a stand taken on a deep philosophical principle.

The most systematic statements of the relation of religious to philosophical (and scientific, natural-philosophical) truth are by those who felt the Averroistic attack most keenly. Aquinas is here pre-eminent. His authority made itself felt, long after his death, especially on the broad question of the distinction between dogmatic theology, whose objects are known 'by the light of divine revelation,' and philosophy, or natural theology, where the *same objects* may be 'known by the light of natural reason.'[10] Philosophy was to perform three tasks for Christian theology: demonstrate the truths at the basis of faith (as that God exists); show analogies common to nature and grace; and fight off attacks on the faith.[11] Philosophy and dogmatic theology were said to differ formally, inasmuch as the *conclusions* of a proof by philosophy might be put at the *beginning* of an exposition of what is known by revelation. (The order of presentation differs, but one can hardly say that the logical orders are simply reversed, since surely the truths known by revelation may in principle be had in any order whatsoever.) More interesting, philosophically, is the idea that they might share their objects.

To say that they share their objects is to discard the Wittgensteinian 'language-game' method of insulation, and to positively invite conflict. One often comes across a quotation from Aristotle's *Metaphysics*, to the effect that many hold conflicting beliefs, and that if reality is as any person's opponents maintain, all will be right in their beliefs.[12] Those who had so much to say against two truths did not need Aristotle. They took his view to be self-evident; but in doing so they missed the point, for they did not ask themselves what anyone could have *meant* by maintaining that there are two truths. The best we can do for them is say that they feared that their opponents, confronted with propositions A and ~ A, might preserve both, using two different sorts of validation. (One might imagine two sorts of assertion sign, $\vdash_f$ for assertion by faith, and $\vdash_s$ for assertion by science; and then one might blithely say $\vdash_f A \ \& \ \vdash_s \ \sim A$.) It is not easy to discuss a social group, however, when its existence has never been proved. Is mine at all a probable medieval solution? Most would have preferred to dodge the issue entirely, of course, but some might have followed another course. For Roger Bacon, for example, the meaning of words is not absolute, but depends on the contexts in which they are uttered. Most impositions are not explicit, but tacit, and even independent of the speaker's conscious actions.[13] Those following in a strongly Neoplatonic tradition (following Augustine, Anselm, and Avicenna on this point), and emphasizing the 'truth of being', would not have been much affected by this Baconian insight; but those in the Aristotelian mould, emphasizing the adequation of mind to being, and insisting that truth is actually and formally present only in the intellect and not in the thing,[14] would not have found it hard to say something like 'There seem to be two truths, but on closer examination of the contexts the ambiguity can be located, and we have in fact not $\vdash A \ \& \ \sim A$ but $\vdash A' \ \& \ \sim A$.'

MILD INCONSISTENCY

Now there are in fact many examples of propositions in medieval philosophy and theology where there are prima facie two truths, but where a meaning equivalence dissolves on closer inspection, thus saving the day. It is a truism that at the heart of the unrest we are discussing there was a consciousness of propositional or conceptual incompatibility; and as I am pointing out, there were two ways of looking at this, the alarmist and the analytical or contextualist, as we might call them. Having said as much, however, there is a danger of trivializing the problem. I will give some examples where there clearly was conceptual incompatibility of a sort, but where this is not deep enough to be seen as a threat to a particular conceptual scheme or outlook:

1. In analysing the upward motion of an air bubble through a tube filled with water that is imagined to reach to the heavens (i.e. beyond air's natural place), Nicole Oresme noted that the air must be said to be by its nature capable of two simple contrary motions.[15] At first sight there is conceptual ambiguity as between Aristotle's and Oresme's 'natural motion'; but the problem is resolved by taking the surrounding elements (in this case the water) into account in the determination of natural motion. It is not that Aristotle survives unscathed, but that an inconsistency has been excised.
2. Jean Buridan considered the possibility that a small body situated between two worlds might have a motion that was natural in one and unnatural in the other.[16] Without pursuing the question, it is obvious that (if this is not taken to militate against the concept of two worlds) Aristotle's notions can be suitably modified (along Newtonian lines, for instance, with natural motion being whatever is produced by the resultant of all component natural motions).
3. Following Aristotle, some motions have a contrary (linear motions), while others (circular motions) do not.[17] It is not that the concepts of motion are incompatible, but that they are not sufficiently specified by the one world. (It is as though we have to use the word 'matter' to speak of lead and gold, and then complain when we found that their densities were different.)
4. According to Aristotle, corruptibles and incorruptibles are not in the same physical genus, for they have different types of potentiality; but in a logical sense they are, for they share the concept of corporeity. Is there an inconsistency in the concept? Avicebron wanted there to be a matter common to all bodies, but Aquinas settled for a higher sort of non-corruptible matter, and by his two-substance theory avoided the inconsistency.[18] His distinction between logical genus and physical genus, like the examples under 1–3 above, can be seen as a form of intellectual pacifism.
5. There are many philosophical models for creation and existence, but speaking very generally one may say that most writers treated existence as a predicate. On the resulting 'property-thing model' the question arises as to what it is we are to take as the 'thing' on which the property is to be tacked. Henry of Ghent was one who described

creation as the attachment of actual existence to an (eternally existing) real essential being. To other theologians this sort of creation did not seem consistent with the concept of 'creation ex nihilo'.[19] By reinterpretation, by changing the model, consistency was preserved. The model I have in mind was that where God himself, as it were, 'holds up the property,' sustaining the thing he has created its existence. Here is another case of a reinterpretation in the interests of preserving intellectual calm.

INVIOLABLE CONCEPTUAL SCHEMES

Examples such as these could no doubt be multiplied many times over. Why was no-one tempted to talk of 'two truths' when confronted with inconsistency? Because by relatively straightforward conceptual refinement, elaboration and specification the difficulties could be removed. Science and philosophy were fair game for those who would modify them in their details. Scholars were only tempted to talk of two truths when there were at stake *two different conceptual schemes* that it was feared might both be viewed as more or less *inviolable*, two 'language games,' so to say, that might clash because they were thought to have an overlap of interpretation. It is the near inviolability that made the scholastic so fearful. The most he could do with Scripture was reinterpret, and there were limits to what was possible in that respect. With Aristotle one might do more; but the fear was that it would not be enough to square philosophy with Scripture.

This problem did not have to involve Scripture as one of the two 'more or less inviolable' schemes. I hardly need go into the details of that well worn theme of the incompatibility supposed to exist between the Ptolemaic and Aristotelian systems of the universe. Ptolemy, as we now know, thought them reconcilable, and was thus prepared to put them on the same epistemological plane.[20] Others thought to save the day by treating Aristotelian cosmology as a higher form of study than a mere geometrical and hypothetical astronomy bereft of a sound metaphysical base. Less well known is Aquinas' way of handling the problem, essentially by claiming that Ptolemy was *unproved*. This tactic deserves further comment.

As Duhem noted in his classic 'ΣΩZEIN TA ΦAINOMENA: Essai sur la notion de théorie physique de Platon à Galilée,' Aquinas took the line advanced by Simplicius: it is not necessary that hypotheses *(sup-*

positiones) invented by the mathematical astronomers be necessarily true, and we should not describe them as such, for one might be able to save the appearances in other ways, not yet worked out by man.[21] This straightforward point of logic, that the truth of a premiss is not guaranteed by that of a conclusion drawn from it, has curiously failed to appeal to many of the great scientists of history; but then, such people have not usually made strong claims for the *necessity* of their principles. The overall tendency of what one might call the Simplicius philosophy is undoubtedly a healthy and progressive one. It is to say, in effect, that no matter how successful science is, one may be forgiven scepticism, and one is always free to try to improve on it. It is a *potentially* progressive philosophy, and that is why Duhem favoured it, of course. What he did not fully appreciate was that it did not have to work this way. He tells us, for example, that the position taken by Thomas allowed astronomers to use Ptolemy's hypotheses, even though their metaphysics might lead them to reject the same. This sounds progressive enough, viewed from an astronomical perspective, until one realises that Thomas' purpose was quite different: he wanted to take the sting out of Ptolemaic criticism of his favourite metaphysician. He wanted Aristotle as a helpmeet in theology, and he wanted to claim those commentators on the *Sentences* of Peter Lombard (II.14, 'de opere secundae diei, qua factum est firmamentum') who had come to think of a much simplified ('crude' would be a more honest description) version of Aristotle's homocentric universe as canonical. Not only was this his point of view, it was the point of view for which he was thanked by later clerics, and it was one that — no doubt unintentionally — tended to devalue experience and experiment. (Fortunately for the development of the sciences there were other forces at work.) Aquinas felt safe with Aristotle, and here achieves intellectual calm in what must surely be thought a curious way: saying that Ptolemy was unproved was, on the analogy of territorial demarcation, as though the Pope were to have said 'I believe that genuine conflict between Spain and Portugal is impossible because, by their very nature, the Portuguese do not wish to make any territorial claims.'

THE ETERNITY DEBATE

Aristotelian and Ptolemaic cosmology differ, as already noted, not only in their geometry but in their modes of justification, which indeed neither shared with the theologian writing as such. Is there any *a priori* reason for supposing that the end-results of their deliberations could not agree; or that a proposition reached by faith should not be expected to agree with one reached by 'rational' processes, in whatsoever form ? Surely such agreement is only possible, indeed can only be *expressed*, if there is an initial agreement that the same language-game is being played out and that epistemological or other motives are alike. Judged by their actions, scholars were generally reluctant to draw a fundamental epistemological distinction between faith and reason so as to insulate one set of propositions from another. One distinction they were prepared to draw in the interests of intellectual harmony was that between *natural law* and *purposive action*, whether of man or of God. This claim is one I shall try to justify at length, in connection with the eternity debate.

At the core of this perennial debate is what at first sight looks like a relatively small group of concepts — creation, existence, conservation in existence, finite and non-finite temporal duration, determinism, possibility, necessity, and divinity in particular. Of all potential conceptual conflicts, two are of especial historical significance. First, was the eternity concept — taken, broadly speaking, in isolation from all ideas of creation — defensible in itself? Second, is it logically possible to combine the concepts of eternity and creation? Thus Bonaventure (not to mention many of his contemporaries) was no friend of Aristotelian eternity, but he would not have denied it meaning, as he would have done the notion of eternal creation. The genesis of this last notion has much to do with Aristotle's concept of a non-creative divinity — a problem for Jewish, Christian, and Islamic philosophers, some of whom tended to look favourably on the solutions offered by Plotinus and other neo-Platonists, involving an eternal creation.

Most of the essential ingredients of the ensuing discussion are already to be found in Plato and Aristotle, and to help assess later history I shall first explain their doctrines in a general way. To say of time that it is potentially infinite — as Aristotle said of it — is to say, in common parlance, that it is eternal; but historically that word has a variety of meanings. It is necessary to establish a few conventions, if we are to avoid confusion. I use the English word 'eternal' to mean 'everlasting,' so that it is the correlate of Aristotle's infinite time; and I shall use it indifferently of past and future. Some philosophers — especially in the Middle Ages — use the word 'sempiternal' in this sense, to avoid confusion with the notion of timelessness, for which they reserve 'eternal'; but there are good reasons for not following their lead. Outside philosophy and theology 'eternal' ('aeternus') most commonly has, and was clearly derived from, the notion of continuing worldly existence; and although many seem to think that 'timeless' is the only alternative meaning, this is far from being the case. I shall begin by eliminating some of the alternatives as irrelevant to what one might call the infinity problem.

Subjects such as mathematics that are held to deal with necessary truths do not need to include in their statements any reference to the time at which those statements are true. They are timeless, in the sense that it is inappropriate to associate them with a time. Time does not enter the game at any point, so to say — it is not a term in the axioms, for instance. (The truths are 'eternal verities' to use a phrase made popular by Descartes, who took the word 'eternal' in a sense I here avoid.) In his early works, the *Meno* and the *Phaedo*, Plato explained *a priori* knowledge in terms of the soul's having lived in an earlier life among timeless unchanging Forms, which are later remembered when the soul enters the world of changing things. Here again one might say that time is inappropriate, that is, in this case, to a description of the world of Forms. One might even have argued that this world can be described as 'timeless' in the same sense as mathematics, had Plato not shifted his ground, first in the *Republic*, and then in the *Timaeus*. There (37E6—38C3) the Forms are unchanging models used by the demiurge who made the changing world. The patterning Form 'has being for all eternity, whereas the Heaven has been and is and shall be perpetually throughout all time.'[22] Time comes into being only with the cyclical movement of the heavens; it moves 'according to number'; and it is not proper to speak of past, present, and future of the Forms, we are told. But how to reconcile this with the image of a craftsman using the models? How can there be any sort of activity in a realm of timelessness? How can God be 'eternal' in this special sense?

If the world of Forms is to be a repository of patterns for the world of sense, what is the pattern for change ? The task seems to be given by Plato to *space* (which was seen as a necessary precondition of the operations of the demiurge), rather than to *time*, which was taken as a product of the demiurge's creation of the moving heavens.[23] It is not explained where the demiurge gets his inspiration for this, much less for other types of change. Aristotle criticised Plato's doctrine of the creation of time,[24] but I am here concerned only with the difficulty of giving a coherent account of timelessness. It surely cannot apply to anything that can be thought of as 'another world' — the demiurge's workshop, for example, on the shelves of which stand his unchanging patterns. Putting Plato's cosmology aside, however, we can see at least four broad meanings that 'timeless' might have:

(1) *Not time-related.* Of a subject-matter, such as that of mathematics.
(2) *Changeless.* In itself incapable of providing an indication of the passage of time, but changeless as judged by related things.
(3) *Monotemporal.* Involving a coalescing of all times into one.
(4) *Extratemporal* Standing outside time. An idea often exploited by theologians speaking of a realm different from ours. If change takes place at all there, it is presumably 'in' some other sort of time than ours.

The third of these meanings was not unknown to Plato, who — according to Cornford he was echoing the words of Parmenides[25] — writes of the One Being; 'It never was nor ever will be, since it is now all at once.' Boethius, in the last chapter of his *De consolatione philosophiae*, talks of eternity as 'the complete possession of eternal life all at once (*totum simul*),' and contrasts it with the everlasting. Augustine had used a similar turn of phrase.[26] It is easy to talk of 'life' in this way, but hard to see what is to be gained by giving all events in a life the same time coordinate, so to speak. But then, I am for the moment only concerned with meanings, and it is enough that I distinguish between 'eternal' (the scholastic 'sempiternal'; the word 'everlasting' is usually thought to carry the additional meaning of 'unchanging in at least some respect') and the four concepts outlined above. Why it is important to do so at the outset is that *none* of the concepts (1)—(4) has anything to do with the infinity of time in the relevant historical senses of infinity. Some writers have allowed themselves to pass nevertheless from one of these four usages into the use of 'eternal,' where questions of infinity do naturally occur.[27]

In his discussion of motion, in Book VIII of the *Physics*, Aristotle takes issue with Plato on the doctrine that time was created with the world. His argument is that time is unthinkable except as a kind of dividing moment, a kind of middle point uniting the beginning of future time and the end of past time. (There *are* other conceptualizations, and the Middle Ages explored some of them, but Aristotle's position was this.) From this he deduced that 'there must always be time'; for there must be times to both 'sides' of any moment. And since it is motion by which we have time, time being 'a kind of affection of motion,' there will always be motion. (This might seem a rather more meaningful statement than that there will always be time; but by the latter he means only that there are no moments flanked by time to one side only.) In the same chapter he draws the same conclusion from the fact that the chain ... cause-effect-cause-effect-cause ... cannot be broken. These are the main reasons supporting the doctrine of the infinity of time already introduced in Book III.

When Aristotle argued that there will always be motion, he need only have said that there will always be *some* sort of motion. In fact his cosmology does not admit of the running down of the celestial part of the universe, which such natural philosophers as Empedocles feared.[28] The heavenly movements are indeed eternal, for Aristotle, and with this fact complications arise in the cosmological account. For imperishable things exist *actually*.[29] They are 'prior in substance to perishable things.' 'If they did not exist, nothing would exist.' They exist of necessity, and thus cannot exist potentially. Eternal movement is one of these imperishables.[30] As he has it in the *Physics*, 'in the case of eternal things, what may be must be.'[31] The material universe is one of those eternal things about which we do not deliberate, he tells us in the *Nichomachean Ethics*, where he adds that in this it is like *mathematics*. The emphasis is not on motion and change but on eternal and necessary existence.

Now bearing in mind Aristotle's hostility to the actually infinite, and what he wrote in Book III of the *Physics*, how are we to reconcile with this the clear implication of the *Metaphysics* that eternal (infinite) movement, and hence eternal (infinite) time exists actually?

Looking back to Book III we find that time is given rather little attention, and is indeed made the basis for explaining Aristotle's views on spatial magnitude. '*There will not be an actual infinite*'[32] In talking of the existence of an infinite, one must use the word 'is' in the way one does when one says 'it is day' or 'it is the games ,' that is, anticipating a recurrence. The infinite is not 'this', as a man or a horse,[33] and is not even potentially so. 'Time indeed and movement are infinite, and also thinking, in the sense that each part that is taken passes in succession out of existence'.[34] In other words, it is as an attribute of certain sequences of events or things — things sharing something, which allows one to say they are of the same sort, but numerically different — that infinity may have whatever actuality it is said to have in the *Metaphysics*. In a sentence: time is actually infinite, but only in a special sense. And when commentators — and Aristotle himself — say that the infinite exists potentially, they mean precisely the same thing, namely that it exists in this peculiar sense. Moreover, there is a good case to be made out for saying that the first statement (time is actually infinite in a certain sense) expresses Aristotle's views more accurately than the second (time exists potentially), although historically it does not play an important role.[35] For most of his commentators, Aristotle was believed to have flatly denied an actual infinity.

ETERNAL CREATION: AL-KINDI VERSUS PLOTINUS

In Aristotle, God and the world were to a great extent separate beings. Plotinus tried to put an end to this dualist view, with his doctrine that all that exists proceeds from (emanates from) a single, incorporeal, first cause, 'the One', that is God. This view was bound to give rise to more problems than it solved, within the three religious groups considered here: as already pointed out, eternal emanation seems hardly compatible with creation *ex nihilo*, and it was one of several doctrines made the subject of a broad attack by Sa'adyâ (d. 942), the first systematic work of the Kalâm movement.[36] A more thorough rejection of eternal emanative creation had been earlier set down by al-Kindî, the ninth century Baghdad philosopher who introduced so much Greek philosophy to the Islamic world, and who is usually seen as opposed to Kalâm. (The two great philosophers al-Fârâbi and Ibn Sînâ (Avicenna), both in the Falsafah tradition of a broadly Greek based philosophy, favoured eternal creation.) Al-Kindî gave priority to scriptural revelation over philosophy, but managed to reconcile the two; that is, he reconciled revealed religion with natural theology and cosmology. In his rejection of eternal emanative creation he drew on arguments already sketched by Philoponus, the sixth century Christian philosopher, but his new presentation is extremely interesting for its quasi-axiomatic form. I will here set out his supposedly self-evident postulates. These concern bodies, in the sense of any extra-mental quantitative entities, and these I denote by 'B'. That a body is finite or infinite I abbreviate by respectively 'f(B)' and 'i(B),' and for the moment I shall say nothing of the meaning of these terms.

The axioms are these, using > and < to mean 'greater than' and 'less than' (negated by being struck through):[37]

A(1) If $B_1 \not< B_2$ and $B_1 \not> B_2$ then $B_1 = B_2$.
A(2) '$B_1 = B_2$' $=_{def}$ 'the dimensions of B_1 and B_2 are equal in actuality and potentiality.'
A(3) f(B) if and only if $\neg$ i(B).
A(4) If $B_1 = B_2$ and $B_3 = B_1 + B_0$ then $B_3 > B_2$.
A(5) If $f(B_1)$ and $f(B_2)$ then $f(B_1 = B_2)$.
A(6) The smaller of two generically related things is inferior to the larger.

A few brief comments on these 'obvious truths' are called for. The first three amount to definitions — although it is not clear as to what is taken as fundamental and what as defined. A(4) is very different, and it is obvious to anyone versed in school algebra that it makes no allowance for negative body, zero body, or infinity's tricks. Al-Kindî seems to *use* the idea of zero body, as will emerge. I am not sure of A(6), which seems to offer only psychological (analogical or metaphorical ?) support to his arguments in an unnecessary way at one point only, so I shall ignore it.[38] In sum, the list of postulates is already looking rather slim.

Al-Kindî's first 'theorem' is that there is no body I such that i(I). The proof begins with his supposing otherwise:

Let i(I), and let B be a finite body, that is, f(B). Let $J =_{def} I - B$.[†] (As before I am using the ordinary '+' and '−' of algebra to represent the addition to and subtraction from body by body. This whole idea is one involving an analogical component, at the stage of interpretation, at least.)

Suppose now that f(J). Then by A(5), f(J + B).

But $J + B = I - B + B = I$, therefore f(I), and by A(3) $\neg$ i(I), which contradicts the opening assumption.

Suppose then that i(J). Let $K =_{def} J + B$.[†]

Now by A(4) we can say that if $J = J$ and $K = J + B$ then $K > J$. (At this stage al-Kindî adds 'or $K = J$,' which will be considered later.) ϕ

The two conditions of the hypothetical are satisfied, thus $K > J$. (Or $K = J$, for al-Kindî.)

Since $i(J)$ therefore $i(K)$. (*Another axiom* is strictly required at this point, namely: If $i(P)$ and $Q > P$, then $i(Q)$. Here is an example of a philosopher accepting 'the obvious' uncritically.)

Al-Kindî now has two alternatives: either $K > J$ or $K = J$. First assume that $K > J$:

J is a portion of K. Let that portion be denoted by K_0, that is, $J = K_0$, wherefore by A(2) the dimensions of K_0 and J are the same. But to talk of such dimensions is to assume that $f(K_0)$ and $f(J)$, because it is to assume limits.[†] Since $i(J)$, this leads to a contradiction.

Suppose then that $K = J$:

By the definition of K, $J + B = J$. (He probably stops here, perceiving a 'whole = part' contradiction. I will finish off the proof from his principles.)

By A(4), if $J = J$ and $J = J + B$, then $J > J$, which is false.

The two alternatives having both been rejected, the supposition $i(J)$ is taken to be at fault. The opening supposition $i(I)$ must be rejected. The *immediate conclusion* is that *there is no infinite body*, that is to say, assuming such, we are led into nothing but contradiction. The *subsequent inferences* are that the universe is spatially and temporally finite. More about them shortly. First, some comments on the rather unusual proof I have just presented here.

The sentence marked thus ✝ is obviously important to the proof. It amounts to a prohibition on the expression of equality between infinites, and could have been tacked on to A(2). It is hard to see the point of al-Kindî's puzzling insertion of the alternative $K = J$ (see ϕ). If $K = J$ then we have added (speaking algebraically) zero B, and at once a loophole in A(4) is revealed. We know that the condition $K > J$ falls foul of the axiom A(2), for it is by invoking this that we dismiss it; but is that the secret purpose, so to speak, of A(2)? As it is expressed, this A(2) seems to rule out too much: it excludes expressions of equality not only between infinites but between what we might loosely call 'ring-like bodies.' (Aristotle has a caveat about these. One may go round a ring endlessly marking off its parts, but only by going over the same ground again and again. It does not count as infinite in the sense of 'that which cannot be gone through'.) But if we are to exclude equalities between infinites, we must exclude inequalities. There is scarcely a line of the 'proof' that al-Kindî can legitimately write down without violating his own axioms. In a nutshell, I am not particularly impressed by his 'axiomatics of infinity', but I know of nothing earlier that is more superficially thorough.

Al-Kindî's 'first proof' is by no means over, but from now on it looks more philosophically conventional. Time is quantitative, therefore there is no actually infinite time. Time is the duration of the body of the universe, which duration is therefore finite. *Proof*: There is no body without time, because time is the number of motion (in Aristotle's phrase). It is necessary only to show that there is body if and only if there is motion. If there is no body, then there is nothing to move; but if there is no motion, does that mean that there is no body? Al-Kindî thought he could show as much, by a process of elimination. Consider the idea of the universe at the instant of its creation, without motion. *Either* the universe was created *ex nihilo* (a form of generation which he says is a type of motion, something Aristotle would not have allowed), *or* the universe is eternal. (Is there no third possibility, such as creation meaning a stirring of activity in a hitherto motionless body? The possibility is not countenanced.) Al-Kindî eliminated the second alternative on the grounds that the eternal does not change, for it is fully actual, cannot become more perfect, cannot move. This tiny little argument is the nub of the whole question. Al-Kindî is rejecting the notion of an actually infinite time, where actuality has the sense of necessary in the *Metaphysics* — 'always in the same state,' as opposed to the possible, which can change and be other than it is.[39] There was no effort to interpret eternity as a potential infinity. As the only alternative worthy of retention we are left with the conclusion that body does not precede motion: if no motion, then no body. There is body if and only if there is motion, and there is motion if and only if there is time. Time is finite, and so, therefore, is the time-duration of body. It all looks more like a conceptual recommendation than a tight argument. It marks the end of al-Kindî's 'First Argument'.[40]

His remaining arguments are derivative and less interesting, but they are worth summarizing, for we meet them again and again. (They seem to have been mainly inspired by Philoponus.)

Time, which is finite, is duration counted by motion; motion is necessary for composition out of matter and form; composition is necessary for there to be bodies; and there are bodies. If time is finite, then motion is finite in duration; therefore so is composition (in duration); therefore so is body (in duration?). This is clearly subsidiary to the 'axiomatic' argument (making time finite). The last step is no less shaky than its foundations.

Before every segment of time there is another, back to the beginning of time, or otherwise any given moment of time could not arise. Suppose past time was infinite. There is a sense in which we *know* from the infinite past to the present, for we have arrived at the present. (This may not be convincing, but it is the best I can do with the argument . . .) If we know it, it is finite. Looking at the matter 'backwards', however, the duration from now to the infinite past is such that we can always go further back in time. It is an infinite duration. But the two periods must be equal, thus an infinite is equal to a finite duration; which is a contradiction. Ergo . . . .

You cannot reach the present time without first reaching a time before it; nor that before you have reached a time before it; and so on. If time were infinite, it would not be possible to arrive at the present, for an infinite time cannot be traversed. The present has indeed arrived. Ergo . . . .

This last argument ends with the corollary that body cannot exist outside time, therefore it too has finite duration. Also future time is got by adding finite time to past time. Therefore future time is finite. (See A(5).)[41]

CREATION AND BEING. GOD'S VOLITION

This long aside on al-Kindî's arguments against eternity and eternal creation has provided us with an example of a philosopher who did not need to fall back on overtly religious argument or authority to achieve his ends. Within the next century and a half, others in the Falsafah tradition were using different tactics from his, to another end. Al-Fârâbî and Avicenna used as a key to the problem a classification of sorts of existence, based on concepts of necessity and possibility. Briefly, one says that God is the only necessary being; possible beings *per se* are items subject to generation and corruption; and the heavenly bodies are possible *per se* but necessary *ab alio*.[42] The universe as a whole comes under the third heading: nature is a totality with a necessity that it draws from God, its first principle. This does not mean that, being derivative from God, it has temporal posteriority; and by glossing 'creation from nothing' as 'creation that is not from something', al-Fârâbî thought he could preserve the Plotinian theory of emanation. The world came, on this view, from the essence of God, that is not a 'something' (such as a coeternal matter would have been). Avicenna took the world to have been brought by God out of a state of non-being into a state of being (whatever that may mean), in a timeless sense, a sense of essential posteriority. Somehow these arguments for eternal emanative creation were presented as the true Aristotle.

Such ruses as theirs did not stem criticism. One of the arguments used against the idea that God created a contingent world in a temporal sense had been that this implied an inconsistency in the concept of God. A necessary, eternal, and *immutable* being would in some sense be *changing* in such a creative act in time.[43] Al-Ghazâlî (d. 1111) seems to have admitted the consistency of the notion of a God eternally bringing a universe into being, but insisted that there was nothing inconsistent in the idea of an eternal agent willing from eternity that the universe come into being at a particular moment. God's 'willing from eternity' does not, he said, imply any change in God.

Now this move by al-Ghazâlî might seem to be a case of a small-scale conceptual patching-up operation, resembling in this respect those I mentioned earlier, from Oresme, Buridan, Aquinas, and others. It is in fact on a much larger scale than those earlier examples. Al-Ghazâlî was a true 'demarcationist' in the sense that he drew a sharp distinction between the science of human volition and that of God's volition, to which second subject the theory of the world's creation belongs. Like Maimonides after him, and even Averroes, he stressed the danger in assuming there to be an analogy between God and man. When philosophers asked what reason God could have had for choosing one moment of time rather than another, to create the world, they were assuming, by analogy with the human case, that volitions were caused by motives. The argument was used in other connections by writers in the Kalâm tradition, when discussing why God should have made one particular choice (say of the number of the stars) rather than another.[44] Al-Ghazâlî accused his opponents of having misunderstood even the human case, and of having tried to combine incompatible philosophical perspectives.[45] The processes of natural causation, granted the right initial conditions, occur necessarily. Ascribing an act to a person, however, we do not consider it to be a natural effect, but the result of choice and desire, an expression of will. The theory of emanative creation, he believed, mistakenly mingled the two sorts of explanation: it gives God a power to act freely, but claims that the result of his creative act is a necessary one. (One could represent this as a consequence of trying to blend Plotinus with Aristotle: for Aristotle, 'that which always is is neither generated nor destructible'. A creative God was not Aristotle's problem.) Maimonides likewise complained about those who would mingle a theory of a world showing purpose and design and the free activity of its creator with that of a world determined according to natural laws:[46]

[Certain recent philosophers] assume that change in [God's] action or will is inadmissible. It is therefore clear that these philosophers abandoned the term 'necessary result', but retained the theory of it . . .

Needless to say, there is in none of this any suggestion of a 'two truths' doctrine in any stronger sense than that there are two theories in play, and one must choose the right one on a particular occasion. There is no suggestion that both work under the same circumstances.

According to Averroes (1126—1198, and thus roughly contemporary with Maimonides, who died in 1204), this was all misconceived. Eternal creation, he believed, is true to Scripture and to Aristotle.[47] The line of demarcation drawn by al-Ghazâlî between voluntary and natural agents was an unfortunate and mistaken one. 'Voluntary' takes on its meaning by contrast with the involuntary or forced sort of act that God is incapable of performing. We are by all means to draw a line between God and man, and their ways of acting; and having drawn it, we shall not be tempted to dismiss eternal creation, just because it does not conform with a theory of human voluntary action. In short, Averroes was a demarcationist, but of another sort, and in consequence able to make a different sort of conceptual recommendation.

THE CLASSIFICATION OF ATTITUDES

After seeing so many variants on doctrinal positions, we are perhaps in danger of losing sight of the main problem, which is that of the status of 'rational' proof in regard to a small number of cosmological theses that are at least in their wording straightforward. Unfortunately, the situation is very much more complicated than it is usually made out to be, and this is true even when we restrict our attention to a 'narrow' problem such as that of the eternity concept alone — overlooking the fact that views on its admissibility were for some coloured by the Plotinian 'combination' problem, that is, of coupling eternity and creation. To give some idea of possible complications in the classification of medieval attitudes towards demonstration, conceivability, and belief, I have represented ten reasonable doctrinal attitudes, on the accompanying diagram. Some of these attitudes might be combined. These positions result from the combination of ten points of view, as

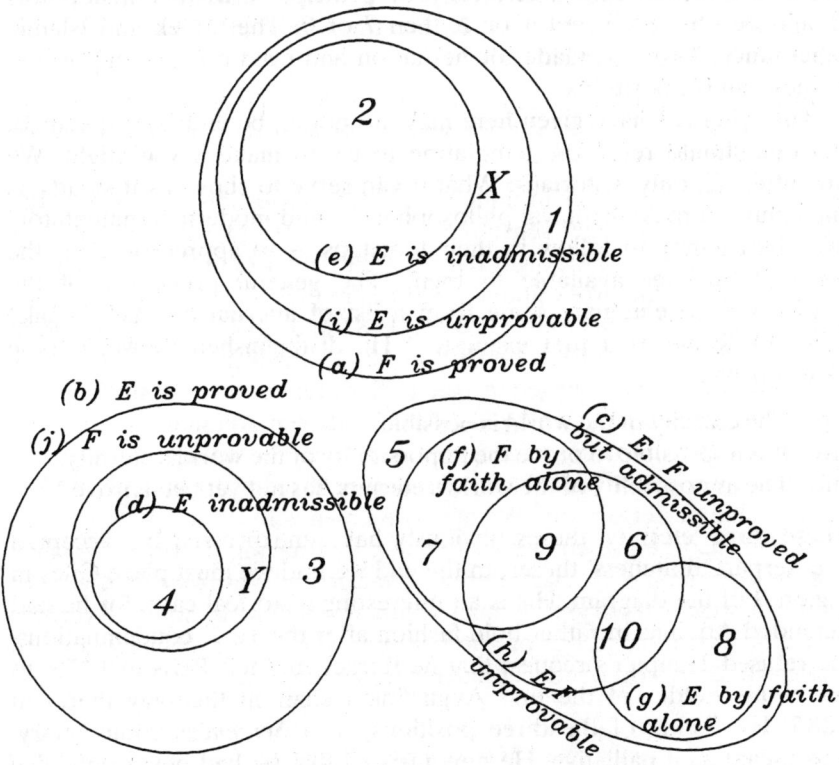

Fig. (a) A rational proof of the finiteness of the world's history (non-eternity; F on the diagram) is already available. (b) A rational proof of the world's eternity (E) is already available. (c) A proof of neither E nor F is as yet available, but both are conceptually admissible. (d) E is conceptually inadmissible. (e) E is conceptually inadmissible. (f) F is held to be true by faith alone.(g) E is held to be true by faith alone. (h) E and F are in principle unprovable by rational non-theological arguments. (i) E is in principle unprovable. (j) F is in principle unprovable.

one might loosely call them. (That there are ten on both occasions is a coincidence.) I will begin with them, lettering them (a) to (j). The ten doctrinal positions are numbered on the accompanying Venn diagrams 1 to 10. (Regions X and Y are ignored as uninteresting.) The overlapping more general 'points of view' are represented by the circular areas, and they may be more fully explained as in the caption of the diagram.

Here F is taken to be in principle unprovable where its contradictory is already proved. I follow the general view of the time, namely that E and F are indeed contradictories. The principal Christian standpoints should be obvious: if not 1 or 2, then 7 or 9. The Greek and Islamic inheritance, however, made for hesitation and even positive opposition to these simple positions.

The scheme I have given here may, no doubt, be endlessly qualified, but one should resist the temptation to try to make it watertight. We are, after all, only historians. What it can serve to show, as it stands, is the failure of most medieval philosophers — and modern commentators are often happy to follow in their footsteps — to appreciate even the main alternatives available to them. The general procedure of the period was to eliminate some from a list of alternatives. Let us take Giles of Rome as a first example.[48] He distinguished between these three positions:

(i) The eternity of the world is possible.
(ii) It is impossible to prove the impossibility of the world's eternity.
(iii) The impossibility of the world's eternity has not yet been proved.

These three 'eternity' theses obviously have unexpressed but accepted counterpart 'finiteness' theses. In the end we find we must place Giles in region 9 of the diagram. His is an interesting historical case, for he had defended Aquinas in rather bold fashion after the 1277 condemnations. He refused Tempier's request that he retract, and left Paris in 1278, to return (as holder of the first Augustinian chair in theology there) in 1285. His listing of the three positions, in a *Sentences* commentary, was meant as a palliative. He now insisted that he had never defended (i) or (ii), whatever impression he might have given previously. It was merely, as he explained, that the reasons so far devised for proving that the world is not from eternity 'do not seem to us to be proofs'. That the world might have been from eternity is a proposition he will be seen admitting only for the sake of argument ('*gratia disputationis*'). He shows, however, that he thought himself capable of refuting the principal arguments in favour of an eternal world — in particular Aristotle's.[49] At first sight, therefore, he seems to be a straightforward sceptic of type 5/7. On the other hand, he was hostile towards Averroes on the grounds that he had attacked doctrines held by those of the Christian faith. Giles obviously ended his career as of type 7, the position of an intellectual who wants to remain above suspicion of unorthodox views. It is *faith* that forces him to drop his flirtation with eternity. When one reads his enunciation of the errors of the philosophers, it is easy to confuse two sorts of error. Thus when he says that such and such a philosopher 'erred in holding to the eternity of motion', one might imagine that he is *asserting* the proposition 'motion began a definite period of time ago,' whereas he is primarily finding fault with the *proof* of the eternity of motion. In the end, however, it is spoken of as an error, *tout court*. But how can he say this, without a counter-argument? Faith gives us the assurance. Thus 'all [Avicenna's] errors arose from the fact that he did not clearly understand the way in which God acts in accordance with the disposition of His wisdom . . .'.[50] Giles has decided not to live dangerously any more.

Another test case might be Godfrey of Fontaines, a close contemporary of Giles, who believed, according to one recent account, not that the eternity of the world is possible, but that its impossibility is not demonstrable.[51] This comment is acceptable as far as it goes. All I wish to point out is that it does not serve to distinguish between five quite different points of view, represented by my 2, X, 6, 9, 10. Of course, consulting Godfrey's writings will allow the elimination of most of these. Here I am only concerned to point out the great complexity of what seems at first sight to be a simple Eternity-Creation debate. (Whether it is permissible to speak of a created but eternal world is one of another range of questions that I put aside for the moment.)

THE VIEWS OF AQUINAS. STRATEGIES OF ARGUMENT

Giles and Godfrey were both writing in the wake of Thomas Aquinas' commentaries and *De æternitate mundi*, and discussion has raged around the precise nature of Thomas' views, from the beginning. It is generally agreed that he was fairly consistent in maintaining that no-one had successfully disproved the world's eternity. (In the historical context, this was to take issue with such scholars as Bonaventure, John Pecham, and Henry of Ghent.) His writings on the subject cover roughly two decades — say 1253—1273. In his commentary on the *Sentences*, from the beginning of this period, he has it that reason can prove (with my present shorthand) neither E nor F. In brief, he falls into region 7 or 9, for F is held by faith alone.[52] In his *Summa contra gentiles* he refutes *particular* arguments for E, but treats arguments previously offered for F as probabilistic. A non-rational proof is possible, he says — that is, arguing on the basis of God's will. (I shall return to this point.) In *De potentia* he disallows proofs directed against E, and of course notes that there is no proof against F. An eternal being distinct in essence from God is said to be a conceptual possibility; but he does not say that such is a possible eternally created effect. In the *Summa theologiae*, to which I return, he repeats that actual arguments for E and F are inconclusive. F rests on faith. This, his greatest work, dates from around 1266 to 1268. A few years later, in his *Compendium theologiae*, he still refrains from saying that the eternal world is possible, but in his final contribution to the subject, his *De aeternitae mundi*, he has reached his boldest position. He finds no contradiction in the concept of eternal creation, and therefore judges it to be a possibility. (As we see, it is dubious whether we should have been putting him in regions 7 or 9 earlier; and even now the matter is open to dispute, thanks to certain textual problems. I follow J. F. Wippel's account for the final change of heart in *De aeternitate mundi*.)[53]

There are three separate matters here of great interest. One is the biographical question: why did Aquinas not defend the possibility of eternal creation earlier? It seems probable, following J. F. Wippel and I. Brady, that he had been stung by an attack by John Pecham.[54] The second point concerns the structure of Aquinas' argument — interesting because it is an attempt at a consistency proof of sorts. And thirdly, Aquinas draws a new and interesting line of demarcation, in the *Summa theologiae* (perhaps elsewhere too, but there it is particularly clear).

In the 'consistency proof', he wishes to show that introducing the concept of eternal creation will not lead to contradiction, from which it will follow that eternal creation is possible. He lists all obstacles he can find to the concept's self-consistency, and removes them. This is not, for us, a proof, but one would hardly expect him to have done more. Having got so far, by what right does he claim to pass from non-contradiction to the *possibility* of eternal creation? Again he lists the obstacles to possibility here. Perhaps eternal creation is impossible because God lacks the necessary power? No! God is infinite. Perhaps because it is *in itself* impossible? Here there are two meanings offered. First there might be no passive potency; second, it might be self-contradictory. He can see no other alternatives. Having eliminated the first alternative (since God can achieve His end without passive potency) we are left with the second. In brief: non-contradiction implies possibility; and he thinks he had removed all obstacles to non-contradiction.

So much for a rather crude summary of a substantial discussion. No less interesting is the earlier *quaestio* from the *Summa theologiae*. As I have said already, he was very much concerned with listing *existing* arguments and dismissing them, or trying and failing to do so. (He even made Aristotle out to be no more than a critic of *actual* arguments against eternity, rather than a positive believer in the idea! Robert Grosseteste had earlier scorned William of Conches for reading Aristotle in this way.[55] In his preamble to the replies to ten arguments listed in the first article of the *quaestio*, he writes:

> It is not necessary for God to will anything but himself, therefore it is not necessary for God to will that the world always existed. It is rather that the world exists as long as God wills it to do so, since its existence depends on [his] will as on its cause. It is not therefore necessary for the world always to exist, wherefore it cannot be demonstrably proved.

The statement is amplified in the responsions that follow, but the drift of his argument is clear, and it applies not only to the thesis E but to the thesis F: neither is provable by the rational arguments of philosophy; neither is provable by any human theory of how God's will operates. But finally, F is accepted on the grounds of faith. There is thus a three-fold division of the theologico-philosophical landscape, into faith, natural philosophy, and a theory of divine volition which is not within our grasp. (We do know a little about God's nature, and we know that he cannot break the law of non-contradiction.)[56]

COMPONENT ARGUMENTS

We have so far been mainly concerned with the strategy of argument, rather than with particular historically influential examples, but we have now seen how even so good a strategist as Aquinas was obliged, in the end, to fight minor battles in his systematic elimination of earlier claims to the truth. There are so many minor variations on actual arguments that it would be foolish to try here to give more than a cursory survey — but there is a good reason for giving it, namely that it illustrates further some of the points I have been making. It would not be desirable to stay entirely within Aquinas' lists, useful as they are.[57] Instead I list, somewhat anonymously, what seem to me to be the commonest arguments for the two main theses (E and F), sketching the reasoning very briefly. In square brackets there will be found a reference to an author where the argument may be found. (This does *not* imply that he agrees with its conclusion.) The key to names is at the end. To make it easier to refer to the arguments later, I give each a name, even though these names can give little idea of the precise nature of the arguments themselves.

The World is Eternal

E1. [*Motion.*] Motion is the fulfilment of the movable insofar as it is movable, and thus necessarily involves the presence of things capable of motion, things that either began or are eternal. The first motion of what began must have been caused. There cannot have been a time when there were only things at rest, unless they were in process of change previously; for rest, the privation of motion, must be caused, and therefore requires a previous change. All processes of change must be preceded by others, thus motion is eternal. (Similar arguments show motion to be imperishable.) And thus what is capable of being in motion is eternal. [A][58]

E2. [*Time.*] Time cannot exist and is unthinkable apart from the moment — a beginning of future time and an end of past time. Any past duration must begin with some moment, which must thus be an end moment for some earlier stretch of time. There must always be time. Now time is the number of motion, or is itself a kind of motion; therefore motion, like time, must be eternal. And thus what is capable of being in motion is eternal. [A][59]

E3. [*Prime matter.*] The prime matter common to the four elements is eternal; for had it come into being — that is, had it been provided with Form — from another substance, it would not have been prime matter, which is by definition a *formless* substance. If prime matter is eternal then the world is eternal. [M]⁶⁰

E4. [*Hylomorphism*] If the world began to be, it came to be out of matter, either without form — which is impossible — or with form — which amounts to a world before the world began, which is nonsense. [Ta]⁶¹

E5. [*Indestructibles.*] There are many indestructible things in the world. The mark of an indestructible is that it does not begin to be. Therefore the world did not entirely begin to be. [Ta]⁶²

E6. [*Possibility.*] If there was a time when the world did not exist, then its existence was necessary, possible, or impossible. Had it been impossible it would never have begun to be. Had it been necessary it could never have been non-existent. Had it been possible that the world be produced, then the world had this possibility before the production took place. One can say that there is a possibility that the substance receive a certain form. One can also say that the agent has a possibility of performing the act of production before actually doing so. In either case there must be in existence something of which the possibility can be predicated. [M]⁶³

E7. [*Void.*] Had the world come into being, where its body is now there was none present previously. But it is impossible that a void should precede the world. [iR]⁶⁴

E8. [*Nunc.*] What starts is not at its end; and what stops is not at its beginning. There is nothing in time but the now (*nunc*), the end of the past and the beginning of the future. Time can neither start nor stop, therefore. Time is the measure of change, and therefore change can neither start nor stop. [iR]⁶⁵

E9. [*Divine cause.*] God is the sufficient cause of the world, its final cause, its exemplar cause, and its efficient cause. Since he is eternal, so is the world. (An eternal action has an eternal effect, and God's action is his substance, which is eternal.) [iS]⁶⁶

E10. [*Perfection.*] God's actions, resulting from his perfect wisdom (his very essence) are perfect. The world must therefore be perfect and — since it cannot be improved — permanent. [M]⁷⁰

E11. [*Divine activity.*] God is not subject to accidents that could bring about a change in his will. It is impossible to imagine him as other than always active, just as it is impossible for us to imagine him as other than always in existence. His activity is to do with the world, which therefore has always existed. [M][68]

The World had a Beginning

F1. [*Efficient cause.*] If the world were eternal, so would be the generation of things (as of children by parents). But the causes of things can be neither an infinite series nor infinitely various in kind; therefore the hypothesis must be rejected. [Ta][69]

F2. [*Days.*] Had the world always existed, an infinity of days would have preceded today. But an infinity cannot, by definition, be traversed. Past time could not have been traversed on this hypothesis, which means that today would not have arrived — but it has clearly arrived. [Ta][70]

F3. [*Planets.*] From the infinite past, Saturn would have revolved an infinite number of times, Jupiter nearly three times as many, the Sun roughly thirty times as many, the Moon more than a dozen times that figure, and the stars in their daily motions more than ten thousand times as many. It is absurd to suppose multiple infinities — a ten thousand times infinity, for example. Thus the world had a beginning. [P][71]

F4. [*Completeness.*] The number of days from the beginning of the world to the present cannot be infinite because it is impossible to add to the infinite, which is by nature complete. [B][72]

F5. [*Succession.*] An actually infinite time is an impossibility, because it would be an infinite given whole whose parts coexisted, and with time this is out of the question. [A] Since God cannot create an infinite in act simultaneously, neither can he do so successively. [A][73]

F6. [*Ordering.*] A series of things cannot be put into an order (which implies a beginning, a middle, and an end) unless it can be found a *first* term. This applies, for example, to stellar revolutions, and the order of the seasons. It would not be possible to find a first term for any such temporal series if it were infinite. The time order must therefore have a beginning. [B][74]

F7. [*Historical.*] The rise and fall of human cultures and arts occur at definite epochs, which would not be the case had the world always existed. [Ta]⁷⁵

F8. [*Souls.*] The world was made for men, and thus never existed without men, who of course have souls that are immortal. Were the world to have existed from the infinite past, an infinite number of human souls would now exist. But this is impossible because the actual infinite is impossible. (The only escape routes from this conclusion are to posit metempsychosis — incompatible with the Aristotelian principle that every form be unique to one determined matter — or that there is one intellect for the whole human species. This is to be rejected as a suppression of individuality, of human responsibility, and of immortality.) [B]⁷⁶

F9. [*Creation.*] That which is made has a beginning to its duration. God produced the world (and holds it in being). Therefore the world began. [Ta]⁷⁷

F10. [*Supremacy.*] If there were no material creation *ex nihilo* at a finite time in the past there would be a *materia praecedens* sharing in God's infinity. But quite apart from this consideration, an eternal world would be God's equal in duration. Both conclusions are to be rejected. [Ta]⁷⁸

F11. [*Intelligences.*] The intelligences responsible for the revolutions of the spheres are finite. They more or less *know* those revolutions. If the latter are infinite, and the intelligences do not forget, then the finite can know the infinite. And there is no escape from this unsatisfactory inference by saying that one may know an infinity of causes or have an infinity of memories by a single idea, for the *effects* of these causes would be infinite. [B]⁷⁹

Key to sources. A — Aristotle; P — Philoponus; iS — Avicenna; iR — Averroes; M — Maimonides; B — Bonaventure; Ta — Aquinas.

COMMENTARY ON THE ARGUMENTS

First, a few general remarks about these arguments. The fact that there are eleven on each side is fortuitous. Some could have been grouped together, and some split into further independent arguments. Six are explicitly theological, but all have theological presuppositions of some sort behind them when they are put forward in the medieval context.

Most are, at least partly, based on Aristotle's own writings, despite the differences of opinion represented. Very occasionally, scholars added new distinctions to Aristotle's. One can hardly do justice to these briefly, but an allusion to general discussions of the infinite is called for. The scholastic infinite *in fieri* is more or less synonymous with Aristotle's potential infinite, and opposed to *infinitum in actu*, or *in facto*. Some spoke of the latter as a unique thing, a maximum, so great that there is no greater, so numerous that there are no more possible, or the like. Taking it in this sense scholars assumed that it was simply nonsensical to speak of differences of scale as between different infinites — as in the *planets* and *completeness* arguments, F3 and F4.

The four arguments with greatest bearing on the concept of eternal creation are E9, E11, F9, F10. I have already commented on E11, and the strong sense of demarcation in Maimonides' treatment of it. The reasoning behind E9 is very odd indeed, to modern eyes: on my previous analogy, it represents the partial breach of a boundary, that between theology and physics. It applies to theology an extension of the cause-effect relation: the quantity of the one must (in some sense) be that of the other; or, when there is no precise quantitative equivalence in the writer's mind, it is at least usually intended that the cause and the effect must be *alike*, whether ostensibly or in some covert way.[80] As Aquinas has it, in his reply to something like E9, 'sequitur effectus secundum exigentiam formae quae est principium actionis' — the principle of action is a form determining the effect, but *not* inevitably making the effect resemble the agent. 'In agentibus ... per voluntatem, quod conceptum est et praedefinitum accipitur ut forma quae est principium actionis' — it is the *concept* that is the operative form. One who sculpts a beautiful woman need not be beautiful, as he might have added by way of explanation.

The argument F9 brings us to the heart of the eternal creation debate. Are the two concepts 'caused by God' and 'always existing' incompatible? We have seen something of the structure of the treatment Aquinas gave to the problem in *De aeternitate mundi*, but not the details.[81] He there maintained that the whole question hinges on two principles: first, that a cause must precede its effect in duration; second, that non-being must precede being in duration, as when we say creation is from nothing. He rejected the first of these, after a short essay analyzing simultaneous and instantaneous causation (sadly lacking in examples!), and the second after a medley of arguments, partly reflecting back on what he has said about simultaneous causation (God being now the cause), and partly from authority (Anselm and Augustine). As I have said already, Aquinas could only justify his belief in the non-eternity of the world by recourse to faith. The world, as he wrote in the *Summa theologiae*, offers no proof in itself (*ex parte ipsius mundi*) of its newness (*novitas*). Of two supporting arguments, the second concerns Revelation, while the first is an extraordinary semantic plea. The specific nature of every object (i.e., the nature as member of a species) abstracts from the here and now (*ab hic et nunc*). On this account, universals are said to be everywhere and always.[82] Therefore it is impossible to show that man, or the heavens, or a stone, did not always exist! (Note the confusion of the timeless and the timed 'always'.) Note that there is not the slightest trace of an empirical argument. Aquinas was dealing with conceptual possibilities, and he thought it important that unbelievers be given no opportunity to laugh at the overenthusiastic faithful, with their impotent proofs.[83]

The *supremacy* argument, F10, is really two arguments rolled into one, and is usually found split into its components. But as Aquinas himself indicated, 'those who hold to the eternity of the world would agree that it was made by God from nothing in the sense that it was not made from anything — not that it was made *after* nothing, which is what *we* understand.'[84] (This, we recall, was al-Fârâbi's stance.) *Ex nihilo* creation might conceivably be sidestepped and the theologians placated, but what of a world sharing in God's eternity? Again the analogical element is of paramount importance. In the historical situation, where God's supremacy had long been associated with his eternity (not to mention his omnipotence, omniscience, and other things), it was natural enough that theologians should continue the tradition uncritically. But even if one accepts eternity as a defining property of God, one cannot offer a defence of the 'supremacy' argument without some further principle relating to what God may and may not share with his creatures. And this has very little to do with the concepts of infinity or eternity as such, at least in their 'potential' senses. Of course Christian philosophers of the Middle Ages had a special doctrine of time, according to which God's eternity is his very essence, and not something serving to measure God's duration.[85] This was generally supposed to be not appropriate to his creatures, and thus the argument F10 could not, strictly speaking even be enunciated. A comparison of man's duration with God's is strictly impossible — so here we have another clear instance of a demarcation between language games. (Where the angels were taken to stand, with their duration measured through *aevum*, a sort of nothing — 'aevum nihil est' wrote the hard-headed William of Ockham —[86] is something I hardly need consider here.)

Despite the general theological suspicion of emanationist philosophy, with its hint of God's eternity shared with his creatures, it was smuggled into the West in the baggage of neo-Platonism, as we have seen, as well as in Avicenna and Averroes, and is well known in the West in John Scotus Erigena and Nicholas of Cusa. The general run of emanationists held that God produced the world necessarily, that the world exists necessarily (as explained), that it is the best of all possible worlds, indeed the only one possible, and even contains all possibilities. Many of the condemnations of 1277 were directed against what were seen as implied limitations on God's power. The trouble was that the sort of necessitarianism that follows on talk of a 'supreme principle' does not fit well with the idea of a personal God for whom goodness and free-will are associated intrinsically. Aquinas was one of the leaders of the attack on the necessitarians, but he did not attack them on the grounds of their eternalism. In short, you could believe in an eternal universe without subscribing to the emanationist view. And yet again, you could be a necessitarian without being an emanationist. (The robot is not free when it creates the motor car, but it creates, in ordinary parlance, at least.)

In connection with the argument E9 we saw something like a union of physics and theology through the concept of cause. In a somewhat similar way we find Duns Scotus trying to bring the two into a relationship in his analysis of F1 — which argues for a beginning to the world from the impossibility of an infinite series of efficient causes. I need not repeat my account of his analysis of such a series, but note only how he argued (not very satisfactorily) for two series of causes, one 'accidentally ordered' and one 'essentially ordered'. He said that the first could be infinite only if we admit that the essentially ordered series has an end.[87] He was thinking of the old father and son series. Such could only be infinite 'granted something permanent which is not part of the succession,' something essentially prior to the series. He might be thought to be trying to distinguish between a hierarchy of laws, or explanatory grounds, for individual causal acts which may, as a concession, be in infinite series. But of course he aims really at finding *God* as the explanatory ground, *outside* the series. On the territorial analogy, there are two realms of discourse, but exchanges can take place across the frontier.

William of Ockham went more or less further down the same road, as far as his conceptualizing of the problem was concerned, although of course he criticized Scotus' proof as a whole. It was Duns Scotus' aim in his first part of the proof of God's existence to find a cause external to the series of contingent beings, a *producing* cause. It was Ockham's aim to make the external cause a conservator, in addition. He simply says that it is *self-evident* that every created being has to be conserved.[88] What of this *conservans*? If not produced by another, then it is the first efficient cause (in what is clearly no longer the traditional meaning of that phrase). If it is produced, then it needs another *conservans*; and so on. Either we stop with a first conserving-efficient-cause, or we accept an infinite series *of conservation*. Now the great advantage of this approach over that of his predecessors is that, according to the ordinary conventions, efficient causes do not *have* to be simultaneous with their effects, so the old argument which Aquinas and Scotus tried to bring, albeit half-heartedly, against simultaneous infinities of efficient causes, is not very convincing. *Conserving* causes, however, *do* have to be simultaneous with what they conserve. A string of such causes of something in existence *now* would be — if infinitely long — a present and actual infinity. This is tacitly accepted to be impossible, for traditional (Aristotelian) reasons. The conservation series must be finite! It must have a first term, namely God the unconserved conservator. Whether that is the sort of God his audience or later theologians wanted is not my concern. I will point out one final thing, however. Ockham has *not* got rid of the possibility of conventional efficient causes (or beings, in causal sequence) stretching back without end, 'to eternity.' He found his God in the present, and not by looking into the distant past.

It is commonly said that the effect of the 1277 condemnations was to reduce the number of topics in theology where rational demonstrations were to be expected. Whatever the evidence, it is interesting to recall how Thomas Bradwardine dealt with the question of the 'endless hierarchy in things,' half a century later. He simply laid down an axiom denying such a hierarchy. When he wrote on God — and the structure of the universe interested him, he said, only inasmuch as it throws light on the being of man and God — he began with these two postulates[89]: (i) God is the highest good, in comparison with whom nothing is better or more perfect; (ii) There cannot be an endless sequence or hierarchy in things and there must be a first cause in the chain of causes. He followed with a forty-part corollary, and argumentation outwardly axiomatic in form, but in reality often impressionistic and weak. Some of his contemporaries, less fastidious than Duns Scotus and William of Ockham, might have thought the second axiom to be a consequence of the first. As for the first, while the axiomatic style might seem to be the very antithesis of analogical argument, its use of concepts like goodness and perfection, of God and (tacitly) of man, leads one to think of God analogically. Bradwardine saw this as a danger to be resisted: there is a distance between God and his creatures, and anthropomorphism can only suggest limits to God's infinity.[90] At first sight, Bradwardine is a conservative theologian, whose dislike of this sort of analogy is symptomatic of a strong form of demarcationism — comparable with that of al-Ghazâli and Maimonides. (We recall their refusal to draw analogies between human and divine volition, for example.) It must be noticed, however, that Bradwardine was not averse to using the sort of rational (axiomatic) *methods* in his theology that he had learned in the more worldly sciences. In this he was at the scholastic end of a long succession that had begun in antiquity. He was like the Eurocrats who want to preserve national boundaries, but who want all national institutions to conform to a single plan.

Those who have searched for a clear statement of a two-truths doctrine have been unsuccessful. My general aim has been to show that they were searching for the impossible, but that to talk loosely of 'two truths' is understandable enough when one is faced with two different but seemingly overlapping conceptual schemes, and especially so when both are very firmly rooted in a single culture. I have not by any means explored all of my material, or indeed any of it as thoroughly as I might have done. I have singled out, though, a number of different ways in which ideological conflict was avoided in the context of the eternity question. Not all ways were wholly respectable. It is easier to be an unprincipled pacifist than a principled belligerent. As we know from the world around us, the ease of practising unprincipled belligerence falls somewhere between the two. I began by setting out a Wittgensteinian standpoint because it has much to recommend it; but I would point out, finally, that it is a defensive philosophy, and that it is not at all clear by what general criteria the frontiers between different realms of discourse are to be drawn — and withdrawn. There are different approaches to the problem of dissolving boundaries — by union, by federation, and by invasion, for example — but boundaries have a habit of reasserting themselves, as human needs change. Perhaps the pragmatic approach to the problem is the correct one. The average nuclear physicist, for example, feels about as strongly for the boundary between Aristotle and Holy Scripture as does the average Chinese about the boundary between ancient Rome and Carthage. Those who feel more strongly are called historians — except by certain other sorts of historian.

Rijksuniversiteit Groningen
Groningen, The Netherlands

NOTES

1. E. Gilson, *History of Christian Philosophy in the Middle Ages* (Sheed & Ward, 1955), p. 402.
2. P. Mandonnet, *Siger de Brabant et l'Averroïsme latin au xiiime siècle*, IIme partie: textes inédits (Louvain, 1908), pp. 175—6. The Biblical references are to I. Cor. i, 19.
3. *Op. cit.*, p. 398.
4. *Ibid.*, p. 725.
5. *Siger de Brabant* (Louvain, 1942), Vol. II, p. 694.
6. For a summary of several modern attitudes to these questions, see van Steenberghen, *op. cit.*, especially pp. 690—700; and for later bibliography the article on Siger in the *Enciclopedia Dantesca*.
7. R. Arnaldez & A. Z. Iskandar, art. 'Ibn Rushd,' *Dictionary of Scientific Biography*, Vol. XII (1975), p. 6.
8. John of Jandun, as quoted by Gilson, *op. cit.*, pp. 523—4, says this in as many words. Gilson detects here a sense of humour, and no doubt John did have a marked sense of irony, although what he says does mirror contemporary values.
9. Gilson, *loc. cit.*
10. *Summa Theologiae*, Ia, 1, 1, ad 2.
11. *Expos, de Trin.*, ii. 3.
12. Ch. 5 (ch. 3 in the medieval Latin versions); 1009 a6—g15.
13. *The Cambridge History of Later Medieval Philosophy*, ed. N. Kretzmann et al., (Cambridge, 1982), pp. 266—7 (Jan Pinborg).
14. As Godfrey of Fontaines said. See J. F. Wippel, *The Metaphysical Thought of Godfrey of Fontaines* (Washington: Catholic University of America Press, 1981), p. 33.
15. See 'Intimations of Cosmic Unity? Fourteenth-century Views on Celestial and Sub-Lunar Motion,' in my *Stars, Minds and Fate* (London: Hambledon Press, 1989),

pp. 301—312. The Oresme reference is *Le Livre du ciel et du monde*, ed. A. D. Menut & A. J. Denomy (Madison: Univ. of Wisconsin Press, 1968), book I, ch. iv, lines 36—8.
16. North, art. cit., note 15.
17. *Ibid.*
18. Note o, p. 38, of Vol. X of the Blackfriars edition of *Summa Theologiae*, ed. W. A. Wallace (corresponding to Ia 66, 2.)
19. Wippel, *op. cit.*, p. 152. I certainly cannot claim to do justice here to Godfrey of Fontaines.
20. See B. Goldstein, 'The Arabic Version of Ptolemy's Planetary Hypotheses', *Trans. of the Amer. Phil. Soc.*, n.s., 57, no. 4 (1967), pp. 3—55. Note that although the Ptolemaic solution was known via other authors (e.g. Alfraganus) to Western astronomers in the Middle Ages, they were unaware of his authorship of the combined system, with each planet's sphere thick enough to hold the necessary geometrical apparatus (epicycle, deferent, etc.).
21. Duhem's work first appeared in the *Annales de philosophie chrétienne*, in five parts. It has often been reprinted, beginning with an edition from Hermann et Fils (Paris, 1908). For Aquinas, see *Expositio super librum de Caelo et Mundo*, II.17. The sentiment is repeated in *Summa Theol.* Ia 32, 1.
22. F. M. Cornford's translation, *Plato's Cosmology* (London, 1937), p. 99.
23. *Timaeus*, 52A.
24. *Phys.*, VIII. 1; 251b 18—28.
25. *Op. cit.*, p. 102.
26. *Confessions*, xi. 13 [xi.XI]: 'In the eternal nothing passes, the whole is present together [*omnes simul stant*].' In xi.15 [xi.XIII] he retales the Platonic doctrine of God as the creator of time (with the creation of heaven and earth).
27. Sense (4) presents a problem of Aristotelian exegesis. At *Phys.* IV.12 (221b 25—222a 9) he explains containment in time (boundedness by times) in a way that suggests that he might have held that something eternal, being unbounded by times is 'outside time'. But so far as I can see, he nowhere draws this conclusion. *Parts* of the history of the eternal thing are, after all, bounded.
28. *De Caelo*, 284a 24—26; *Met.* 1050b 24.
29. This statement is a version of what A. Lovejoy called the principle of plenitude, various equivalent forms of which have been explained by J. Hintikka in an article reprinted in modified form as ch. 5 of his *Time and Necessity* (Oxford Univ. Press, 1973). The basic version is: *Each possibility must be realised at some moment of time*. Hintikka, arguing that Aristotle has two notions of possibility, distinguishes them as *possibility* and *contingency* ('contingent' meaning 'possible' but not necessary,' that is, 'possible to be and possible not to be'). From the basic principle it follows that *what is eternal is so by necessity*. (If it were other than necessary, that is, possibly nonexistent, then by the first principle there would be an actual moment when it did not exist.) Other derived principles are: *nothing eternal is contingent* and *that which never is is impossible*.
30. These statement are all taken from *Met.* 1050b.
31. *Phys.* III.4; 203b 29.
32. 206a 21.

33. 206a 20—35.
34. 208a 20.
35. J. Hintikka puts the case very clearly in another article, reprinted in virtually its original state as ch. 6 of his *Time and Necessity*.
36. F. E. Peters, *Aristotle and the Arabs* (New York Univ. Press, 1968), pp. 152—3.
37. W. L. Craig, *The Kalām Cosmological Argument* (Macmillan, 1979), p. 23. I have introduced my own compromise logical notation.
38. *Ibid.*, p. 24, 2(c) (i) (a) (ii).
39. 1015a 33—36; 1026b 27—29.
40. Craig, *loc. cit.*, omits the half of the argument from motion to body saying that he is omitting the other half! See his n.67. Although the axiomatic approach is uncommon, there are philosophers who approximate to it, but less closely than al-Kindī, and of them Maimonides is worth mentioning. In Part II of his *Guide for the Perplexed* he set out 26 propositions employed by other philosophers, propositions that he accepted provisionally in order to see what could be deduced from them. In fact only the last of them was for him a matter of hypothesis: 'Time and motion are eternal, constant, and in actual existence.' He thought it admissible but not demonstrable ('as the Aristotelian commentators assert'), nor, on the other hand, impossible ('as the Mutakallimūn say'). See p. 172 of the translation of *The Guide* by M. Friedländer (London, 1904), where Maimonides recommends to his readers to follow Moses' teaching of creation *ex nihilo*. He is going to show that whether or not we believe in creation *ex nihilo* or eternity, he can prove the existence of God (ed. cit., pp. 149—54). As he notes, the Mutakallimūn assume 'as an axiom' that it is impossible to conceive how an infinite number of things could come into existence, even successively.
41. I am not concerned here with al-Kindī's Plotinian-style arguments for God the True One. He assumed that the universe must be caused. (It could not cause itself to exist; therefore it and the multiplicity in it have a Creator-cause . . .)
42. For further bibliographical references on this subject, see S. Feldman, 'The theory of eternal creation in Hasdai Crescas and some of his predecessors,' *Viator, 11* (1980), pp. 291—3.
43. Feldman, *ibid.*, p. 294.
44. *Ibid.*, pp. 296—7.
45. *Ibid.*, p. 297.
46. Maimonides, *op. cit.*, p. 191.
47. S. van den Bergh, *Averroës' Tahāfut-al-Tahāfut* (London, 1954), discussion 1, paras. 65—8 and discussion 3, paras. 147—72.
48. J. F. Wippel, 'Did Thomas Aquinas defend the possibility of an eternally created world?,' *Journal of the History of philosophy, 19* (1981), pp. 21—37, especially p. 22 (for Giles of Rome).
49. Giles of Rome, *Errores Philosophorum*, ed. J. Koch; trans. J. O. Riedl (Marquette Univ. Press, 1944), p. 15 etc.
50. *Ibid.*, p. 29.
51. Wippel, *op. cit.*, (n. 14 *supra*), ch. 3.
52. For further bibliography on commentaries and opinions on Aquinas, see Wippel, art. cit. (n.48 *supra*). The reference to the commentary on the *Sentences* is Book II,

d.1, q.1, art.5. The references for the remainder of my paragraph are: *Summa contra gentiles*, II. cap. 31—8; *De potentia*, q.3 art. 17; *Summa theol.*, Ia 46, 1—2; *Compendium theologiae*, cap. 98—9.
53. Wippel, *ibid.*, passim.
54. *Ibid.*, p. 37.
55. Roberti Grosseteste ... *Commentarius in VIII libros Physicorum Aristotelis*, ed. R. C. Dales (Univ. of Colorado Press, 1963), p. XX. (Maimonides — p. 176, *ed. cit.*, namely at II. xv — claims that Aristotle was well aware that he had not proved the eternity of the world!) In *Summa theol.*, Ia 46, 1, reply, Aquinas refers to *Physics*, VIII.1, *De Caelo*, I.10, and *Topics*, I.9. For some of the vast literature on Aquinas see Wippel (n. 48 *supra*), n. 3.
56. Wippel, *ibid.*, n. 45.
57. In *Summa theol.* Ia 46, 1—2, Aquinas presents two sets of arguments and the replies to them. Of the first ten arguments (for the thesis that the universe of creatures has always existed) the first eight have a look of Aristotelian physics about them, and the last two of schools' theology. They are mostly aprioristic: taking Aristotelian concepts for granted, they tend to show that an assumption of non-eternity leads to an analytically false conclusion. The same goes for the replies; and for the following claim that the world's beginning is demonstrable (eight arguments); and for his replies to those; and so on.
58. 251a 8—251b 9.
59. 251b 10—28.
60. Friedländer trans. (n. 40, *supra*), p. 174, no. 2.
61. *Summa theol.*, Ia 46, 1, 1.
62. Ia 46, 1, 2.
63. *Op. cit.*, no. 4.
64. In *Phys.* VIII. 11.
65. *Ibid.; cf. Summa theol.* Ia 46, 1,7.
66. *Met.* 9.1, 9 and 10. *Cf. Summa, theol.* Ia 46, 1, 9.
67. *Op. cit.*, no. 7.
68. *Ibid.*, no. 6. I have added the last sentence, which seems to be implied.
69. Ia 46, 2, 7. *Cf.* Aristotle's *Met.*, 994a 1—.
70. Ia 46, 2, 6.
71. Simplicius *In Phys.* 1179, 15.
72. Bonaventure, *Comm. Sent.*, Bk. II, dist. i, p. 1.
73. Aristotle, *Physics*. Book VIII. 5.
74. *Op. cit.* (his second argument).
75. *Summa theol.*, Ia 46, 2, 4.
76. Bonaventure, *op. cit.* (his fifth argument).
77. Ia 46, 2, 1.
78. Ia 46, 2, 2.
79. *Op. cit.* (his fourth argument).
80. *Met.*, Book 8 (theta).
81. *De aeternitate mundi*; a translation from the text of J. Perrier (1949) is included in Appendix 2 of *Summa Theologiae*, vol 8, ed. T. Gilby, London, 1967. See pp. 152—157.
82. *Post. An.*, I. 31 (87b 33).

83. *Summa theol.*, Ia 46, 2 ad 1.
84. *Ibid.*, Ia 46 2 ad 2.
85. L. Baudry, *Lexique philosophique de Guillaume d'Occam* (Paris, 1957), p. 19.
86. *Ibid.*, p. 20.

    Within the cosmology of Thomas Aquinas there were certain actions that could take place without motion on the part of the agent. The sun, for instance, a corporeal thing, the measure of time — but itself unmeasurable by time was supposed responsible for a flux of light without first being unable to act and later being able to act. [*De subst. sep.*, cap. IX, sec. 51—2.] God, the supreme Being, was supposed to infuse *esse* to pure intellective (angelic) creatures in the same way, to the extent, that is, that God is unchanging and unmoving and not to be measured by our time. It did not seem possible to suppose that this influx might be measured by our sort of duration. There must, thought Aquinas and others, be a type of duration proper to the angelic essence, midway between time (our time) and eternity (God's time).

    It is not clear that Aquinas considered this threefold measure of time enough, however. Matter has passive potency. God has active potency. The angels must be somewhere between the two extremes. Might they not approximate to God's actuality in varying degrees? In fact Aquinas graded his angels in such a way that those closest to God, with the least capacity to receive additional perfection, that is, the least capacity for change, participate to a higher degree in God's eternity than those nearer to man. For the subject generally, see H. P. Kainz, *Active and Passive Potency in Thomistic Angelology* (The Hague, 1972).
87. John Duns Scotus, *Philosophical Writings — A Selection*, tr. A. Wolter (Indianapolis: Bobbs-Merrill, 1975), (First edn., 1962, Nelson), p. 47. *Cf. De primo rerum pincipio*, chs. 1 and 2.
88. D. Webering, *Theory of Demonstration According to William Ockham* (New York: St. Bonaventure, 1953), pp. 103—6 for precise references.
89. *De Causa Dei.* see H. A. Oberman, *Archbishop Thomas Bradwardine, a Fourteenth-Century Augustinian* (Utrecht, 1958), pp. 50 ff.
90. Oberman, *loc. cit.* Note how strongly analogical is this 'divine activity' argument, E 11. It would in any case have been thought fallacious, surely, by any serious commentator (I have seen only one commentary on it), since a concealed premiss is that God's activity *only* concerns the world. Maimonides fastens on the doctrine of the Active Intellect, the incorporeal being that acts (according to the Aristotelians) not continually, but only at certain times. The argument E 11 is hardly one we need consider further, since here we have the testimony of two groups of philosopher-theologians, one of which can conceive God interrupting his worldly activity, while the other cannot. Beyond this point the theologians must take over.

    Bradwardine made a number of points with a bearing on the sort of conservation question raised by Ockham, and one wonders whether perhaps Leibniz had read the passage where Bradwardine asks whether God leaves the world to run on, after creation, without interfering in secondary causes. I ought to add that Leibniz was one of the intermediaries between that long series of scholastic writers on the eternity question, and Kant — whose First Antinomy retraces much conventional ground.

MICHAEL SEGRE

SCIENCE AT THE TUSCAN COURT, 1642—1667

INTRODUCTION

The scientific revolution of the seventeenth century was rooted in the Italian Renaissance, and was undoubtedly initiated by Galileo. Yet, after Galileo's death, Italy ceased to be the main centre of this revolution and the focus of scientific activity moved to the other side of the Alps.

What was the cause of this scientific decline? The impression one gets from reading the classical literature is that Italian science was paralyzed by the restrictions imposed by the Church after Galileo's trial.[1] This view is by no means false, though relatively little attention has been given to the work of Galileo's followers in Italy after his death, and to the social and intellectual context of their work. Whereas the picture of Galileo's trial is, by now, more or less clear, little has been added concerning the effects of this trial on the generations of Italian scientists who came after him and should have extended and spread his work.[2] Galileo's direct collaborators and pupils, such as Cavalieri, Torricelli, Viviani and Borelli, left a large amount of published or unpublished works which have not been thoroughly studied. Can these works reveal anything new on the decline of science in Italy during the second half of the 17th century? Do they strengthen or weaken the view that it was the Church which was mainly responsible for the decline of Italian science in Italy after Galileo's trial?[3]

In this article I propose to concentrate on Tuscany, Galileo's homeland, where a number of his followers pursued the research program of their teacher and succeeded, for a period of twenty-five years, in making important contributions to seventeenth century science. This work was supported by the Tuscan ruling family, the Medici, — an enterprise which undoubtedly required a certain degree of diplomacy when handling issues that might arouse the objections of the Church. I propose to outline and analyze briefly the salient points in the history of Tuscan science in this period. I wish to support the view that the role of the Church in causing the decline of science in Italy has been exaggerated, and to argue that in Tuscany the Medici — contrary to what is generally claimed — were at least partially responsible for this decline. I do not intend to claim that the ecclesiastical suppression had no prime role, but only to point out that after Galileo's trial the Church, practically speaking, stopped interfering with the work of the Tuscan scientists, and the control of scientists was enforced by the Medici, who restrained their work in a way that was sometimes exaggerated.

My claim is to a certain degree conjectural, but relies on the reading of much of the available material concerning the scientific interests of the Medici after Galileo's death. A large part of this material is contained in the Galilean collection of MSS in the National Library in Florence, in particular the twelve — still unpublished — volumes of correspondence of the *Accademia del Cimento*, the Tuscan official scientific academy active between 1657 and 1667.[4]

IN THE WAKE OF GALILEO

Galileo's sentence, pronounced by the Roman Inquisition on June 22, 1633, condemned him for having violated Cardinal Bellarmine's injunction of 1616 ordering him 'to relinquish altogether the said opinion that the Sun is the center of the world and immovable and that the Earth moves.' The sentence was to be sent to all Apostolic Nuncios and Inquisitors, 'especially the Inquisitor of Florence, who shall read the sentence in full assembly and in the presence of most of those who profess the mathematical art.'[5]

Galileo's trial, however, did not put an end to his scientific work. On the contrary: whereas in the three preceeding decades he occupied himself mainly with politics of science, Galileo's nine last years were devoted to pure scientific research, and the publication in 1638 of his *Two New Sciences* marked the climax of his scientific career. Galileo was then under house arrest, over seventy years old, ill and blind, and was able to complete his last remarkable achievement, at least in part, thanks to the help of such devoted followers as Vincenzio Renieri — the professor of mathematics at the University of Pisa — and his two young assistants, Evangelista Torricelli and Vincenzio Viviani. The Medici contributed to his work by continuing to support him and supply all his needs.[6]

Scientific activity under the patronage of the Tuscan Court went on also after Galileo's death in 1642. The Tuscan Grand Duke, Ferdinand

II, appointed Torricelli to the post of court mathematician and encouraged his work. In the five years during which Torricelli served at the Tuscan court, he made some important mathematical discoveries, and in 1644, with Viviani's help, performed the famous barometer experiment. In general, neither Galileo nor his followers were free to do what they wanted; but with due care — e.g. by refraining from upholding the Copernican view in print — they could continue working freely, even in astronomy. Both Torricelli and Renieri, at least, occupied themselves undisturbed with astronomy.[7]

Galileo's followers, of course, had to be prudent, and neither Torricelli nor Renieri did, in fact, publish their work in astronomy. Nor did Torricelli publish the description of the barometer experiment, which contradicted the Aristotelian theory of the plenum, and — by confirming the existence of the vacuum and implicitly confirming the existence of atoms — could have had far-reaching theological implications relating to transubstantiation, thereby arousing serious objections. But even these obstacles could be overcome: instead of publishing his experiment, Torricelli sent a letter with its description to Michelangelo Ricci, a follower of Galileo in Rome and a future cardinal whom he could trust and who corresponded regularly with many scientists in Europe. Ricci knew how to deal with such a delicate case and could spread the news.[8]

Thus, at least in the five years that followed Galileo's death, science in Tuscany under the Medici patronage continued to produce important results. The Medici, for their part, continued not interfering with the work of their scientists, with one possible exception, related to Torricelli's official title. Whereas Galileo was appointed 'Mathematician *and* Philosopher' to the Grand Duke, there is no document to show that Torricelli was officially given a title. True, he may have held the title of 'Mathematician,' as appears on the inner front page of his *Opera geometrica* — his only work published during his lifetime. Although we do not know what Torricelli wanted and eventually asked for, nor what the court's pleasure was, this seemingly negligible semantic detail cannot be overlooked when one considers how strongly Galileo had insisted on his title. Paolo Galluzzi suggests that the Grand Duke may have withheld the title of Court Philosopher to pretend that he was separating mathematics from philosophy, so as to meet the instrumentalistic request of the Church that a mathematical theory cannot describe physical reality. If this was interference, it was a minor formality, dictated by the special conditions of those days, most probably to protect Torricelli. But, as we shall see, the interference grew beyond proportion in the years that followed Torricelli's death in 1647.[9]

THE ACCADEMIA DEL CIMENTO

Ten years after Torricelli's death, in 1657, the Medici — Grand Duke Ferdinand II and his brother Prince Leopold — founded the *Accademia del Cimento*, to promote scientific investigation in Tuscany. The *Accademia* was active for ten years, run by the hard-working Prince Leopold, and investigated the fields that occupied Galileo: physics and astronomy. It soon became very famous and was regarded as being on the same level as the French *Academie Royale des Sciences* and the English Royal Society of London. It is surprising, therefore, to see that some of its aspects were unusual for an institution of this type, specifically its formal structure, its membership and above all, its official publication, the *Saggi di naturali esperienze* (here briefly labelled *Saggi*), published in 1667.[10]

The *Accademia del Cimento*, in fact, had no official statutes or regulations, calling to mind the non-existence of a written agreement on Torricelli's official status. The Lincean Academy, which was active in Rome during the first half of the century, already had a detailed declaration of principles which regulated the conduct of its members down to such minute details as the length of their meetings.[11]

There is no official list of the members of the *Accademia del Cimento*. We know some of the names from the Academy diaries, which are still unpublished, but there may have been other members who are not mentioned. The *Accademia del Cimento* used to meet in a private and informal way in different places, usually depending on the whereabouts of the prince, and its activity was interrupted for long periods. The massive interference of Prince Leopold — himself an amateur scientist — in its work at all levels and throughout its existence, is amply testified by correspondence. Little room for any decision-making was left to the scientists.[12]

The best mind and the driving spirit of the Academy was one of Galileo's leading followers, Giovanni Alfonso Borelli. Borelli came from the South of Italy and by the fifties of the century was regarded, both in Italy and abroad, as the best scientist in the peninsula. In 1656 he was offered the chair of mathematics at the University of Pisa, and invited to join the *Accademia del Cimento*.

Together with Borelli, one finds in the Academy Galilean scientists like Viviani and Candido del Buono, physicians like Francesco Redi and Antonio Oliva, Aristotelian philosophers like Alessandro Marsili and Carlo Rinaldini, and probably even a classicist, Carlo Dati, who had little to contribute to natural sciences. One wonders how these widely different intellectuals could have collaborated. And, indeed, the relations in the *Accademia del Cimento* became tense and often degenerated to personal quarrels.[13]

Probably the strangest thing about the *Accademia del Cimento* is the official account of its work, the *Saggi*. This book, written by the secretary of the Academy, Lorenzo Magalotti, under the strict supervision of Prince Leopold and purporting to summarize ten years of the Academy's work, is in many ways an unusual 'scientific' publication, even by the standards of those days. It presents a rather restricted sample of experiments, considering that the work of the Academy in physics and astronomy fills no less than 34 volumes of manuscripts, mostly still unpublished.[14] All but one of the experiments are in physics (the last experiment described being on the digestion of some animals), with very little theory — at most a reference to, or a quotation from, a philosopher (Plato, Plutarch, Gassendi, Galileo, Boyle and a few others). There is no indication by whom and when the experiments were carried out, and almost no attempt to comment on the results. The *Saggi* also totally ignore any work in astronomy, although the *Accademia* obtained some important results, confirming, *inter alia*, the hypothesis that Saturn was surrounded by a ring or that comets are not sublunar events.[15]

The *Saggi* seems to have been published first and foremost to impress the reader with form rather than with content. Even for the Baroque period the book is grossly overdecorated and gave the publisher an enormous amount of work. Targioni Tozzetti, the eighteenth century Florentine historian of science who republished this work, comments, not without scorn, that 'the woodcut initials, the decorations and the tailpiece are too large, suitable for a choir book and unsuitable to the text of the page.' Henry Oldenburg, the secretary of the Royal Society, on receiving a copy of the *Saggi*, remarked in a letter to Boyle that the book was "pompous" (in the 17th century "pompous" usually meant magnificent), adding that there was nothing new in it except for a few experiments on amber.[16]

Why did the Medici gather together such a heterogeneous group of intellectuals? Why no formal structure? Why did the Academy publish only a partial description of its work? And why such a bizarre publication? Was it because of the pressure exerted by the Church?

SCIENCE, THE CHURCH AND THE TUSCAN COURT

The Medici may of course have acted the way they did to avoid a controversy with the Church. Prince Leopold might have been trying to disguise the Academy's Galilean approach by introducing into the Academy scientists with different tendencies. Within the Academy, Marsili and Rinaldini, the two Aristotelian philosophers, appear to have formed a pole opposite to that of the Galilean Borelli. In a way, the *Accademia del Cimento* was structured rather like Galileo's *Dialogue*, in which one of the interlocutors, Simplicio, was an Aristotelian philosopher; perhaps here, however, the Aristotelians had more consolidated positions.[17]

The *Saggi* were probably so strange and partial because committing controversial topics, such as astronomy, to writing would have been asking for trouble. A description of 'neutral' experiments, detached from any theory or possible controversy, was much safer. This painful choice prevented the members of the *Accademia* from receiving some of the credit they deserved, but it did eliminate the danger of a clash with the Church and permitted the *Accademia del Cimento* to carry on its work *sub rosa*. All in all, had it not been for this policy, Galileo's disciples might have been persecuted, and from this point of view the prince, who ran the Academy, could be regarded as an enlightened supporter of Galilean science who not only contributed to its encouragement but also had the ability to preserve it.

All this may be true; yet the interaction between the Church and Galileo's followers was much more complex than the way it is generally presented. The Church is generally presented as being an anti-Galilean monolith, and it is seldom pointed out that most of Galileo's followers were ecclesiastics. Castelli, Galileo's major follower was a Benedictine; Cavalieri, a Jesuate (not a Jesuit); Vincenzio Renieri, an Olivetan monk; Ricci became a Cardinal. (Torricelli and Viviani were not ecclesiastics, but had been trained by the Jesuits.) Hence it was not the Church as a body, but more specifically some components of it that may have interfered in the work of Galileo's followers. Galileo's main opponents were either Jesuits or members of the Inquisition orders, the Dominicans and the Franciscans, and very little research has so far been conducted on the archives of these orders, so that one cannot say to what extent they indeed interfered in the work of Galileo's followers. The remaining evidence — that Galileo's followers were intimidated by the 'Church' — is rather restricted. Let us have a look at the type of evidence that is available.[18]

There is one documented case of plain persecution of a follower of Galileo, that of Viviani's tutor, the mathematician Father Clemente Settimi, of the *scuole pie*. Settimi had been imprisoned for a short time in 1641 by the Roman Inquisition because he was suspected of having read Galileo's forbidden *Dialogue*.[19]

The other evidence I have been able to find is rather shaky. There is, for instance, a letter written in 1665 by Borelli to Prince Leopold de' Medici, proposing the publication in France of some observations that the former had made on the comet that appeared at the beginning of the same year, 'so that people on this side of the Alps [meaning Italy] may see the free way of speaking in assemblies of Jesuits and other men of letters, and how everybody speaks of the Pythagorean [i.e., Copernican] system, so that the sentence may become acceptable and less frightening.' Although this letter indicates that in the sixties of the century some Italian scientists were afraid, it also implies that in Borelli's view there was no real danger.[20]

Another instance, related by Targioni Tozzetti, concerns Viviani who, 'being rightly afraid of some search, kept all the writings of Galileo, of his disciples and of his correspondents in a pit in his house.' Yet Antonio Favaro, who collected Galileo's works and papers, says that it is not at all clear that fear was the reason for placing the writings in the pit. Targioni Tozzetti, in his great admiration for Galileo's struggle for Copernicanism, may have exaggerated a little.[21]

Perhaps somewhat sounder evidence comes from a letter written in 1658 by Carlo Rinaldini, the Aristotelian member of the *Accademia del Cimento*, to Prince Leopold warning that in view of the publication of the *Saggi* 'the Jesuits are making a din before the time; they say that if that book of natural observations contains anything concerning some of them, they will have the right people to answer.' Did this, however, relate to the Inquisition and the 'forbidden science' or were the Jesuits just a competing school? Paolo Galluzzi, who published part of this document, suggests that Rinaldini, who was basically anti-Galilean, may have written this letter, *inter alia*, in an attempt to hinder the work of the Academy.[22]

Admittedly there is evidence of growing objections to Galileanism all over Italy toward the end of the century. In 1671, the mathematician and astronomer Honoré Fabri (himself an Inquisitor) was tried by the Roman Inquisition and sentenced to fifty days imprisonment for having suggested that should a proof of the earth's motion be found, the relevant passages in the Bible could be interpreted more symbolically. Later, between 1688 and 1697, some members of the Academy of Investigators and Atomists in Naples were tried as atheists. And in 1693, Antonio Baldigiana, Consultor of the Holy Office, informed Viviani that congregations were held in Rome against modern physics. This wave of repression did not spare Tuscany, but again, it was enforced by the Medici: it was Grand Duke Cosimo III, son of Ferdinand II, who prohibited private teaching 'in writing or in voice' of the Democritean or atomist philosophy, though it is true that much the same might have happened without Cosimo's intervention. The fact remains that, as far as Tuscany before 1667 is concerned, there seems to be no evidence that Galileo's followers were persecuted or intimidated.[23]

The list of instances that I have presented may not be complete, but the fact that I was unable to find any additional ones could indicate that the danger was in fact smaller than is generally assumed. To what extent, then, was the policy of self-censorship imposed by Prince Leopold justified?

Borelli's published writings indicate that Leopold's caution was to a great extent unnecessary. Nothing, as a matter of fact, happened to Borelli when he published the astronomical observations he made during his stay in Tuscany, which were part and parcel of the work of the *Accademia del Cimento*. He published two works, one on the comet of 1664–65, and the second on Jupiter's moons. The former, published under a pseudonym, was based on observations on the comet made by him and other academicians. The latter purports to be a physical account of the motion of Jupiter's moons, but is in reality a disguised (and outstanding) analysis of the Copernican System.[24] It seems evident, therefore, that by paying a little attention to the *manner* of presenting one's material, the *Accademia* could have allowed much more to be published than it did. In fact, as a note, probably written by the secretary of the *Accademia*, Lorenzo Magalotti, to Prince Leopold, testifies, the Academy planned to publish some of its astronomical works concerning Saturn, provided that 'Borelli, to avoid difficulty, content himself with keeping his proofs outside the Copernican system."[25] These writings, however, were not published. Why?

I suggest that the main concern of the Medici was not specifically to prevent a clash with the Inquisition, but to fulfill their aspiration to appear magnificent. They tried to satisfy all the institutions, and individual intellectuals, concerned with Galilean and non-Galilean science, including, of course, the Inquisition. As Borelli said on one occasion: 'These princes try to avoid clamorous appearance that might arouse malevolence and clamour, and in short [they see] that true philosophy spreads in a pleasant way and soft manner.'[26] This approach led them, perhaps involuntarily, to force into the *Accademia del Cimento* different scientific approaches, so that they could present themselves as the patrons of one 'universal' scientific culture. Experimental physics served this purpose in the best way. It did not involve theoretical discussions or controversies and could satisfy both Galilean and non-Galilean scientists.

The *Saggi* is an outcome of this approach: it does not argue against anybody, and does not mention the name of any experimenter. Moreover, before publishing the *Saggi*, Leopold, to be on the safe side, sent drafts of the work to the representatives of three approaches to science: the Galilean Borelli, the Aristotelian Rinaldini, and the ecclesiastical Ricci. Only after these three intellectuals were satisfied were the *Saggi* sent to press.[27]

From the point of view of pleasing everybody, Leopold's caution may not have been excessive; after all, he had much more to lose than his scientists. But from the point of view of science, the effect of this policy was not a positive one, and in the long run it was the main cause for the cessation of the work of the *Accademia del Cimento*. Scientists of opposing views did not succeed in collaborating constructively, and personal relations in the Academy became tense. In 1667, Borelli, evidently having lost all patience with the policy of compromise and self-censorship and with the continuous quarrels, abandoned Tuscany. The *Accademia* without Borelli ceased to exist. It was at this point, and not soon after Galileo's trial, that scientific activity in physical science in Tuscany collapsed.

Scientific research at the Tuscan Court could have produced many more praiseworthy results, had it not been for the hesitant policy of the late Medici. By trying to satisfy Galilean and non-Galilean intellectuals simultaneously, they displeased everybody and, to a certain extent, also discredited their own scientists. Also, by trying to force science into an empirical frame, they helped create an image of Galilean science more empirical than it probably was.[28]

My conclusion, namely that the interference of the Medici hindered the work of the *Accademia del Cimento*, is given additional support by the research carried out in life sciences at the Tuscan Court, in parallel with the work of the *Accademia del Cimento*. This research was led by Court Physician Francesco Redi, whose most important result was an experiment to refute the Aristotelian belief that worms generate spontaneously. This experiment could be considered as controversial as Torricelli's experiment, but in this case, Ferdinand II and Leopold, having had little interest in life sciences, did not interfere in the research carried out in this field.[29]

THE MEDICI ATTITUDE TOWARDS PRESERVATION OF SCIENCE RECONSIDERED

One of the great merits of the Medici, from the earliest days of their reign in Tuscany, was to understand the value of Renaissance culture and to encourage it; they rightly appear in history as the dynasty which promoted the Renaissance. At the beginning of the seventeenth century, Grand Duke Cosimo II, following the example of his predecessors, was clever enough to understand the importance of the new science and to support Galileo. He was also wise enough not to interfere with Galileo's work.

After Galileo's trial the Medici continued not to interfere in his work and that of his successors, *inter alia* permitting Torricelli to make some important contributions to seventeenth century science. But when, after Torricelli's death, they began interfering with research in the physical sciences, this work was stopped. It is true that without the court's support scientists like Borelli would not have been able to pursue some of their work and that the *Accademia del Cimento* would have never existed. But it is also true that Prince Leopold de' Medici unnecessarily forced Galilean scientists to 'collaborate' with Aristotelian philosophers, imposed secrecy on the work of the Academy and censorship on the publications of its works, and created a climate that hindered the work of the Academy. The result was that the Academy dissolved, research in the physical sciences at the Tuscan Court stopped, and Galileo's followers were given much less credit than they deserved; even today much of their scientific work is still unpublished.

This, of course, still gives rise to a variety of questions. A first group of questions concerns the exact responsibility of the Church in the repression of science, and the interaction between religious and secular policies towards science at that time. Is the responsibility of the Church a matter of historical truth, or is it, to a larger extent than usually thought, the result of an image developed in the wake of the Enlightenment and nineteenth century positivism? Can Leopold be excused by saying that he acted the way he did in order to try and preserve the results of Galilean science through difficult times?

A second group of questions more directly concerns the decline of science in Tuscany in the second half of the seventeenth century. What happened to Tuscan science *outside* the Tuscan Court? Further investigation in this direction may help to clarify whether and to what extent Tuscan science was already by that time mortally sick from other causes, thus minimizing the effect, either positive or negative, of the Medici's patronage.

A last group of questions, not dealt with in this paper, involves the sociology of science: must a scientific argument inevitably influence the interpersonal relations among the scientists? What are the limits of the political support which can be "offered" to a scientist?

Universität München
München, Germany

NOTES

Abbreviations

OG: *Le Opere di Galileo Galilei. Edizione nazionale*, edited by Antonio Favaro, 20 vols. (Florence: Giunti Barbèra, 1890–1909. Reprinted, 1929–1939 and 1968).

TT: Giovanni Targioni Tozzetti, *Notizie degli aggrandimenti delle scienze fisiche accaduti in Toscana nel corso di anni LX del secolo XVII*, 3 vols. (Florence, 1780. Reprinted, Bologna: Forni, 1967).

1. One can find this view in many works on the popular and scholarly level alike. Let me quote three very different ones: *i. TT. ii.* Indro Montanelli and Roberto Gervaso, *L'Italia del Seicento* (Milan: Rizzoli, 1969); this is a popular book of Italian history enjoying very large sales and emphasising the responsibility of the Church (see p. 310). *iii.* The article written by A. Natucci on Vincenzo Viviani in the *Dictionary of Scientific Biography*, *14*: 48—50 (New York: Charles Scribner's Sons, 1976).
2. A documentary history of Galileo's trial, presenting the translation into English of the pertinent documents, was recently published by Maurice A. Finocchiaro under the title *The Galileo Affair: A Documentary History* (Berkeley, etc.: Univ. of California Press, 1989).
3. In addition to *TT*, an encompassing study of Tuscan science after Galileo is Raffaello Caverni, *Storia del metodo sperimentale in Italia*, 6 vols. (Florence, 1891—1900. Reprinted, Bologna: Forni, 1970). Other useful works are: Angelo Fabroni, *Lettere inedite di uomini illustri*, 2 vols. (Florence, 1773—75); Giorgio Abetti and Pietro Pagnini (eds.), *Le Opere dei discepoli di Galileo Galilei, Edizione nazionale*, Vol. I: *L'Accademia del Cimento, Parte prima* (Florence: S. A. G. Barbèra, 1942); in the same series: Paolo Galluzzi e Maurizio Torrini, *Carteggio*, 2 vols. (Florence: Giunti Barbèra, 1975—84). *Celebrazione dell'Accademia del Cimento nel tricentenario della fondazione* (Pisa: Domus Galilaeana, 1958); W. E. Knowles Middleton, *The Experimenters: A Study of the Accademia del Cimento* (Baltimore: Johns Hopkins Univ. Press, 1971); Gino Arrighi *et al.*, *La scuola galileiana: prospettive di ricerca* (Florence: La Nuova Italia, 1979); Paolo Galluzzi, 'L'Accademia del Cimento: 'gusti' del principe, filosofia e ideologia dell' esperimento,' *Quaderni storici*, N. 48, Anno 16, Fasc. 3, (1981), pp. 788—844.
4. Gal. MSS 275—286.
5. *OG 19*: 322, 283; the sentence: 402—406. Translated by Giorgio de Santillana, *The Crime of Galileo* (Chicago: The Univ. of Chicago Press, 1955), respectively pp. 126, 293, 306—10.
6. Galileo Galilei, *Discorsi e dimostrazioni matematiche intorno a' due nuove scienze* ... (Leyden, 1638). *OG 8*. There are several translations into English. I used Henry Crew and Alfonso de Salvio, *Dialogues Concerning Two New Sciences* (New York: Macmillan, 1914). On the relations between Galileo and the Medici, see my 'Galileo as a Politician,' *Sudhoffs Archiv 72* (1988), pp. 69—82.
7. Renieri was entrusted by Galileo with the task of drawing up the tables of Jupiter's moons, and King Louis's counsellor, Baldassar de Monconys, records in his diary in 1646, after having visited Torricelli, that Torricelli occupied himself with astronomy and cosmology. See *Opere di Evangelista Torricelli*, edited by Gino Loria and Giuseppe Vassura, 4 vols., Vols. I—III (Faenza: Montanari, 1919), Vol. IV (Faenza: Lega, 1944), 4: 84—85.
8. Galluzzi & Torrini, *Carteggio 1*: 122—3. The theological implications of Galileo's atomistic views are presented by Pietro Redondi, *Galileo Heretic*, translated by Raymond Rosenthal (Princeton: Princeton Univ. Press, 1987). Redondi claims that Galileo's theory of matter, as presented in his *Assayer* (1623), — rather than his campaign for Copernicanism — was the main cause of his future misfortune, since it brought into question the sacrament of the Eucharist.

9. Evangelista Torricelli, *Opera geometrica* (Florence, 1644). On the inner front page of this work, the title is given as *Magni Ducis Mathematico*. Paolo Galluzzi, 'Vecchie e nuove prospettive torricelliane,' in G. Arrighi *et al.*, *La scuola galileiana*, 13—51, p. 46. On Galileo's title, see my 'Galileo as a Politician' p. 75.
10. *Saggi di naturali esperienze fatte nell'Accademia del Cimento sotto la protezione del Serenissimo Principe Leopoldo di Toscana e descritte dal segretario di essa Accademia* (Florence, 1667). Translated into English by Middleton in *The Experimenters*.
11. On the Lincean see Stillman Drake, 'The Accademia dei Lincei,' *Science*, 151 (1966), pp. 1194—1200.
12. The diaries of the academy, Gal. MSS 260, 261, 262. On Leopold's role in the Academy, see Middleton, *The Experimenters*, pp. 56—61.
13. If Dati was not a member, he at least suggested a number of experiments to the Academy and also published the description of the Torricellian experiment under the pseudonym of Timauro Antiate, in a work entitled *Lettera a Filaleti* (Florence, 1663). On the tense relations in the Academy, see Galluzzi, 'L'Accademia del Cimento,' pp. 819—23. The prince acknowledged that the antagonism in the Academy hindered its work: Middleton, 'The Experimenters,' p. 316.
14. Gal. MSS 259 to 274 and 290 to 307.
15. Gal. MSS 271—274. This work is outlined in *TT 1*: 382—404 and documented in *TT 2*—2, pp. 737—800; *cf.* also Middleton, *The Experimenters*, pp. 256—62; Albert Van Helden, "The Accademia del Cimento and Saturn's Rings," *Physis* 15 (1973), pp. 237—59.
16. *TT 1*: 418; *The Correspondence of Henry Oldenburg*, Edited and Translated by A. Rupert Hall and Marie Boas Hall, Vol. IV (Madison, Milwaukee and London: Univ. of Wisconsin Press, 1967), p. 248.
17. Galileo Galilei, *Dialogo ... sopra i due massimi sistemi del mondo tolemaico e copernicano* (Florence, 1632); *OG, 7*. The most recent translation into English I know of is by Stillman Drake, under the title *Dialogue Concerning the Two Chief World Systems — Ptolemaic and Copernican* (Berkeley: Univ. of California Press, 1962).
18. To the best of my knowledge the only research carried out so far in this direction is by Adriano Prosperi, 'L'Inquisizione fiorentina al tempo di Galilei,' in Paolo Galluzzi (ed.), *Novità celesti e crisi del sapere, atti del convegno internazionale di studi galileiani, Supplemento agli Annali dell'Istituto e Museo di Storia della Scienza*, Fasc. 2, Monografia 7 (1983), pp. 316—325. Prosperi consulted the archives of the Florentine Inquisition and found that many documents which might have been relevant to Galileo had disappeared. The remaining material concerning Galileo is too slight to allow any conclusions to be drawn.
19. *OG 18*: 372.
20. *TT 1*: 399.
21. *TT 1*: 124. Antonio Favaro, 'Documenti inediti per la storia dei manoscritti Galileiani nella Biblioteca Nazionale di Firenze,' *Bullettino di Bibliografia e di Storia delle Scienze Matematiche e Fisiche 18* (1885), pp. 1—112, 151—230, pp. 56—57.
22. Galluzzi, 'L'Accademia del Cimento,' p. 823.

23. A brief outline of the anti-Galilean wave at the end of the century is presented by Redondi, *Galileo Heretic*, pp. 319—20.
24. Giovanni Alfonso Borelli, *Lettera del movimento della cometa apparsa il mese di decembre del 1664* (Pisa, 1665), and *Theoricae mediceorum planetarum ex causis physicis deducta* (Florence, 1666). The first work was published under the pseudonym of Pier Maria Mutolo.
25. *TT 1*: 385.
26. In a letter published by Tullio Derenzini in 'Alcune lettere di Giovanni Alfonso Borelli ad Alessandro Marchetti,' *Physis 1* (1959), pp. 224—43, at p. 233.
27. All the existing drafts of the *Saggi*, with the notes of the referees, have been published in Abetti & Pagnini, (eds.), *Le Opere dei discepoli di G. Galilei*.
28. The view presenting Galileo as an empiricist was challenged by Alexandre Koyré in his *Études Galiléennes* (Paris: Hermann, 1939, 1966), translated into English by John Mepham under the title *Galileo Studies* (Hassocks, Sussex: Harvester Press, 1978). I discussed the role of experiment in Galileo's work in my 'The Role of Experiment in Galileo's Physics,' *Archive for History of Exact Sciences 23* (1980), pp. 227—52.
29. Redi's result was presented in his *Esperienze intorno alla generazione degl'insetti* (Florence, 1668). Redi still thought that there was spontaneous generation in oak galls. It was left to Malpighi to disprove this three decades later.

NOTES ON CONTRIBUTORS

The Editor has asked the contributors to the volume to furnish him with their own brief description for this section of the book. The materials he received varied from one-liners to full-fledged encomia. He saw himself forced, therefore, to exert his editorial prerogative in order to bring the various contributions in line with reasonable expectations.

S.U.

GAD FREUDENTHAL is Chargé de recherche at the Centre national de la recherche scientifique (C.N.R.S.) in Paris. He is the editor of *Étudés sur / Studies on Hélène Metzger* (Leyden: Brill, 1990) and of a volume of essays in the sociology of science by the late Joseph Ben-David (University of California Press, 1990). His main research interests are medieval science in Hebrew and ancient and medieval theories of matter. He has recently completed a book on Aristotle's "chemistry" and its aftermath.

JOHN GLUCKER is Professor of Classical Philology and Philosophy, Tel-Aviv University. He is author of *Antiochus and the Late Academy* (Göttingen, 1978), two books on Greek philosophy in Hebrew, and numerous articles in learned periodicals.

BERNARD R. GOLDSTEIN is Professor of Religious Studies and History of Science at the University of Pittsburgh (U.S.A.). He is the author of, among many other publications in the field of medieval astronomy, a number of books and articles on the scientific work of Levi ben Gerson. Most recently he has published, with David Pingree, in *Transactions of the American Philosophical Society*, vol. 80.6 (1990), *Levi ben Gerson's Prognostication for the Conjunction of 1345*.

EDWARD GRANT is Distinguished Professor of History and Philosophy of Science and Professor of History at Indiana University, Bloomington, U.S.A.. He has published seven volumes and over 50 articles on medieval science, with emphasis on mathematics, physics and cosmology. A book on medieval cosmology has just been completed.

NOTES ON CONTRIBUTORS

FRITZ KRAFFT is currently Professor and Head of the Institute of the History of Pharmacy at the Philipps University in Marburg. Trained in classics, philosophy and physics, he served as Professor of History of Science at the University of Mainz between 1970-1988. His main fields of research are science from classical antiquity to the eighteenth century, with special emphasis on Greek science and its role in the rise of the new sciences; other research interests are the history of astronomy, cosmology, atomism, Copernicus, Kepler, Guericke, and nuclear fission. His books include *Otto von Guericke* (1978) and *Das Selbstverständnis der Physik im Wandel der Zeit* (1982), Italian edition (1990).

Y. TZVI LANGERMANN is a member of the staff of the Institute of Microfilmed Hebrew Manuscripts, Jewish National and University Library, Jerusalem, where he works on Hebrew and Arabic texts that concern science and philosophy. His dissertation, an edition and translation of Ibn al-Haytham's *On the Configuration of the World*, has recently been published by Garland Publishing. Another recent book is *The Jews of Yemen and the Exact Sciences* (Jerusalem, 1987).

DAVID LINDBERG is Evjue-Bascom Professor of the History of Science, and Director of the Institute for Research in the Humanities, at the University of Wisconsin. His recent publications include "Science as Handmaiden: Roger Bacon and the Patristic Tradition," *Isis*, **78** (1978), 518–36; *God and Nature: Historical Essays on the Encounter between Christianity and Science* (University of California Press, 1986), ed. with Ronald L. Humbers; and *Reappraisals of the Scientific Revolution* (Cambridge University Press, 1990), ed. with Robert S. Westman. He is presently completing a book entitled *The Origins of Scientific Thought: Western Science in Philosophical, Religious, and Institutional Context, 600 B.C. to A.D. 1500* (University of Chicago Press).

JOHN NORTH has since 1977 been Professor of the History of Philosophy and the Exact Science at the University of Groningen, The Netherlands. He writes on the history of cosmological ideas, and recently published *Chaucer's Universe* (1988, 1990), a study of the astronomical element in Chaucer's poetry. Another recent publication is *Stars, Minds and Fate* (London: Hambledon Press, 1989).

NOTES ON CONTRIBUTORS

MICHAEL SEGRE was trained as a historian of science at the Hebrew University of Jerusalem and has taught for a number of years at Tel-Aviv University. He is a specialist on Galileo and teaches presently at the Institute for History of Science, Munich University. His book on Galileo's followers, *In the Wake of Galileo*, is in print.

EDITH SYLLA is a Professor of History at North Carolina State University, Raleigh, U.S.A., and a specialist on late medieval and early modern natural philosophy, mathematics, and logic. She also serves as Assistant Dean for Research and Graduate Programs. She has published widely in her areas of expertise and has recently been working on Jacob Beruoulli and the origins of the mathematical theory of probability and its applications.

SABETAI UNGURU, presently a Fellow of the Wissenschaftskolleg zu Berlin, is Professor of the History of Science and incoming Director of the Cohn Institute for the History and Philosophy of Science and Ideas at Tel-Aviv University. His research areas are the history of ancient and medieval mathematics (the Greek and Latin traditions) and the history of medieval optics. Among his recent publications are two volumes on the history of mathematics in Hebrew and a forthcoming critical Latin edition with translation, introduction, notes and commentaries of Books II and III of Witelo's *Perspectiva* (*Studia Copernicana*, XXVIII, 1991).

ABRAHAM WASSERSTEIN, born in Germany in 1921, has been Professor of Greek (now Emeritus) at the Hebrew University of Jerusalem since 1969. Before that he was Professor of Classics at the University of Leicester, England (1960-1969) and lecturer in Greek at the University of Glasgow (1951-1960). He was elected a Fellow of the Royal Astronomical Society in 1961. He is interested in Greek and Latin literature, in Greek philosophy and science, and in the transmission of science from ancient Greece to medieval and modern Europe, all areas in which he has published extensively.

NAME INDEX

This index was prepared with the assistance of Maria A. Gowans, my secretary, during the tenure of my fellowship at the Wissenschaftskolleg zu Berlin. The numbers in parentheses indicate notes.

S.U.

Abetti, G., 306(3)
Achena, M., 69(29)
Adrastus *of Aphrodisias*, 192, 219(14)
Aegidius Romanus (*Giles of Rome*), 115, 116, 121(6), 257, 274, 276–277, 291(48–49)
Aeschines, 5
Aeschylus, 8, 17(8)
Aetius, 42(2), 44(30)
Aiton, E. J., 96(3), 219(13), 220(22), 224(60), 225(70), 226(80, 87)
Albert *of Saxony*, 121(8)
Albertus (*Magnus*), 62, 64(65), 72(65), 255
Albinus, 5, 6, 11, 16(6), 17(13–14, 23)
Albo, J., 69(28)
Alexander IV (*Pope*), 253
Alexander of *Aphrodisias*, 43(16), 200, 220(16), 225(70)
Alfraganus, 290(20)
Alhazen (*cf.* also Haytham, Ibn-al), 167–169, 174–176, 191
Almosnino, M., 96(3)
Althoff, M., 246(1)
Alvarus, T., 155(52)
Alverny, M.-Th. d', 248(44)
Ambrose, *Saint*, 112
Amicus, B., 106–111, 115, 119, 122(16, 18–19), 123(22–25, 27, 29); 124(35, 37–41), 125(51), 127(75)
Amis, K., 177
Anaxagoras, 22, 33
Anaximander, 19(4), 67(15), 188
Anselm, *Saint*, 260, 285
Apollonius *of Perga*, 187, 219(14)
Apuleius, 11, 16(10)
Aquinas, Thomas, *see*: Thomas, *Saint*

Aramah, R. I., 91
Arcesilaus, 9
Archimedes, 42(6), 75–76
Aristarchus, 19(1), 42(1), 192
Aristotle, 3, 6, 8, 11–14, 17(18), 22, 26, 29–30, 34–36, 42(5), 43(9, 18)–44, 47–51, 55–56, 58–61, 63–64, 66–67(10), 68(19), 69(30), 70(39), 72(62, 66), 78, 86, 88, 94–96(11), 101–102, 104–105, 110, 112, 118–120(1–2), 122(13), 130, 132–139, 141, 146–148, 151–152, 164–165, 186–191, 194–196, 203, 218(4–5), 219(11), 220(21), 223(47), 229–235, 237–238, 240, 242–243, 246(3), 247(7, 12), 256–268; 270–271, 273–274, 276, 279–280, 282, 284, 289, 290(29), 291(55), 292(69, 73)
Armstrong, A. H., 236, 247(21–23, 26–27), 248(30, 32–40)
Arnaldez, R., 289(7)
Arrighi, G., 306(3), 307(9)
Aschkenazi, 96(3)
Augustine, *Saint*, 44(30), 109, 120(1), 123(29), 239, 260, 266, 285
Avempace, 138, 147–148
Averroes (*cf.* also Rushd Ibn), 88, 97(18), 103, 123(26), 134–140, 195, 253, 258–259, 273–274, 277, 281, 286
Aversa, R., 111, 112–115, 118, 123(29–30), 124(44–46), 125(47–51), 127(68)
Avicebron, 261
Avicenn[a]e, *see*: Sînâ, Ibn

Bacon, F., 163

Bacon, R., 163—166, 178(10), 229, 239—240, 243, 248(46), 250(76), 260
Baihan, Abu-r, 164
Bailey, C., 246(3)
Baldigiana, A., 302
Balneolis, Leo *de, see*: Levi, ben Gerson
Baneth, 89
Barker, P., 77, 80(n)
Barnes, J., 70(39)
Barocas, V., 246(1)
Basil, *Saint*, 112, 119—120(1)
Bate, H., 220(22)
Baudry, J., 66(3), 67(8), 68(17—18)
Baudry, L., 292(85—86)
Baur, L., 121(4)
Behler, E., 66(3)
Bellarmine (*Cardinal*), 296
Bellutus, B., 125(51)
Bergh, S. *van den*, 291(47)
Bernard, *Saint; of Clairvaux*, 256
Berthelot, M., 71(52)
Bienkowska, B., 96(6)
Bierce, A., 177
Birkenmajer, A., 221(29)
Bîrunî, al-, 53, 68(23)
Bisliches, M. L., 69(33)
Bitruji, al-, 86, 96(10), 97(14—15), 225(70)
Björnbo, A. A., 248(45)
Bode, 224(63)
Böhm, W., 120(1)
Boethius *of Dacia*, 257—258, 266
Bonaventure, *Saint*, 257, 264, 277, 282—283, 292(72, 76), 293(88)
Borelli, G. A., 295, 298—304, 308(24, 26)
Botarel, Moses Farissol, 75
Bouyges, M., 97(17—18)
Boyer, C. B., 179(26)
Boyle, R., 299
Bradwardine, T., 129, 130, 131, 146—149, 151—152, 155(39), 287—288, 293(90)
Brady, I., 278

Brahe, T., vii, 80, 81(n), 105, 108—109, 112, 114, 119, 124(32, 41), 191, 199, 203—205, 207, 211—213, 220(24—25), 222(46), 225(70)
Brewer, J. S., 178(9, 12)
Bridges, J. H., 178(11), 248(50)
Bruno, G., 193
Buono, C. *del*, 299
Buridan[us], J., 102—104, 121(5—6, 8, 10), 187, 219(9), 261, 273
Burley, W., 152
Burnet, J., 44(24)
Buzon, C. *de, see*: Chevalley, C.

Caesar, 43(16)
Callippus, 44(18), 186—187, 225(70)
Campanus, 136, 138, 142
Campbell, 9
Cantor, G., 250(76)
Cardano, G., 105
Carlebach, J., 78, 80(n)
Caroti, S., 155(52)
Caspar, M., 76, 81(n), 223(48, 50), 224(60), 225(66), 249(54)
Cassius Longinus, 6—7, 33—34
Castelli, 300
Cavalieri, 295, 300
Caverni, R., 306(3)
Chalcidius, 14
Chandler, B., 153(6)
Charpa, U., 219(13)
Chase, F. H., 120(1)
Cherniss, H., 9
Chesterton, G. K., 177
Chevalley, C., 249(54, 60—66)
Cicero, 11, 44(30—32)
Clagett, M., 154(37—38)
Clavius, Chr., 111, 112, 124(41—43), 125(52)
Cohen, J., 75
Colson, F. H., 68(17)
Commandino, 75—76, 80
Cooperman, B. D., 95(1)
Copernicus, N., 42(1), 86, 96(6), 97(19), 129, 130, 154(12), 187,

NAME INDEX

189—195, 197, 204—205, 207—211, 213—214, 219(13), 220(21, 23, 27), 221(29, 30)
Cornaeus, M., 115, 117—118, 127(69—71)
Cornford, F. M., 266, 290(22)
Cosimo III (*Grand Duke*), 302, 304
Cousin, 15(5), 17(13)
Craig, W. L., 290(37), 291(38—40)
Crantor, 3
Crescas, R. Hasdai, 69(28)
Crew, H., 306(6)
Critias, 12
Crombie, I. M., 9, 17(22), 72(69), 167—168, 172—173, 178(16), 179(26), 181(50—54)
Crosby, H. L. Jr., 154(39), 155(40)
Curtze, M., 78, 80(n)
Czechowic, M., 96(7)

Dales, R. C., 291(55)
Damascene, J., 120(1)
Damascius, 6
Damian, P., 256
Dante, A., 64, 258
Dati, C. (*pseud.: Timauro Antiate*), 299, 307(13)
Davidson, H. A., 67(3), 68(24), 69(28), 72(67), 94—95(1), 98(24, 27)
Delacrut, M., 96(3)
Democritus, 37, 230, 246(3)
Demosthenes, 5
Denomy, A. J., 289(15)
Derenzini, T., 308(26)
Descartes, R., 265
Dicaearchus *of Messene*, 8
Dicks, D. R., 43(13)
Diels, H., 15(4), 16(10)
Diemer, A., 220(27)
Dieterici, F., 68(23)
Dillon, J., 6, 15(2), 16(10)
Dilts, 5
Diogenes *Laertius*, 3, 5—6, 11, 16(6, 10), 17(13—14, 16, 20, 24), 246(4)
Dobryzycki, J., 96(6)

Donahue, W. H., 122(14—15), 126(63), 127(73—74)
Dodds, E. R., 16(7)
Drake, S., 109, 124(33—34, 36), 307(11, 17)
Dreyer, J. L. E., 43(13), 222(46)
Dubnow, S., 96(4, 7—8)
Düring, I., 67(10)
Duhem, P., 67(4, 12), 68(17, 23, 25), 70(37), 72(65), 262—263, 290(21)
Dumbleton, J., viii, 129—153(10), 154(11, 20), 155(46, 50—51)
Duncan, A. M., 224(60)
Duns, *John Scotus*, 123(20), 286—288, 293(87)
Duran, Shime^c on ben Zehmah, 69(28)
Dyck, Walther *von*, 249(54)

Easterling, H. H., 218(3)
Ecphantus, 192
Edel, A., 247(10—11)
Einstein, A., 43(9)
Elliott, J. H., 153(2)
Empedocles, 33, 42(5), 231, 246(6), 267
Epicurus, 36, 37(30—31), 38(32), 39(33), 230
Euclid, 44(19), 136, 138, 142, 151, 169, 248(45)
Eudemus of *Rhodes*, 189
Eudoxus, 24(13), 27(17), 28(18), 44(18), 186—188, 210, 218(5), 220(15), 225(70)
Euripides, 5
Eusebius *of Caesarea*, 4, 16(7)
Evoy, J. Mc, see: McEvoy, J

Fabricius, D., 200—202, 214, 222(46), 223(53)
Fabri, H., 302
Fabricius, J., 109
Fabroni, A., 306(3)
Fakhry, M., 98(25)
Fârâbî, al-, 55, 66, 268, 272, 285
Farrington, B., 44(25)

Favaro, A., 301, 305, 306(5), 307(21)
Feldman, S., 69(28), 291(42—45)
Ferdinand II (*Arch/Grand/duke*), 213, 296—300, 302, 304
Ficino, M., 240, 242—243, 249(53)
Field, G. C., 18(26)
Finocchiaro, M. A., 306(2)
Finzi, M., 75
Fischer, K. A. F., 98(22)
Fisher, N. W., 178(10, 13)
Forbes, R. J., 71(51)
Fotinis, A. P., 247(17)
Fracastoro, G., 225(70)
Freudenthal, G., vii, 67(15)
Friedländer, M., 291(40), 292(60)
Friedlander, I., 96(4)
Frisch, Chr. 225(74)
Fritz, K. *von*, 246(3)
Furth, M., 71(48)

Gadamer, H.-G., 66(3)
Gaius, 6
Galen, 34, 66
Galilei, G., viii, 108—109, 119, 121(5, 8), 123(26), 124(36), 125(52), 129, 187, 193—194, 215, 219(10), 262, 295—305, 306(2—3, 6—8), 307(9, 17—18), 308(28)
Galluzzi, P., 297, 302, 306(3—4, 8), 307(9, 13, 18, 22)
Gandz, S., 98(20—21)
Gaon, Sa^{c}adya, 84
Gassendi, 299
Gershom, R. ben Shlomo, 70(36)
Gersonides, see: Levi, ben Gerson
Gervaso, R., 306(1)
Gesner, C., 60, 71(53)
Ghazâlî, al-, 53, 68(23), 273—274, 288
Gilbert, W., 195—198, 200, 202—203, 212—214, 216, 221(35—36), 222(37), 222(38—39, 46), 225(74)
Gilby, T., 292(81)
Giles *of Rome, see*: Aegidius Romanus
Gilson, E., 258, 289(1, 8—9)
Ginzburg, M., 96(3)
Glucker, J., vii, 3

Godfrey *of Fontaines*, 277, 289(14), 290(19)
Goes, E. *de*, 122(17)
Goldstein, B. R., vii, 76—80(n, n); 81(n, n, n, n, n, n, n, n); 96(10), 97(15), 290(20)
Gossen, H., 34, 220(25)
Grant, E., viii, 97(13), 121(5), 121(6—7), 130, 153(3—5), 154(12, 19), 155(52)
Gregory *of Nyssa*, 112
Grosseteste, R. (Lincoln?), 120(4), 135, 141, 154(21), 177, 239, 248(31), 279, 291(55)
Grube, G. M. A., 17(15, 22)
Günther, S., 221(32)
Guericke, O. *von*, 220(25), 224(56)
Guerlac, H., 220(21), 221(29)
Guthrie, W. K., 42(2), 43(7), 44(21, 24, 27), 67(13)

Haas, A. E., 246(1)
Haase, R., 226(103)
Hahm, D. E., 67(16)
Hall, A. R. and Boas Hall, M., 153(2), 307(16)
Hall, T. S., 67(15)
Hammer, F., 249(54)
Harrassowitz, O., 71(50)
Harriot, T., 77, 250(75)
Hartner, W., 98(26)
Haytham, Ibn -al (*cf.* also Alhazen), 78, 81(n), 97(12, 17)
Heath, Sir Thomas, 43(13), 44(22)
Heiberg, J. L., 220(15)
Helden, A. *Van*, 307(15)
Hellman, D., 96(3), 124(31—32), 220(24)
Hellman, G., 221(33)
Henry *of Ghent*, 261, 277
Heraclitus, 33
Hermann, C. F., 16(10), 17(13, 16—17, 23)
Hermias, 6
Hermocrates, 12
Heron *of Alexandria*, 42(6)

NAME INDEX

Herophilus, 33—34
Herwart *von Hohenburg, see*: Hohenburg
Hett, W. S., 247(7, 14—15)
Heytesbury, W., 129, 151—152
Hicetas, 192
Hicks, R. D., 246(4)
Hidâyat, Husain, 71(58)
Hintikka, J., 290(29, 35)
Hipparchus, 109, 187
Hippocrates, 24, 44(26), 66
Hippolytus, 5—6, 11, 16(10), 42(2)
Hiyya, A. bar, 94
Hobbes, Th., 5
Hofer, J., 123(28)
Hohenburg, H. *von*, 199—200, 213, 218, 226(77)
Holmyard, E. J., 58, 70(37, 41—42), 72(63—64)
Hooke, R., 195
Hudry, F., 248(44)
Hume, D., 5

Ideler, 33
Immanuel b. Jacob Bonfils *of Tarascon*, 75
Immanuel *of Rome*, 66, 72(71)
Isidore *of Seville*, 164
Iskandar, A. Z., 289(7)
Israeli, I. (*the younger*), 69(28), 84, 86, 96(10)
Israeli, (*the elder*), 84
Isserles, Moses; Rabbi; *of Cracow*, viii, 83—95(1, 3) 96, 98(24—25, 27)

Jâbir, Hayyân Ibn, 61
Jacob *of Bezyce*, 96(7)
Jamil, A., 68(23)
Jessen, J., 174
Joel, M., 69(28)
John XXI (*Pope; Peter of Spain*), 257
John *of Jandun*, 103—104, 123(25), 259, 289(8)
Jones, H. L., 68(21)
Jones, W. H. S., 44(26)

Kainz, H. P., 293(86)
Kant, I., 47(1), 66(1), 254—257, 293(90)
Katzenellenbogen, M., (*the Rabbi of Padua*), 85, 96(5)
Kellner, M. M., 76, 81(n)
Kenna, S. Mac, *see*: MacKenna, S.
Kenny, A., 153(1)
Kepler, J., vii, viii, 25, 76, 80, 81(n, n); 173—175, 185—186, 188, 191, 193, 195—196, 198—218(1), 219(11), 220(26), 221(31—32), 222(37, 40, 43, 46), 223(47, 50, 53), 224(54—55, 57—60, 63); 225(64—73, 75—76, 78); 226(80, 82, 87—100, 102—103); 229—230, 240—246, 247(20), 248(46), 249(53—68), 250(69—72, 74—77)
Kindî, al-, 238, 240, 247(18), 248(44—45), 268—272, 291(40—41)
Kircher, A., 222(39), 226(95)
Kirk, G. S., 42(2), 44(24)
Klaus, G., 221(29)
Koch, J., 291(49)
Koenigsberger, H. G., 153(2, 3)
Koyré, A., 18(25); 173, 176, 181(57, 63), 250(73), 308(28)
Kraemer, J., 96(9)
Krafft, F., viii, 218(2), 219(8, 12—13), 220(15, 20, 25—27), 221(28, 30—31), 224(60, 62), 225(72), 226(75, 95, 103)
Kraus, P., 71(55)
Kretzmann, N., 153(1), 289(12)
Kristeller, P. O., 15(1)
Kühn, 34
Kunitzsch, P., 98(22)
Kurland, S., 69(27)

Lactantius, 43(9)
Langermann, Y. T., viii, 98(22)
Lapidge, M., 67(16)
Laplace, 66
Lasserre, F., 220(15)
Lattes, I. *de*, 75
Leclerc, I., 249(67)

Lee, H. D. P., 72(62)
Leibniz, 293(90)
Lejeune, A., 168, 179(27)
Lemendone, 96(7)
Leopold (*Prince de' Medici*), 298—305, 307(12)
Leucippus, 230
Levi, ben Gerson, vii, 69(28), 75—78, 79—81(n, n), 84, 94, 97(11)
Lindberg, D. C., viii, 78, 81(n, n,), 97(13), 168, 173—175, 179(26—27), 181(58—61), 246(1, 3, 6), 247(7, 19, 24), 248(31, 44—45, 47—52), 250(76)
Lippmann, E. O. *von*, 69(28)
Lisker, Ch., 95(3)
Locke, J., 5
Loeb, J. (*Maharal of Prague*), 94
Lohr, C. H., 123(20)
Lombard, P., 263
Long, A. A., 44(33)
Long, H. L., 16(10)
Longomontanus, 214
Loria, G., 306(7)
Louis (*King*), 306(7)
Lovejoy, A., 290(29)
Lucretius, 22(11), 43(9), 44(30), 68(18), 231

Machir, Jacob ben, 84
MacKenna, S., 248(38, 40—42)
Maddison, R. E. W., 250(73)
Maestlin, M., 204—207, 209, 214, 225(71), 226(77)
Magalotti, L., 299, 303
Maier, A., 249(53)
Maimonides, 56, 68(24), 84, 86—87, 89—90, 94, 96(9), 98(21, 23, 25), 273—274, 281—282, 284, 288, 291(40, 46, 55), 293(90)
Malcutius, O., 226(93)
Malpighi, 308(29)
Mandeville, D. C., 58, 70(37, 41—42), 72(63—64)
Mandonnet, P., 289(2—4)
Marchetti, A., 308(26)
Marchia, F. *de*, 195

Maricourt, P., 163
Marsili, A., 299—300
Martyr, J., 5
Massé, H., 69(29)
Mastrius, B., 125(51)
Maurach, G., 220(25)
Maurolico, 250(76)
McEvoy, J., 248(31)
Medici, *the [de]*, 295—298, 300—305, 306(6)
Meer, M., 75
Meniel, Chr., 71(50)
Melissus, 135
McMullin, E., 249(67)
Mendelsohn, E., 96(9), 155(39)
Menut, A. D., 289(15)
Mepham, J., 308(28)
Meyer, K., 220(26)
Michaud-Quantin, P., 248(47)
Middleton, K. W. E., 306(3), 307(10, 12—13, 15)
Miechowicz, M., 96(7)
Mittelstraß, J., 219(13), 220(19)
Moerbeke, W. of, 120(1)
Molland, A. G., 155(39)
Monconys, B. *de*, 306(7)
Montanelli, I., 306(1)
Moody, E. A., 121(5)
Mordecai Finza *of Mantua*, 75
Moreaux, 3
Mullin, E. Mc, *see*: McMullin, E.
Murdoch, J., 155(52)
Mutoli, P. M., 308(24)

Natucci, A., 306(1)
Nazzam, -al, 98(25)
Neubauer, 95(3), 98(25)
Neugebauer, O., 45(13), 89, 97(19), 98(21)
Newton, I., 120, 151, 155(39), 195, 218, 229
Nic[h]olas *of Cusa*, 164, 192, 194, 286
Norman, R., 196—197, 221(33)
North, J. D., viii, 131, 151, 153(6), 289(16—17)
Numenius, 6

NAME INDEX

Obadiah, D. ben, 89—90
Oberman, H. A., 293(89—90)
Ockham, W. of, 115, 121(5), 129, 151, 240, 248(53), 249(53), 286—288, 293(90)
Oded, B., 81(n, n)
Oehler, K., 67(13)
Oldenburg, H., 299
Oliva, A., 299
Olympiodorus, 5—6
Oresme, N., 154(37), 155(39), 261, 273, 289(15)
Orsini, F., 75
Osiander, A., 190, 194, 205, 219(13)
Oviedo, F. de, 106, 119, 122(17—18), 127(75—77)

Page, B. S., 248(38)
Pagnini, P., 306(3)
Pais, A., 43(9)
Parmenides, 13, 135, 266
Patrizi, F., 243
Paul of Venice, 152, 155(52)
Pecham, J., 277—278
Peck, A. L., 67(9)
Pedersen, O., 97(14)
Pelner Cosman, M., 153(6)
Peregrinus, P., 163
Perrier, J., 292(81)
Peters, F. E., 290(36)
Petraitis, C., 72(62)
Petz, H., 76, 81(n)
Peurbach, G., 83, 85, 88, 92—93, 96(3), 97(14—15), 98(19)
Phaedrus, 7—8
Philebus, 6
Philip of Opus, 7(16), 17(16)
Philo, 50, 68(17, 19—20)
Philolaus, 192, 206
Philoponus, J., 66(3), 98(24), 105, 119, 120(1), 268, 271, 282
Pinborg, J., 153(1)
Pines, S., 72(69), 96(9), 98(23)
Plato[n], vii, 3(5), 4—10(23, 25), 11—12(26—27), 13—15(5), 16(6—8), 17(10, 16—17), 18(28), 23—27, 47, 49(5), 66, 105, 112, 119, 133, 135, 185, 188—189, 194—196, 205—207, 209—210, 214, 225(66), 237, 246(6), 262, 264—266, 299
Platter, F., 174
Pliny, 109, 123(29)
Plotinus, 6, 195, 229, 234—238, 240, 242, 244—245, 264, 268, 273
Plutarch[us], 15(2), 194, 299
Pohlenz, M., 67(16), 68(17)
Polemarchus, 44(18)
Poncius, J. (Punch, J.), 106, 108, 119, 123(20—21), 127(75)
Porta, G., 196—198, 221(34), 226(82)
Pos[e]idonius, 52, 68(21), 188
Proclus, 4, 6(11—12), 7(13), 9(21—22), 15(5), 17(13, 20), 44(23), 206
Prosperi, A., 307(17)
Ptolemy, 24(14), 25—26, 43(14), 76, 86, 89, 97(19), 168, 186, 189—193, 204, 207, 209—211, 219(14), 220(17), 262—263, 290(20)
Pythagoras, 56, 189

Quintilian, 4
Qurrah, Thabit bin, 88

Ramus, P., 225(70)
Raphael, 13
Raven, J. E., 42(2), 44(24)
Râzî, -al, 61
Redi, F., 299, 304, 308(29)
Redondi, P., 306(8), 308(23)
Regenbogen, O., 68(17)
Regiomontanus, 76
Reinhardt, K., 68(21)
Renan, E., 75, 81(n)
Renieri, V., 296—297, 300, 306(7)
Rex, F., 71(56, 57)
Rh[a]eticus, G. J., 204, 208
Rhodes, G. de, 115, 118, 126(68)
Ricci, M., 297, 300, 303
Riccioli, G. B., 115—118, 125(57), 126(59—62, 64—65, 67—68), 127(72—73), 219(11)
Riedl, J. O., 291(49—50)
Rinaldini, C., 299—303
Rist, J. M., 67(16)

Ritter, 9
Roche, J. J., 75—76, 81(n)
Ronchi, V., 246(1), 249(66)
Rose, P. L., 75—76, 81(n)
Rosen, E., 222(40), 224(54), 226(75)
Rosenthal, J., 96(7)
Rosenthal, R., 306(8)
Ross, W. D., 247(9, 11—12)
Rothmann, Chr., 222(46), 225(70)
Rubio, A., 119
Rudolph II(*Emperor*), 213
Runia, D. T., 16(6)
Rushd, Ibn (*cf.* also Averroes), 54—56, 64—65, 69(28), 72(67), 88, 96(11), 97(18), 258, 289(7)
Ruska, J., 71(52—54, 59)
Russell, J. L., 226(103)

Sa^cadyâ, 98(24), 268
Sabra, A. I., 78, 81(n), 96(9), 97(17—18)
Safâ, Ikhwân, al-, 53, 62, 68(23)
Saffrey, D., 17(11)
Salvio, A. *de*, 306(6)
Sambursky, S., 120(1)
Santillana, G. *de*, 306(5)
Sasson, (H. H.) Y. ben, 95(1), 96(7—8)
Scaliger, J. C., 225(66)
Scheiner, Chr., 109, 215
Schiaparelli, G., 43(13)
Schimank, H., 220(25)
Schramm, M., 220(16—17)
Schubart, W., 15(4)
Schultz, 5
Seck, F., 222(43)
Segre, M., viii
Serbellonus, S., 125(50)
Settimi, *Father Clements*, 301
Sextus, 17(24)
Shalom, A., 69(28)
Shatzmiller, J., 75, 81(n, n)
Shehaby, N., 97(17)
Shorey, P., 9
Siev, A., 95(1, 3), 96(5), 98(27)
Siger *of Brabant*, 256—258, 289(6)
Simplicius, 23, 26, 43(13, 17—18)—44, 66(3), 120(1), 123(26), 188, 190—191, 195, 220(16), 225(70), 262—263, 292(71), 300
Sînâ, Ibn (Avicenna), viii, 47—48, 55(30), 56—59, 62—65, 66(71), 69(29), 70(34, 37), 72(60, 65, 67, 69, 71), 260, 268, 272, 277, 281, 286
Smith, J. A., 247(9, 13)
Sobol, P., 121(4)
Socrates, 9, 10(24), 11—12
Solmsen, F., 67(5—6, 9—10, 13—14)
Sophocles, 8
Sosigenes, 27(16), 43(18)—44, 189—191, 220(22)
Spica, 123(29)
Staff, J., 75—77
Stapleton, H. E., 71(68)
Staub, J. J., 76, 81(n)
Steenberghen, F. *van*, 258, 289(6)
Steinmetz, P., 68(19)
Steinschneider, M., 69(31—32), 96(3)
Stephanus, 4
Sticker, B., 220(26)
Strabo, 52, 68(21—22)
Straker, S. M., 76, 78, 80, 81(n, n), 249(66), 250(71)
Swerdlow, N. M., 96(3), 97(19)
Swineshead, R., 129—130, 152 155(52)
Sylla, E., viii, 153(1), 155(39, 41)

Tachau, K. H., 248(53)
Tal, Shlomo, 95(1)
Telesio, B., 105
Tempier, É., 256—260, 276
Theaetetus, 44(20)
Theon *of Smyrna*, 192, 219(14)
Theophrastus, 48, 50—51, 63, 65, 68(17)
Themistius, 105
Thomas, *Saint*, 121(6), 123(26), 131, 163—164, 178(7), 190—191, 219(11), 225(70), 257—259, 261—263, 273, 276—287, 290(21), 291(48, 52), 292(55—56, 86)
Thrasyllus, 7
Tibbon, Samuel b. Judah Ibn, *see:* Tibbon, Shmuel b. Yehudah Ibn

Tibbon, Shmuel b. Yehudah Ibn, 56, 57(34), 64—65, 69(32—33), 70(34, 36), 72(65, 71)
Tideus, 248(45)
Tigerstedt, E. N., 18(28)
Timaeus, 10(24), 12—13, 135
Titius, 224(63)
Torricelli, E., 295—298, 300, 304, 306(7), 307(9)
Torrini, M., 306(3, 8)
Touati, Ch., 69(28), 75, 82(n)
Tozzetti, T., 299, 301, 305, 306(1), 307(20—21), 308(25)

Underwood, E. A., 246(3)
Unguru, S., 178(10, 13)

Vajda, G., 70(34—35), 72(67)
Vallesius, 126(68)
Vassura, G., 306(7)
Vinci, Léonard de, 68(17)
Viviani, V., 295—297, 299—302, 306(1)
Vlastos, G., 12, 13(28), 18(28)
Vogl, S., 248(45)

Wadding, L., 123(20)
Wallace, W. A., 121(5, 8), 123(26), 125(52), 179(26), 290(18)
Walters, B., 81(n)
Walther, B., 75
Wasserstein, A., vii, 43(10, 12), 44(19—20, 30)

Webering, D., 293(88)
Weinberg, J. R., 248(43)
Weisheipl, J. A., 153(8), 249(67)
Westerink, L. G., 6, 15(3), 16(10), 17(11—12, 16—17, 23)
Wiedemann, E., 78, 82(n)
Wieland, W., 66(3)
William of Conches, 279
Wilson, C. A., 154(19), 226(87, 103)
Wippel, J. F., 278, 289(14), 290(19), 291(48, 51—54), 292(55—56)
Witelo, viii, 163—164, 166—174, 176—177, 181(54—56)
Wittgenstein, L., 254—255
Witt, R. E., 16(10)
Wojcieh (Albertus) of Brudzewo, 85, 96(6)
Wolff, M., 219(8)
Wolfson, H. A., 68(19, 24), 98(24—25), 219(11)
Wolter, A., 293(87)
Wyckoff, D., 72(65—66)

Xenocrates, 3
Xenophanes, 13, 19(2—3), 20, 28, 42(2)
Xenophon, 12

Yarden, Dov, 73(71)

Zeno of Elea, 15(5)
Zenon of Citium, 50, 68(17)

BOSTON STUDIES IN THE PHILOSOPHY OF SCIENCE

Editors:

ROBERT S. COHEN
(Boston University)

1. Marx W. Wartofsky (ed.), *Proceedings of the Boston Colloquium for the Philosophy of Science 1961–1962.* 1963.
2. Robert S. Cohen and Marx W. Wartofsky (eds.), *In Honor of Philipp Frank.* 1965.
3. Robert S. Cohen and Marx W. Wartofsky (eds.), *Proceedings of the Boston Colloquium for the Philosophy of Science 1964–1966. In Memory of Norwood Russell Hanson.* 1967.
4. Robert S. Cohen and Marx W. Wartofsky (eds.), *Proceedings of the Boston Colloquium for the Philosophy of Science 1966–1968.* 1969.
5. Robert S. Cohen and Marx W. Wartofsky (eds.), *Proceedings of the Boston Colloquium for the Philosophy of Science 1966–1968.* 1969.
6. Robert S. Cohen and Raymond J. Seeger (eds.), *Ernst Mach: Physicist and Philosopher.* 1970.
7. Milic Čapek, *Bergson and Modern Physics.* 1971.
8. Roger C. Buck and Robert S. Cohen (eds.), *PSA 1970. In Memory of Rudolf Carnap.* 1971.
9. A. A. Zinov'ev, *Foundations of the Logical Theory of Scientific Knowledge (Complex Logic).* (Revised and enlarged English edition with an appendix by G. A. Smirnov, E. A. Sidorenka, A. M. Fedina, and L. A. Bobrova). 1973.
10. Ladislav Tondl, *Scientific Procedures.* 1973.
11. R. J. Seeger and Robert S. Cohen (eds.), *Philosophical Foundations of Science.* 1974.
12. Adolf Grünbaum, *Philosophical Problems of Space and Time.* (Second, enlarged edition). 1973.
13. Robert S. Cohen and Marx W. Wartofsky (eds.), *Logical and Epistemological Studies in Contemporary Physics.* 1973.
14. Robert S. Cohen and Marx W. Wartofsky (eds.), *Methodological and Historical Essays in the Natural and Social Sciences. Proceedings of the Boston Colloquium for the Philosophy of Science 1969–1972.* 1974.
15. Robert S. Cohen, J. J. Stachel, and Marx W. Wartofsky (eds.), *For Dirk Struik. Scientific, Historical and Political Essays in Honor of Dirk Struik.* 1974.
16. Norman Geschwind, *Selected Papers on Language and the Brain.* 1974
17. B. G. Kuznetsov, *Reason and Being: Studies in Classical Rationalism and Non-Classical Science.* 1987
18. Peter Mittelstaedt, *Philosophical Problems of Modern Physics.* 1976
19. Henry Mehlberg, *Time, Causality, and the Quantum Theory* (2 vols.). 1980.

20. Kenneth F. Schaffner and Robert S. Cohen (eds.), *Proceedings of the 1972 Biennial Meeting, Philosophy of Science Association.* 1974
21. R. S. Cohen and J. J. Stachel (eds.), *Selected Papers of Léon Rosenfeld.* 1978.
22. Milic Čapek (ed.), *The Concepts of Space and Time. Their Structure and Their Development.* 1976.
23. Marjorie Grene, *The Understanding of Nature, Essays in the Philosophy of Biology.* 1974.
24. Don Ihde, *Technics and Praxis. A Philosophy of Technology.* 1978.
25. Jaakko Hintikka and Unto Remes, *The Method of Analysis. Its Geometrical Origin and Its General Significance.* 1974.
26. John Emery Murdoch and Edith Dudley Sylla, *The Cultural Context of Medieval Learning.* 1975.
27. Marjorie Grene and Everett Mendelsohn (eds.), *Topics in the Philosophy of Biology.* 1976.
28. Joseph Agassi, *Science in Flux.* 1975.
29. Jerzy J. Wiatr (ed.), *Polish Essays in the Methodology of the Social Sciences.* 1979.
30. Peter Janich, *Protophysics of Time.* 1985.
31. Robert S. Cohen and Marx W. Wartofsky (eds.), *Language, Logic and Method.* 1983.
32. R. S. Cohen, C. A. Hooker, A. C. Michalos, and J. W. van Evra (eds.), *PSA 1974: Proceedings of the 1974 Biennial Meeting of the Philosophy of Science Association.* 1976.
33. Gerald Holton and William Blanpied (eds.), *Science and Its Public: The Changing Relationship.* 1976.
34. Mirko D. Grmek (ed.), *On Scientific Discovery.* 1980.
35. Stefan Amsterdamski, *Between Experience and Metaphysics. Philosophical Problems of the Evolution of Science.* 1975.
36. Mihailo Marković and Gajo Petrović (eds.), *Praxis, Yugoslav Essays in the Philosophy and Methodology of the Social Sciences.* 1979.
37. Hermann von Helmholtz, *Epistemological Writings. The Paul Hertz/Moritz Schlick Centenary Edition of 1921 with Notes and Commentary by the Editors.* (Newly translated by Malcolm F. Lowe. Edited, with an Introduction and Bibliography, by Robert S. Cohen and Yehuda Elkana). 1977.
38. R. M. Martin, *Pragmatics, Truth, and Language.* 1979.
39. R. S. Cohen, P. K. Feyerabend, and M. W. Wartofsky (eds.), *Essays in Memory of Imre Lakatos.* 1976.
40. B. M. Kedrov and V. Sadovsky. *Current Soviet Studies in the Philosophy of Science.* Forthcoming.
41. M. Raphael, *Theorie des Geistigen Schaffens auf Marxistischer Grundlage.* Forthcoming.
42. Humberto R. Maturana and Francisco J. Varela, *Autopoiesis and Cognition. The Realization of the Living.* 1980.
43. A. Kasher (ed.), *Language in Focus: Foundations, Methods and Systems. Essays Dedicated to Yehoshua Bar-Hillel.* 1976.
44. Trân Duc Thao, *Investigations into the Origin of Language and Consciousness.* (Translated by Daniel J. Herman and Robert L. Armstrong; edited by Carolyn

R. Fawcett and Robert S. Cohen). 1984.
45. A. Ishimoto (ed.), *Japanese Studies in the History and Philosophy of Science.*
46. Peter L. Kapitza, *Experiment, Theory, Practice.* 1980.
47. Maria L. Dalla Chiara (ed.), *Italian Studies in the Philosophy of Science.* 1980.
48. Marx W. Wartofsky, *Models: Representation and the Scientific Understanding.* 1979.
49. Trân Duc Thao, *Phenomenology and Dialectical Materialism.* 1985.
50. Yehuda Fried and Joseph Agassi, *Paranoia: A Study in Diagnosis.* 1976.
51. Kurt H. Wolff, *Surrender and Catch: Experience and Inquiry Today.* 1976.
52. Karel Kosik, *Dialectics of the Concrete.* 1976.
53. Nelson Goodman, *The Structure of Appearance.* (Third edition). 1977.
54. Herbert A. Simon, *Models of Discovery and Other Topics in the Methods of Science.* 1977.
55. Morris Lazerowitz, *The Language of Philosophy. Freud and Wittgenstein.* 1977.
56. Thomas Nickles (ed.), *Scientific Discovery, Logic, and Rationality.* 1980.
57. Joseph Margolis, *Persons and Minds. The Prospects of Nonreductive Materialism.* 1977.
58. G. Radnitzky and G. Andersson (eds.), *Progress and Rationality in Science,* 1978.
59. Gerard Radnitzky and Gunnar Andersson (eds.), *The Structure and Development of Science.* 1979.
60. Thomas Nickles (ed.), *Scientific Discovery: Case Studies.* 1980.
61. Maurice A. Finocchiaro, *Galileo and the Art of Reasoning.* 1980.
62. William A. Wallace, *Prelude to Galileo.* 1981.
63. Friedrich Rapp, *Analytical Philosophy of Technology.* 1981.
64. Robert S. Cohen and Marx W. Wartofsky (eds.), *Hegel and the Sciences.* 1984.
65. Joseph Agassi, *Science and Society.* 1981.
66. Ladislav Tondl, *Problems of Semantics.* 1981.
67. Joseph Agassi and Robert S. Cohen (eds.), *Scientific Philosophy Today.* 1982.
68. Władysław Krajewski (ed.), *Polish Essays in the Philosophy of the Natural Sciences.* 1982.
69. James H. Fetzer, *Scientific Knowledge.* 1981.
70. Stephen Grossberg, *Studies of Mind and Brain.* 1982.
71. Robert S. Cohen and Marx W. Wartofsky (eds.), *Epistemology, Methodology, and the Social Sciences.* 1983.
72. Karel Berka, *Measurement.* 1983.
73. G. L. Pandit, *The Structure and Growth of Scientific Knowledge.* 1983.
74. A. A. Zinov'ev, *Logical Physics.* 1983.
75. Gilles-Gaston Granger, *Formal Thought and the Sciences of Man.* 1983.
76. R. S. Cohen and L. Laudan (eds.), *Physics, Philosophy and Psychoanalysis.* 1983.
77. G. Böhme et al., *Finalization in Science,* ed. by W. Schäfer. 1983.
78. D. Shapere, *Reason and the Search for Knowledge.* 1983.
79. G. Andersson, *Rationality in Science and Politics.* 1984.
80. P. T. Durbin and F. Rapp, *Philosophy and Technology.* 1984.
81. M. Marković, *Dialectical Theory of Meaning.* 1984.

82. R. S. Cohen and M. W. Wartofsky, *Physical Sciences and History of Physics*. 1984.
83. E. Meyerson, *The Relativistic Deduction*. 1985.
84. R. S. Cohen and M. W. Wartofsky, *Methodology, Metaphysics and the History of Sciences*. 1984.
85. György Tamás, *The Logic of Categories*. 1985.
86. Sergio L. de C. Fernandes, *Foundations of Objective Knowledge*. 1985.
87. Robert S. Cohen and Thomas Schnelle (eds.), *Cognition and Fact*. 1985.
88. Gideon Freudenthal, *Atom and Individual in the Age of Newton*. 1985.
89. A. Donagan, A. N. Perovich, Jr., and M. V. Wedin (eds.), *Human Nature and Natural Knowledge*. 1985.
90. C. Mitcham and A. Huning (eds.), *Philosophy and Technology II*. 1986.
91. M. Grene and D. Nails (eds.), *Spinoza and the Sciences*. 1986.
92. S. P. Turner, *The Search for a Methodology of Social Science*. 1986.
93. I. C. Jarvie, *Thinking about Society: Theory and Practice*. 1986.
94. Edna Ullmann-Margalit (ed.), *The Kaleidoscope of Science*. 1986.
95. Edna Ullmann-Margalit (ed.), *The Prism of Science*. 1986.
96. G. Markus, *Language and Production*. 1986.
97. F. Amrine, F. J. Zucker, and H. Wheeler (eds.), *Goethe and the Sciences: A Reappraisal*. 1987.
98. Joseph C. Pitt and Marcella Pera (eds.), *Rational Changes in Science*. 1987.
99. O. Costa de Beauregard, *Time, the Physical Magnitude*. 1987.
100. Abner Shimony and Debra Nails (eds.), *Naturalistic Epistemology: A Symposium of Two Decades*. 1987.
101. Nathan Rotenstreich, *Time and Meaning in History*. 1987.
102. David B. Zilberman (ed.), *The Birth of Meaning in Hindu Thought*. 1987.
103. Thomas F. Glick (ed.), *The Comparative Reception of Relativity*. 1987.
104. Zellig Harris et al., *The Form of Information in Science*. 1987
105. Frederick Burwick, *Approaches to Organic Form: Permutations in Science and Culture*. 1987.
106. M. Almási, *Philosophy of Appearances*. Forthcoming.
107. S. Hook, W. L. O'Neill, and R. O'Toole, *Philosophy, History and Social Action. Essays in Honor of Lewis Feuer*. 1988.
108. I. Hronszky, M. Fehér, and B. Dajka (eds.), *Scientific Knowledge Socialized. Selected Proceedings of the Fifth Joint International Conference on History and Philosophy of Science Organized by the IUHPS, Veszprém, 1984*. Forthcoming.
109. P. Tillers and E. D. Green (eds.), *Probability and Inference in the Law of Evidence. The Uses and Limits of Bayesianism*. 1988.
110. E. Ullmann-Margalit (ed.), *Science in Reflection. The Israel Colloquium: Studies in History, Philosophy, and Sociology of Science*. 1988.
111. K. Gavroglu, Y. Goudaroulis, and P. Nicolacopoulos (eds.), *Imre Lakatos and Theories of Scientific Change*. 1989.
112. Barry Glassner and Jonathan D. Moreno (eds.), *The Qualitative-Quantitative Distinction in the Social Sciences*. 1989.
113. K. Arens, *Structures of Knowing: Psychologies of the Nineteenth Century*. 1989.

114. A. Janik, *Style, Politics and the Future of Philosophy*. 1989.
115. F. Amrine (ed.), *Literature and Science as Modes of Expression*. 1989.
116. James Robert Brown and Jürgen Mittelstrass (eds.), *An Intimate Relation: Studies in the History and Philosophy of Science Presented to Robert E. Butts on his 60th Birthday*. 1989.
117. F. D'Agostino and I. C. Jarvie (eds.), *Freedom and Rationality: Essays in Honor of John Watkins*. 1989.
118. D. Zolo, *Reflective Epistemology: The Philosophical Legacy of Otto Neurath*. 1989.
119. Michael Kearn, Bernard S. Phillips and Robert S. Cohen (eds.), *George Simmel and Contemporary Sociology*. 1989.
120. Trevor H. Levere and William R. Shea (eds.), *Nature, Experiment, and the Sciences: Essays on Galileo and the History of Science in Honour of Stillman Drake*. 1989.
121. P. Nicolacopoulos (ed.), *Greek Studies in the Philosophy and History of Science*. 1990.
122. R. Cooke and D. Costantini (eds.), *Statistics in Science. The Foundations of Statistical Methods in Biology, Physics and Economics*. 1990
123. P. Duhem (ed.), *The Origins of Statics*. 1991.
124. K. Gavroglu and Y. Goudaroulis (eds.), Heike Kamerlingh Onnes, *Through Measurement to Knowledge. The Selected Papers of Heike Kamerlingh Onnes 1853–1926*. 1991.
125. M. Capek, *The New Aspects of Time. Its Continuity and Novelties. Selected Papers in the Philosophy of Science*. 1991.
126. S. Unguru, *Physics, Cosmology and Astronomy, 1300–1700: Tension and Accommodation*. 1991.